代码整洁之道
Clean Code
A Handbook of Agile Software Craftsmanship

[美] 罗伯特·C. 马丁（Robert C. Martin） 著
韩磊 译

人民邮电出版社
北京

图书在版编目（CIP）数据

代码整洁之道 /（美）罗伯特・C. 马丁
(Robert C. Martin) 著；韩磊译. -- 2版. -- 北京：
人民邮电出版社，2020.2
书名原文：Clean Code: A Handbook of Agile
Software Craftsmanship
ISBN 978-7-115-52413-3

Ⅰ. ①代… Ⅱ. ①罗… ②韩… Ⅲ. ①软件开发
Ⅳ. ①TP311.52

中国版本图书馆CIP数据核字(2019)第240396号

内 容 提 要

软件质量，不但依赖架构及项目管理，而且与代码质量紧密相关。这一点，无论是敏捷开发流派还是传统开发流派，都不得不承认。

本书提出一种观点：代码质量与其整洁度成正比。干净的代码，既在质量上较为可靠，也为后期维护、升级奠定了良好基础。作为编程领域的佼佼者，本书作者给出了一系列行之有效的整洁代码操作实践。这些实践在本书中体现为一条条规则（或称"启示"），并辅以来自实际项目的正、反两面的范例。只要遵循这些规则，就能编写出干净的代码，从而有效提升代码质量。

本书阅读对象为一切有志于改善代码质量的程序员及技术经理。书中介绍的规则均来自作者多年的实践经验，涵盖从命名到重构的多个编程方面，虽为一"家"之言，然诚有可资借鉴的价值。

- ◆ 著　　[美] 罗伯特・C. 马丁（Robert C. Martin）
  译　　韩　磊
  责任编辑　杨海玲
  责任印制　王　郁　焦志炜
- ◆ 人民邮电出版社出版发行　北京市丰台区成寿寺路11号
  邮编　100164　电子邮件　315@ptpress.com.cn
  网址　https://www.ptpress.com.cn
  大厂回族自治县聚鑫印刷有限责任公司印刷
- ◆ 开本：800×1000　1/16
  印张：25.5　　　　　　　　　　2020年2月第2版
  字数：557千字　　　　　　　　2024年11月河北第19次印刷
  著作权合同登记号　图字：01-2008-5467号

定价：99.00元

读者服务热线：(010)81055410　印装质量热线：(010)81055316
反盗版热线：(010)81055315
广告经营许可证：京东市监广登字20170147号

版权声明

Authorized translation from the English language edition, entitled CLEAN CODE: A HANDBOOK OF AGILE SOFTWARE CRAFTSMANSHIP, 1st Edition, ISBN: 0132350882 by MARTIN, ROBERT C., published by Pearson Education, Inc, Copyright © 2009 by Pearson Education, Inc.

All rights reserved. No part of this book may be reproduced or transmitted in any form or by any means, electronic or mechanical, including photocopying, recording or by any information storage retrieval system, without permission from Pearson Education, Inc.

CHINESE SIMPLIFIED language edition published by POSTS & TELECOM PRESS, Copyright © 2020.

本书中文简体字版由 Pearson Education Inc 授权人民邮电出版社独家出版。未经出版者书面许可，不得以任何方式复制或抄袭本书内容。

本书封面贴有 Pearson Education（培生教育出版集团）激光防伪标签，无标签者不得销售。

版权所有，侵权必究。

代码猴子与童子军军规
（译者序）

2007年3月，我在SD West 2007技术大会上聆听了Robert C. Martin（"鲍勃大叔"）的主题演讲"Craftsmanship and the Problem of Productivity: Secrets for Going Fast without Making a Mess"。一身休闲打扮的"鲍勃大叔"，以一曲嘲笑低水平编码者的Code Monkey（代码猴子）开场。

是的，我们就是一群代码猴子，上蹿下跳，自以为领略了编程的真谛。可惜，当我们抓着几个酸桃子，得意扬扬地坐到树枝上，却对自己造成的混乱熟视无睹。那堆"可以运行"的乱麻程序，就在我们的眼皮底下慢慢腐坏。

从听到那场以TDD为主题的演讲之后，我就一直关注"鲍勃大叔"，还有他在TDD和整洁代码方面的言论。2008年，人民邮电出版社（计算机分社）的编辑拿一本书给我看，封面上赫然写着Robert C. Martin的大名。看完原书序和前言，我已经按捺不住，接下了翻译此书的任务。这本书名为 Clean Code，乃是Object Mentor（"鲍勃大叔"开办的技术咨询和培训公司）一干大牛在编程方面的经验累积。按"鲍勃大叔"的话来说，就是"Object Mentor整洁代码派"的说明。

正如Coplien在序中所言，宏大建筑中最细小的部分，比如关不紧的门、有点儿没铺平的地板，甚至是凌乱的桌面，都会将整个大局的魅力破坏殆尽。这就是整洁代码之所系。Coplien列举了许多谚语，证明整洁的价值，中国也有修身齐家治国平天下之语。整洁代码的重要性毋庸置疑，问题是如何写出真正整洁的代码。

本书既是整洁代码的定义，亦是如何写出整洁代码的指南。"鲍勃大叔"认为，"写整洁代码，需要遵循大量的小技巧，贯彻刻苦习得的'整洁感'。这种'代码感'就是关键所在……它不仅让我们看到代码的优劣，还予我们以借戒规之力化劣为优的攻略。"作者阐述了在命名、函数、注释、代码格式、对象和数据结构、错误处理、边界问题、单元测试、类、系统、并发编程等方面如何做到整洁的经验与最佳实践。长期遵照这些经验编写代码，所谓"代码感"也就自然而然滋生出来。更有价值的部分是"鲍勃大叔"本人对3个Java项目的剖析与改进过程的实操记录。通过这多达3章的重构记录，"鲍勃大叔"充分地证明了童子军军规在编程领域同样适用：离开时要比发现时更整洁。为了向读者呈现代码的原始状态，所有代码注释都不做翻译。

接触软件开发技术二十多年以来，我见过许多对于代码整洁性缺乏足够重视的开发者，

II　代码猴子与童子军军规（译者序）

不算过分地说，这体现了他们职业素养与基本功的双重缺陷。我翻译《C#编程风格》（*The Elements of C# Style*）和本书，实在也是希望在这方面看到开发者在重视度和实际应用方面的提升。我于 2009 年翻译了这本书，如今 10 年过去了，开发者们越来越认同"整洁代码非常重要"。在本书英文版出版后，作者对原版做了许多修正和勘误，这些修正和勘误也体现在了这一中文修订版中。这也正体现了"整洁代码派"的基本理念：不断迭代。

在本书的结束语中，"鲍勃大叔"提到别人给他的一条腕带，上面的字样是"Test Obsessed"（沉迷测试）。"鲍勃大叔""发现自己无法取下腕带。不仅因为腕带很紧，而且那也是条精神上的紧箍咒……它一直提醒我，我做了写出整洁代码的承诺。"有了这条腕带，代码猴子成了模范童子军。我想，每位开发者都需要这样一条腕带吧。

韩　磊

2019 年 12 月

序

乐嚼（Ga-Jol）是在丹麦最受欢迎的糖果品种之一，它浓郁的甘草味道，完美地弥补了此地潮湿且时常寒冷的天气。对于我们这些丹麦人，乐嚼的妙处还在于包装盒顶上印的哲言慧语。今早我买了一包两件装，在其包装盒上发现这句丹麦谚语：

> Ærlighed i små ting er ikke nogen lille ting.

"小处诚实非小事。"这句话正好是我想在这里说的。以小见大。本书写到了一些价值殊胜的小主题。

"神在细节之中。"建筑师 Ludwig mies van der Rohe（路德维希·密斯·凡·德·罗）[1]如是说。这句话引发了有关软件开发、特别是敏捷软件开发中架构所处地位的若干争论。鲍勃（Bob）[2]和我时常发现自己沉湎于此类对话中。没错，Ludwig mies van der Rohe 的确专注于效用和基于宏伟架构之上的永恒建筑形式。然而，他也为自己设计的每所房屋挑选每个门把手。为什么？因为小处见大。

就 TDD[3] 话题展开目前仍在继续的"辩论"时，鲍勃和我认识到，我们均同意软件架构在开发中占据重要地位，但就其确切意义而言，我们之间还有分歧。然而，这种矛与盾孰利的讨论相对而言并不重要，因为在项目开始之时，我们理所当然应该让专业人士投入些许时间去思考及规划。20 世纪 90 年代末期有关仅以测试和代码驱动设计的概念已一去不返。与任何宏伟愿景相比，对细节的关注反而是更为关键的专业性基础。首先，开发者通过小型实践获得可用于大型实践的技能和信用度。其次，宏大建筑中最细小的部分，比如关不紧的门、有点儿没铺平的地板，甚至是凌乱的桌面，都会将整个大局的魅力破坏殆尽。这就是整洁代码之所系。

架构只是软件开发用到的借喻之一，主要用在那种等同于建筑师交付毛坯房一般交付初始软件产品的场合。在 Scrum 和敏捷（Agile）的日子里，人们关注的是快速将产品推向市场。我们要求工厂全速运转、生产软件。这就是人类工厂：懂思考、会感受的编码人，他们由产品备忘或用户故事开始创造产品。来自制造业的借喻在这种场合大行其道。例如，Scrum 就从装配线式的日本汽车生产方式中获益良多。

---

[1] 20 世纪中期著名现代建筑大师，秉承"少即是多"的建筑设计哲学，缔造了玻璃幕墙等现代建筑结构。——译者注
[2] 本书主要作者 Robert C. Martin 绰号 Uncle Bob，这里的"鲍勃"及后文的"鲍勃大叔"就是指 Robert C. Martin。——译者注
[3] Test Driven Development，测试驱动开发。——译者注

即便是在汽车工业里，大量工作也并不在于生产而在于维护，或避免维护。对于软件而言，百分之八十或更多的工作量集中在我们美其名曰"维护"的事情上：其实就是修修补补。与其接受西方关于生产好软件的传统看法，不如将其看作建筑工业中的房屋修理工，或者汽车领域的汽修工。日本式管理对于这种事是怎么说的呢？

大约在 1951 年，一种名为"全员生产维护"（Total Productive Maintenance，TPM）的质量保证手段在日本出现。它关注维护甚于关注生产。TPM 的主要支柱之一是所谓的 5S 原则体系。5S 是一套规程，用"规程"这个词，是为了便于读者理解。5S 原则其实是精益（lean）——西方视野中的一个时髦词，也是在软件领域渐领风骚的时髦词——的基石所在。正如"鲍勃大叔"（Uncle Bob）在前言中写到的，良好的软件实践遵循这些规程：专注、镇定和思考。这并非总指有关实作，有关推动工厂设备以最高速度运转。5S 哲学包括以下概念。

- 整理（Seiri）①，或谓组织（想想英语中的 sort 一词）。搞清楚事物之所在——通过恰当地命名之类的手段——至关重要。觉得命名标识无关紧要？读读后面的章节吧。
- 整顿（Seiton），或谓整齐（想想英文中的 systematize 一词）。有句美国老话说：物皆有其位，而后物尽归其位（A place for everything, and everything in its place）。每段代码都该在你希望它在的地方——如果不在那里，就需要重构了。
- 清楚（Seiso），或谓清洁（想想英文中的 shine 一词）。就像清理工作地的拉线、油污和边角废料，对于那种四处遗弃的带注释的代码及反映过往或期望的无注释代码，本书作者怎么说的来着？除之而后快。
- 清洁（Seiketsu），或谓标准化。有关如何保持工作地清洁的组内共识。本书有没有提到在开发组内使用一贯的代码风格和实践手段？这些标准从哪里来？读读看。
- 身美（Shitsuke）②，或谓纪律（自律）。在实践中贯彻规程，并时时体现在个人工作中，而且要乐于改进。

如果你接受挑战——没错，就是挑战，阅读并应用本书，你就会理解和赞赏上述概念的最后一条。我们最终是在驶向一种负责任的专业精神之根源所在，这种专业性隶属于一个关注产品生命周期的专业领域。在我们遵循 TPM 来维护机动车和其他机械时，停机维护——等待缺陷显现出来——并不常见。我们更上一层楼：每天检查机械，在磨损机件停止工作之前就换掉它，或者按常例每 1000 英里（约 1609.3 km）就更换润滑油以防止磨损和开裂。对于代码，应无情地做重构。还可以更进一步，就像 TPM 运动在数十年前的创新：一开始就打造更易维护的机械。写出可读的代码，其重要程度不亚于写出可执行的代码。1960 年左右，围绕 TPM 引入的终极实践，关注用全新机械替代旧机械。诚如 Fred Brooks 所言，我们或许应该每 7 年就重做一次软件的主要模块，清理缓慢陈腐的代码。也许我们该把重构周期从以年计缩短到以周、以天甚至以小时计。那便是细节所在了。

---

① 这些概念最初出现于日本，5 个概念的日文罗马字拼音首字母正好都是 S，所以这里也保留了日文罗马字拼音写法。中译本以日文汉字直接译出，读者留意，不可直接对应其中文意思。——译者注

② 中文意为"素养、教养"。——译者注

细节中自有天地，而在生活中应用此类手段时也有微言大义，就像我们一成不变地对那些源自日本的做法寄予厚望一般。这并非只是东方的生活观，英美民间也遍是这类警句。上引"整顿"二字就曾出现在某位俄亥俄州牧师的笔下，他把齐整看作是"荡涤种种罪恶之良方"。"清楚"又如何呢？整洁近乎虔诚（Cleanliness is next to godliness.）。一张脏乱的桌子足以夺去一所丽宅的光彩。老话怎么说"身美"的？守小节者不亏大节（He who is faithful in little is faithful in much.）。对于时时准备在恰当时机做重构，为未来的"大"决定夯实基础，而不是置诸脑后，有什么说法吗？及时一针省九针（A stitch in time saves nine.）。早起的鸟儿有虫吃（The early bird catches the worm.）。今日事今日毕（Don't put off until tomorrow what you can do today.）。在精益实践落入软件咨询师之手前，这就是其所谓"最后时机"的本义所在。如何摆正单项工作在整体中的位置呢？巨木生于树籽（Mighty oaks from little acorns grow.）。如何在日常生活中做好简单的防备性工作呢？防病好过治病（An ounce of prevention is worth a pound of cure.）。一天一苹果，医生远离我（An apple a day keeps the doctor away.）。整洁代码以其对细节的关注，使深埋于我们或现有、或曾有、或该有的壮丽文化之下的智慧根源获得荣耀。

即便是在宏伟的建筑作品中，我们也听到关注细节的回响。想想 Ludwig mies van der Rohe 的门把手吧，那正是整理。认真对待每个变量名。你当用为自己第一个孩子命名般的谨慎来给变量命名。

正如每位房主所知，此类照料和修葺永无休止。建筑师 Christopher Alexander——模式与模式语言之父——把每个设计动作看作是较小的局部修复动作。他认为，设计良好结构才是建筑师的本职所在，而更大的建筑形态则当留给模式及居住者搬进的家私来完成。设计始终在持续进行，不只是在新建一个房间时，也在我们重新粉刷墙面、更换旧地毯或者换厨房水槽时。大多数艺术门类也持类似主张。在寻找其他推崇细节的人时，我们发现，19 世纪法国作家古斯塔夫·福楼拜（Gustav Flaubert）名列其中。法国诗人保尔·瓦雷里（Paul Valery）认为，每首诗歌都无写完之时，得持续重写，直至放弃为止。全心倾注于细节，屡见于追求卓越的行为之中。虽然这无甚新意，但阅读本书对读者仍是一种挑战，你要重拾久已弃置脑后的良好规则，自发自主，"响应改变"。

不幸的是，我们往往见不到人们把对细节的关注当作编程艺术的基础要件。我们过早地放弃了在代码上的工作，并不是因为它业已完成，而是因为我们的价值体系关注外在表现甚于关注要交付之物的本质。疏忽最终结出了恶果：坏东西一再出现。无论是在行业里还是学术领域，研究者都很重视代码的整洁问题。供职于贝尔软件生产研究实验室（Bell Labs Software Production Research）——没错，就是生产！——时，我们有些不太严密的发现，认为前后一致的缩进风格明显标志了较低的缺陷率。我们原指望将质量归因于架构、编程语言或者其他高级概念；将我们的专业能力归功于对工具的掌握和各种高高在上的设计方法，没想到那些安置于厂区的机器，那些编码者，他们居然通过简单地保持一致缩进风格创造了价值，这简直是一种侮辱。我在许多年前就在书中写过，这种风格远不止是一种单纯的能力那么简单。日本式的世界观深知日常工作者的价值，而且，还深知工作者简单的日常行为所锻造的开发系统的价值。质量是上百万次全心投入的结果，而非仅归功于任何来自天堂的伟大方法。这些行为简单却不简陋，也不意味着简易。相反，它们是人力所能达的不仅伟大而且美丽的造物。忽略它们，就不能成为完整的人。

当然，我仍然提倡放宽思路，也推崇根植于深厚领域知识和软件可用性的各种架构手法的价值，但本书与此无关——至少，没有明显关系。本书精妙之处，其意义之深远，不该无人赏识。它正与 Peter Sommerlad、Kevlin Henny 及 Giovanni Asproni 等真正写代码的人现今所持的观念相吻合。他们鼓吹"代码即设计"和"简单代码"。我们要谨记，界面就是程序，而且其结构也极大地反映出程序结构，但也理应始终谦逊地承认设计存在于代码中，这至关紧要。制造上的返工导致成本上升，但重做设计却创造出价值。我们应当视代码为设计——作为过程而非终点的设计——这种高尚行为的漂亮体现。耦合与内聚的架构韵律在代码中脉动。Larry Constantine 以代码的形式——而不是用 UML 那种高高在上的抽象概念——来描述耦合与内聚。Richard Garbriel 在"Abstraction Descant"（抽象刍议）一文中告诉我们，抽象即恶。代码除恶，而整洁的代码则大抵是圣洁的。

回到我那个小小的乐嚼包装盒，我想要重点提一下，那句丹麦谚语不只是教我们重视小处，更教我们小处要诚实。这意味着对代码诚实、对同僚坦承代码现状，最重要的是在代码问题上不自欺。是否已尽全力"把露营地清理得比来时还干净"？签入代码前是否已做重构？这可不是皮毛小事，它正高卧于敏捷价值的正中位置。Scrum 有一种建议的实践，主张重构是"完成"（Done）概念的一部分。无论是架构还是代码都不强求完美，只求竭诚尽力而已。人孰无过，神亦容之（To err is human; to forgive, divine.）。在 Scrum 中，我们使一切可见。我们晾出脏衣服。我们坦承代码状态，因为它永不完美。我们日渐成为完整的人，配得起神的眷顾，也越来越接近细节中的伟大之处。

在自己的专业领域中，我们亟需能得到的一切帮助。假使干净的地板能减少事故发生，假使归置到位的工具能提升生产力，我也会倾力做到。至于本书，在我看过的有关将精益原则应用于软件的印刷品中，是最具实用性的。那班求索者多年来并肩奋斗，不但是为求一己之进步，更将他们的知识通过和你手上正在做的事一般的工作贡献给这个行业。看过"鲍勃大叔"寄来的原稿之后，我发现，世界竟略有改善了。

对高瞻远瞩的练习业已结束，我要去清理自己的书桌了。

James O. Coplien
于丹麦默尔鲁普

前言

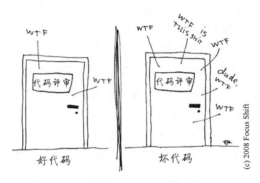

（经 Thom Holwerda 允许复制上图）

你的代码在哪道门后面？你的团队或公司在哪道门后面？为什么会在那里？只是一次普通的代码评审，还是产品面世后才发现一连串严重问题？我们是否在战战兢兢地调试自己错以为没问题的代码？客户是否在流失？经理们是否把我们盯得如芒刺在背？当事态变得严重起来时，如何保证我们在那道正确的门后做补救工作？答案是：技艺（craftsmanship）。

习艺之要有二：知和行。你应当习得有关原则、模式和实践的知识，穷尽应知之事，并且要对其了如指掌，通过刻苦实践掌握它。

我可以教你骑自行车的物理学原理。实际上，经典数学的表达方式相对而言确实简洁明了。重力、摩擦力、角动量、质心等，用一页写满方程式的纸就能说明白。有了这些方程式，我可以为你证明出骑车完全可行，而且还可以告诉你骑车所需的全部知识。即便如此，你在初次骑车时还是会跌倒在地。

编码亦同此理。我们可以写下整洁代码的所有"感觉良好"的原则，放手让你去干（换言之，让你从自行车上摔下来）。那样的话，我们算是哪门子老师？而你又会成为怎样的学生呢？

不！本书可不会这么做。

学写整洁代码很难。它可不止于要求你掌握原则和模式，你得在这上面花工夫。你须自行实践，且体验自己的失败。你须观察他人的实践与失败。你须看看别人是怎样蹒跚学步，再转头研究他们的路数。你须看看别人是如何绞尽脑汁做出决策,又是如何为错误决策付出代价的。

阅读本书要多用心思。这可不是那种降落前就能读完的"感觉不错"的"飞机书"。本书要让你用功,而且是非常用功。如何用功?阅读代码——大量代码。而且你要去琢磨某段代码好在什么地方、坏在什么地方。在我们分解然后组合模块时,你得亦步亦趋地跟上。这会花些工夫,不过值得一试。

本书大致可分为3个部分。第一部分介绍编写整洁代码的原则、模式和实践。这部分有相当多的示例代码,读起来颇具挑战性。读完这几章,就为阅读第二部分做好了准备。如果你就此止步,只能祝你好运啦!

第二部分最需要花工夫。这部分包括几个复杂性不断增加的案例研究。每个案例都清理一些代码——把有问题的代码转化为问题少一些的代码。这部分极为详细。你的思维要在讲解和代码段之间跳来跳去。你得分析和理解那些代码,琢磨每次修改的来龙去脉。

你付出的劳动将在第三部分得到回报。这部分只有一章,列出从上述案例研究中得到的启示和灵感。在遍览和清理案例中的代码时,我们把每个操作理由记录为一种启示或灵感。我们尝试去理解自己对阅读和修改代码的反应,尽力了解为什么会有这样的感受、为什么会如此行事。结果得到了一套描述在编写、阅读、清理代码时思维方式的知识库。

如果你在阅读第二部分的案例研究时没有好好用功,那么这套知识库对你来说可能所值无几。在这些案例研究中,每次修改都仔细注明了相关启示的标号。这些标号用方括号标出,如[H22]。由此你可以看到这些启示在何种环境下被应用和编写。启示本身没有价值,启示与案例研究中清理代码的具体决策之间的关系才有价值。

如果你跳过案例研究部分,只阅读了第一部分和第三部分,那就不过是又看了一本关于写出好软件的"感觉不错"的书。但如果你肯花时间琢磨那些案例,亦步亦趋——站在作者的角度,迫使自己以作者的思维路径考虑问题,就能更深刻地理解这些原则、模式、实践和启示。这样的话,就像一个人熟练地掌握了骑车的技术后,自行车就如同其身体的延伸部分那样;对你来说,本书所介绍的整洁代码的原则、模式、实践和启示就成为你本身具有的技艺,而不再是"感觉不错"的知识。

致谢

插图

感谢两位艺术家 Jennifer Kohnke 和 Angela Brooks。Jennifer 绘制了每章起始处创意新颖、效果惊人的插图,以及 Kent Beck、Ward Cunningham、Bjarne Stroustrup、Ron Jeffries、Grady Booch、Dave Thomas、Michael Feathers 和我本人的肖像。

Angela 绘制了文中那些精致的插图。这些年她为我画了一些画,包括《敏捷软件开发:原则、模式与实践》(*Agile Software Development: Principles, Patterns, and Practices*)一书中的大量插图。她是我的长女,常给我带来极大的愉悦。

资源与支持

本书由异步社区出品，社区（https://www.epubit.com/）为您提供相关资源和后续服务。

提交勘误

作者和编辑尽最大努力来确保书中内容的准确性，但难免会存在疏漏。欢迎您将发现的问题反馈给我们，帮助我们提升图书的质量。

当您发现错误时，请登录异步社区，按书名搜索，进入本书页面，单击"提交勘误"，输入勘误信息，单击"提交"按钮即可。本书的作者和编辑会对您提交的勘误进行审核，确认并接受后，您将获赠异步社区的 100 积分。积分可用于在异步社区兑换优惠券、样书或奖品。

扫码关注本书

扫描下方二维码，您将会在异步社区微信服务号中看到本书信息及相关的服务提示。

与我们联系

我们的联系邮箱是 contact@epubit.com.cn。

如果您对本书有任何疑问或建议,请您发邮件给我们,并请在邮件标题中注明本书书名,以便我们更高效地做出反馈。

如果您有兴趣出版图书、录制教学视频,或者参与图书翻译、技术审校等工作,可以发邮件给我们;有意出版图书的作者也可以到异步社区在线提交投稿(直接访问 www.epubit.com/selfpublish/submission 即可)。

如果您来自学校、培训机构或企业,想批量购买本书或异步社区出版的其他图书,也可以发邮件给我们。

如果您在网上发现有针对异步社区出品图书的各种形式的盗版行为,包括对图书全部或部分内容的非授权传播,请您将怀疑有侵权行为的链接发邮件给我们。您的这一举动是对作者权益的保护,也是我们持续为您提供有价值的内容的动力之源。

关于异步社区和异步图书

"**异步社区**"是人民邮电出版社旗下 IT 专业图书社区,致力于出版精品 IT 技术图书和相关学习产品,为作译者提供优质出版服务。异步社区创办于 2015 年 8 月,提供大量精品 IT 技术图书和电子书,以及高品质技术文章和视频课程。更多详情请访问异步社区官网 https://www.epubit.com。

"**异步图书**"是由异步社区编辑团队策划出版的精品 IT 专业图书的品牌,依托于人民邮电出版社近 30 年的计算机图书出版积累和专业编辑团队,相关图书在封面上印有异步图书的 LOGO。异步图书的出版领域包括软件开发、大数据、AI、测试、前端、网络技术等。

异步社区

微信服务号

关于封面

本书封面的图片是 M104：草帽星系（The Sombrero Galaxy）。M104 坐落于处女座（Virgo），距地球仅 3000 万光年，其核心是一个质量超大的黑洞，有 100 万个太阳那么重。

这幅图是否让你想起了 Klingon 星球（克林贡）[①]的卫星 Praxis（普拉西斯）爆炸的事？我清楚地记得，在《星舰迷航 VI》中，大爆炸之后碎片四溅，飞舞出一个赤道光环的场景。至此，光环就成为科幻电影中爆炸场景的必然产物了，甚至就在《星舰迷航》系列电影的后续情节中，Alderaan（阿尔德然）的爆炸也有类似场景出现。

环绕 M104 的光环是什么造成的？它为何会有如此巨大的膨胀率和如此明亮而微小的内核？在我看来，仿佛那位于中心位置的黑洞勃然大怒，向星系的中心扔出了一个 3 万光年大的洞一般。在这场宇宙大崩塌所及范围之内的居民全都大难临头了。

超大质量的黑洞以星体为食，将星体的相当部分质量转换为能量。方程式 $E = MC^2$ 已经足够体现杠杆作用了，但当 M 有一颗星体那么大的质量时，看吧！在那巨兽酒足饭饱之前，有多少星体会一头撞进它的胃里？核心部分空洞的大小，是否说明了什么呢？

封面上的 M104 图片，是用来自哈勃望远镜的那幅著名的可见光相片（上图）和 Spitzer（斯比泽）轨道探测器最新的红外影像（下图）组合而成。

在红外影像中，光环中的热粒子闪耀着穿过了中心膨胀体。这两幅影像组合起来，显现出我们从未见过的景象，展示了久远之前曾熊熊燃烧的火海。

封面图片：来自斯比泽太空望远镜

[①] 系列剧《星舰迷航》（Star Trek）中的故事情节，Praxis 星爆炸，由此导致联邦和 Klingon 达成首次和平协议。

目录

第 1 章 整洁代码 ... 1
- 1.1 要有代码 ... 2
- 1.2 糟糕的代码 ... 2
- 1.3 混乱的代价 ... 3
  - 1.3.1 华丽新设计 ... 4
  - 1.3.2 态度 ... 4
  - 1.3.3 谜题 ... 5
  - 1.3.4 整洁代码的艺术 ... 5
  - 1.3.5 什么是整洁代码 ... 6
- 1.4 思想流派 ... 10
- 1.5 我们是作者 ... 11
- 1.6 童子军军规 ... 12
- 1.7 前传与原则 ... 12
- 1.8 小结 ... 13
- 1.9 文献 ... 13

第 2 章 有意义的命名 ... 14
- 2.1 介绍 ... 14
- 2.2 名副其实 ... 15
- 2.3 避免误导 ... 16
- 2.4 做有意义的区分 ... 17
- 2.5 使用读得出来的名称 ... 18
- 2.6 使用可搜索的名称 ... 19
- 2.7 避免使用编码 ... 20
  - 2.7.1 匈牙利语标记法 ... 20
  - 2.7.2 成员前缀 ... 21
  - 2.7.3 接口和实现 ... 21
- 2.8 避免思维映射 ... 21
- 2.9 类名 ... 22
- 2.10 方法名 ... 22
- 2.11 别抖机灵 ... 22
- 2.12 每个概念对应一个词 ... 23
- 2.13 别用双关语 ... 23
- 2.14 使用解决方案领域名称 ... 24
- 2.15 使用源自所涉问题领域的名称 ... 24
- 2.16 添加有意义的语境 ... 24
- 2.17 不要添加没用的语境 ... 26
- 2.18 最后的话 ... 27

第 3 章 函数 ... 28
- 3.1 短小 ... 31
- 3.2 只做一件事 ... 32
- 3.3 每个函数一个抽象层级 ... 33
- 3.4 switch 语句 ... 34
- 3.5 使用具有描述性的名称 ... 35
- 3.6 函数参数 ... 36
  - 3.6.1 单参数函数的普遍形式 ... 37
  - 3.6.2 标识参数 ... 37
  - 3.6.3 双参数函数 ... 38
  - 3.6.4 三参数函数 ... 38
  - 3.6.5 参数对象 ... 39
  - 3.6.6 参数列表 ... 39
  - 3.6.7 动词与关键字 ... 39
- 3.7 无副作用 ... 40
- 3.8 分隔指令与询问 ... 41
- 3.9 使用异常替代返回错误码 ... 42
  - 3.9.1 抽离 try/catch 代码块 ... 42
  - 3.9.2 错误处理就是一件事 ... 43
  - 3.9.3 Error.java 依赖磁铁 ... 43

目录

- 3.10 别重复自己 ········· 44
- 3.11 结构化编程 ········· 44
- 3.12 如何写出这样的函数 ········· 45
- 3.13 小结 ········· 45
- 3.14 SetupTeardownIncluder 程序 ········· 45
- 3.15 文献 ········· 48

第 4 章 注释 ········· 49
- 4.1 注释不能美化糟糕的代码 ········· 50
- 4.2 用代码来阐述 ········· 51
- 4.3 好注释 ········· 51
  - 4.3.1 法律信息 ········· 51
  - 4.3.2 提供信息的注释 ········· 51
  - 4.3.3 对意图的解释 ········· 52
  - 4.3.4 阐释 ········· 53
  - 4.3.5 警示 ········· 53
  - 4.3.6 TODO 注释 ········· 54
  - 4.3.7 放大 ········· 55
  - 4.3.8 公共 API 中的 Javadoc ········· 55
- 4.4 坏注释 ········· 55
  - 4.4.1 喃喃自语 ········· 55
  - 4.4.2 多余的注释 ········· 56
  - 4.4.3 误导性注释 ········· 58
  - 4.4.4 循规式注释 ········· 59
  - 4.4.5 日志式注释 ········· 59
  - 4.4.6 废话注释 ········· 60
  - 4.4.7 可怕的废话 ········· 62
  - 4.4.8 能用函数或变量时就别用注释 ········· 62
  - 4.4.9 位置标记 ········· 62
  - 4.4.10 括号后面的注释 ········· 63
  - 4.4.11 归属与署名 ········· 63
  - 4.4.12 注释掉的代码 ········· 64
  - 4.4.13 HTML 注释 ········· 64
  - 4.4.14 非本地信息 ········· 65
  - 4.4.15 信息过多 ········· 65
  - 4.4.16 不明显的联系 ········· 66
  - 4.4.17 函数头 ········· 66
  - 4.4.18 非公共代码中的 Javadoc ········· 66
  - 4.4.19 范例 ········· 66
- 4.5 文献 ········· 70

第 5 章 格式 ········· 71
- 5.1 格式的目的 ········· 72
- 5.2 垂直格式 ········· 72
  - 5.2.1 向报纸学习 ········· 73
  - 5.2.2 概念间垂直方向上的区隔 ········· 73
  - 5.2.3 垂直方向上的靠近 ········· 74
  - 5.2.4 垂直距离 ········· 75
  - 5.2.5 垂直顺序 ········· 79
- 5.3 横向格式 ········· 80
  - 5.3.1 水平方向上的区隔与靠近 ········· 81
  - 5.3.2 水平对齐 ········· 82
  - 5.3.3 缩进 ········· 83
  - 5.3.4 空范围 ········· 84
- 5.4 团队规则 ········· 85
- 5.5 "鲍勃大叔"的格式规则 ········· 85

第 6 章 对象和数据结构 ········· 88
- 6.1 数据抽象 ········· 88
- 6.2 数据、对象的反对称性 ········· 90
- 6.3 得墨忒耳律 ········· 92
  - 6.3.1 火车失事 ········· 92
  - 6.3.2 混杂 ········· 93
  - 6.3.3 隐藏结构 ········· 93
- 6.4 数据传送对象 ········· 94
- 6.5 小结 ········· 95
- 6.6 文献 ········· 96

第 7 章 错误处理 ········· 97
- 7.1 使用异常而非返回码 ········· 98

7.2 先写 try-catch-finally 语句 ········· 99
7.3 使用未检异常 ························· 100
7.4 给出异常发生的环境说明 ········ 101
7.5 依调用者需要定义异常类 ········ 101
7.6 定义常规流程 ························· 103
7.7 别返回 null 值 ························ 104
7.8 别传递 null 值 ························ 105
7.9 小结 ······································ 106
7.10 文献 ···································· 106

第 8 章 边界 ·· 107
8.1 使用第三方代码 ····················· 108
8.2 浏览和学习边界 ····················· 109
8.3 学习 log4j ······························ 110
8.4 学习性测试的好处不只是
    免费 ······································ 112
8.5 使用尚不存在的代码 ·············· 112
8.6 整洁的边界 ··························· 113
8.7 文献 ······································ 114

第 9 章 单元测试 ································ 115
9.1 TDD 三定律 ·························· 116
9.2 保持测试整洁 ························ 117
9.3 整洁的测试 ··························· 118
   9.3.1 面向特定领域的测试
         语言 ························· 120
   9.3.2 双重标准 ···················· 121
9.4 每个测试一个断言 ················· 123
9.5 F.I.R.S.T. ······························ 125
9.6 小结 ······································ 125
9.7 文献 ······································ 126

第 10 章 类 ··· 127
10.1 类的组织 ······························ 128
10.2 类应该短小 ·························· 128
   10.2.1 单一权责原则 ············ 130
   10.2.2 内聚 ························· 131
   10.2.3 保持内聚性就会得到
          许多短小的类 ············· 132
10.3 为了修改而组织 ··················· 138
10.4 文献 ···································· 141

第 11 章 系统 ····································· 142
11.1 如何建造一个城市 ················ 143
11.2 将系统的构造与使用分开 ····· 143
   11.2.1 分解 main ················· 144
   11.2.2 工厂 ·························· 145
   11.2.3 依赖注入 ··················· 145
11.3 扩容 ···································· 146
11.4 Java 代理 ····························· 149
11.5 纯 Java AOP 框架 ················ 151
11.6 AspectJ 的方面 ···················· 154
11.7 测试驱动系统架构 ················ 154
11.8 优化决策 ······························ 155
11.9 明智使用添加了可论证价值的
     标准 ···································· 155
11.10 系统需要领域特定语言 ······· 156
11.11 小结 ································· 156
11.12 文献 ································· 156

第 12 章 迭进 ····································· 158
12.1 通过迭进设计达到整洁
     目的 ···································· 158
12.2 简单设计规则 1：运行所有
     测试 ···································· 159
12.3 简单设计规则 2~4：重构 ····· 159
12.4 不可重复 ······························ 160
12.5 表达力 ································· 162
12.6 尽可能少的类和方法 ············ 163
12.7 小结 ···································· 163
12.8 文献 ···································· 163

第 13 章 并发编程 ······························ 164
13.1 为什么要并发 ······················· 165
13.2 挑战 ···································· 166
13.3 并发防御原则 ······················· 167

13.3.1	单一职责原则	167
13.3.2	推论：限制数据作用域	167
13.3.3	推论：使用数据副本	168
13.3.4	推论：线程应尽可能地独立	168
13.4	了解 Java 库	168
13.5	了解执行模型	169
13.5.1	生产者-消费者模型	170
13.5.2	读者-作者模型	170
13.5.3	宴席哲学家	170
13.6	警惕同步方法之间的依赖	170
13.7	保持同步区域微小	171
13.8	很难编写正确的关闭代码	171
13.9	测试线程代码	172
13.9.1	将伪失败看作可能的线程问题	172
13.9.2	先使非线程代码可工作	172
13.9.3	编写可插拔的线程代码	173
13.9.4	编写可调整的线程代码	173
13.9.5	运行多于处理器数量的线程	173
13.9.6	在不同平台上运行	173
13.9.7	装置试错代码	174
13.9.8	硬编码	174
13.9.9	自动化	175
13.10	小结	176
13.11	文献	176
第14章	逐步改进	177
14.1	Args 的实现	178
14.2	Args：草稿	185
14.2.1	所以我暂停了	196
14.2.2	渐进	197
14.3	字符串类型参数	199
14.4	小结	236
第15章	JUnit 内幕	237
15.1	JUnit 框架	238
15.2	小结	251
第16章	重构 SerialDate	252
16.1	首先，让它能工作	253
16.2	让它做对	255
16.3	小结	268
16.4	文献	268
第17章	味道与启发	269
17.1	注释	270
17.2	环境	271
17.3	函数	271
17.4	一般性问题	272
17.5	Java	288
17.6	名称	291
17.7	测试	295
17.8	小结	296
17.9	文献	296
附录 A	并发编程 II	297
附录 B	org.jfree.date.SerialDate	326
结束语		388

第 1 章

整洁代码

阅读本书有两种原因：第一，你是个程序员；第二，你想成为更好的程序员。很好。我们需要更好的程序员。

这是一本有关编写好程序的书。它充斥着代码。我们要从各个方向来考察这些代码。从顶向下，从底往上，从里而外。读完后，就能知道许多关于代码的事了。而且，我们还能说出好代码和糟糕的代码之间的差异。我们将了解到如何写出好代码。我们也会知道，如何将糟糕的代码改成好代码。

1.1 要有代码

有人也许会以为，关于代码的书有点儿落后于时代——代码不再是问题，我们应当关注模型和需求。确实，有人说过我们正在临近代码的终结点。很快，代码就会自动产生出来，不需要再人工编写。程序员完全没用了，因为业务人员可以从规约直接生成程序。

这是不可能的！我们永远抛不掉代码，因为代码呈现了需求的细节。在某些层面上，这些细节无法被忽略或抽象，必须明确之。将需求明确到机器可以执行的细节程度，就是编程要做的事。而这种规约正是代码。

我期望语言的抽象程度继续提升。我也期望领域特定语言的数量继续增加。那会是好事一桩，但那终结不了代码。实际上，在较高层次上用领域特定语言撰写的规约也将是代码！它也得严谨、精确、规范和详细，好让机器理解和执行。

那帮以为代码终将消失的伙计，就像是巴望着发现一种无规范数学的数学家们一般。他们巴望着，总有一天能创造出某种机器，我们只要动动念头、嘴都不用张，就能叫它依计行事。那机器要能透彻理解我们，只有这样，它才能把含混不清的需求翻译为可完美执行的程序，精确满足需求。

这种事永远不会发生。即便是人类，倾其全部的直觉和创造力，也造不出满足客户模糊感觉的成功系统来。如果说需求规约原则教给了我们什么，那就是归置良好的需求就像代码一样正式，也能作为代码的可执行测试来使用。

记住，代码确然是我们最终用来表达需求的那种语言。我们可以创造各种与需求接近的语言。我们可以创造帮助把需求解析和汇整为正式结构的各种工具。然而，我们永远无法抛弃必要的精确性——所以代码永存。

1.2 糟糕的代码

最近我在读 Kent Beck 著《实现模式》（*Implementation Patterns*）[①]一书的序。他这样写道："……本书基于一种不太牢靠的前提：好代码的确重要……"这前提不牢靠？我反对！我

---

① [Beck07]。

认为这是该领域最强固、最受支持、最被强调的前提了（我想 Kent 也知道）。我们知道好代码重要，是因为其短缺实在困扰了我们太久。

20 世纪 80 年代末，有家公司写了一个很流行的杀手应用，许多专业人士都买来用。然后，发布周期开始拉长。缺陷总是不能修复。装载时间越来越久，崩溃的概率也越来越大。至今我还记得自己在某天沮丧地关掉那个程序，从此再不用它。在那之后不久，该公司就关门大吉了。

20 年后，我见到那家公司的一位早期雇员，问他当年发生了什么事。他的回答令我愈发恐惧起来。原来，当时他们赶着推出产品，代码写得乱七八糟。特性越加越多，代码也越来越烂，最后再也没法管理这些代码了。是糟糕的代码毁了这家公司。

你是否曾为糟糕的代码所深深困扰？如果你是位有点儿经验的程序员，必定多次遇到过这类困境。我们有专用来形容这事的词：沼泽（wading）。我们蹚过代码的水域。我们穿过灌木密布、瀑布暗藏的沼泽地。我们拼命想找到出路，期望有点儿什么线索能揭示到底发生了什么事；但目光所及，只是越来越多死气沉沉的代码。

你当然曾为糟糕的代码所困扰过。那么——为什么要写糟糕的代码呢？

是想快点儿完成吗？是要赶时间吗？有可能。或许你觉得自己要干好而所需的时间不够；假使花时间清理代码，老板就会大发雷霆。或许你只是不耐烦再搞这套程序，期望早点儿结束。或许你看了看自己承诺要做的其他事，意识到得赶紧弄完手上的东西，好接着做下一件工作。这种事我们都干过。

我们都曾经瞟一眼自己亲手造成的混乱，决定弃之而不顾，走向新一天。我们都曾经看到自己的烂程序居然能运行，然后断言能运行的烂程序总比什么都没有强。我们都曾经说过有朝一日再回头清理。当然，在那些日子里，我们都没听过勒布朗（LeBlanc）法则：稍后等于永不（Later equals never.）。

1.3　混乱的代价

只要你干过两三年编程，就有可能曾被某人的糟糕的代码绊倒过。如果你编程不止两三年，也有可能被这种代码拖过后腿，进度延缓的情况会很严重。有些团队在项目初期进展迅速，但有那么一两年的时间却慢如蜗行。对代码的每次修改都影响到其他两三处代码，修改无小事。每次添加或修改代码，都得对那堆扭纹柴了然于心，这样才能往上扔更多的扭纹柴。这团乱麻越来越大，再也无法理清，最后束手无策。

随着混乱的增加，团队生产力也持续下降，以致趋向于零。当生产力下降时，管理层就只有一件事可做了：增加更多人手到项目中，期望提升生产力。可是新人并不熟悉系统的设计。他们搞不清楚什么样的修改符合设计意图，什么样的修改违背设计意图。而且，他们以及团队中的其他人都背负着提升生产力的可怕压力。于是，他们只会制造更多的混乱，驱动生产力向零那端不断下降。如图 1-1 所示。

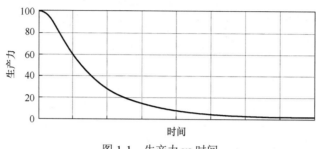

图 1-1　生产力 vs.时间

1.3.1　华丽新设计

最后，开发团队造反了，他们告诉管理层，再也无法在这令人生厌的代码基础上做开发了。他们要求做全新的设计。管理层不愿意投入资源完全重起炉灶，但他们也不能否认生产力低得可怕。他们只好同意开发者的要求，授权去做一套看上去很美的华丽新设计。

于是就组建了一支新军。谁都想加入这个团队，因为它是张白纸。他们可以重新来过，搞出点真正漂亮的东西来。但只有最优秀、最聪明的家伙被选中，其余人则继续维护现有系统。

现在有两支队伍在竞赛了。新团队必须搭建一套新系统，新系统要实现旧系统的所有功能，另外，还得跟上对旧系统的持续改动。在新系统功能足以抗衡旧系统之前，管理层不会替换掉旧系统。

竞赛可能会持续极长时间。我就见过延续了十年之久的。到了完成的时候，新团队的老成员早已不知去向，而现有成员则要求重新设计一套新系统，因为这套系统太烂了。

假使你经历过哪怕是一小段我谈到的这种事，那么你一定知道，花时间保持代码整洁不但关乎效率，还关乎生存。

1.3.2　态度

你可曾遇到过某种严重到要花数个星期来做本来只需数小时即可完成的事的混乱状况？你可曾见过本来只需做一行修改，结果却涉及上百个模块的情况？这种事太常见了。

怎么会发生这种事？为什么好代码会这么快就变质成糟糕的代码？理由多得很。我们抱怨需求变化背离了初期设计。我们哀叹进度太紧张，没法干好活。我们把问题归咎于那些愚蠢的经理、苛求的用户、没用的营销手段和那些电话消毒剂。不过，亲爱的呆伯特

（Dilbert）[1]，我们是自作自受[2]。我们太不专业了。

这话可不太中听。怎么会是自作自受呢？难道不关需求的事？难道不关进度的事？难道不关那些蠢经理和没用的营销手段的事？难道他们就不该负点责吗？

不。经理和营销人员指望从我们这里得到必需的信息，然后才能做出承诺和保证；即便他们没开口问，我们也不该羞于告知自己的想法。用户指望我们验证需求是否都在系统中实现了。项目经理指望我们遵守进度。我们与项目的规划脱不了干系，对失败负有极大的责任；特别是当失败与糟糕的代码有关时尤为如此！

"且慢！"你说。"不听经理的，我就会被炒鱿鱼。"多半不会。多数经理想要知道实情，即便他们看起来不喜欢实情。多数经理想要好代码，即便他们总是痴缠于进度。他们会奋力卫护进度和需求；那是他们该干的。你则当以同等的热情卫护代码。

再说明白些，假使你是位医生，病人请求你在给他做手术前别洗手，因为那会花太多时间，你会照办吗？[3]本该是病人说了算；但医生却绝对应该拒绝遵从。为什么？因为医生比病人更了解疾病和感染的风险。医生如果按病人说的办，就是一种不专业的态度（更别说是犯罪了）。

同理，程序员遵从不了解混乱风险的经理的意愿，也是不专业的做法。

1.3.3 谜题

程序员面临着一道基础价值谜题。有几年经验的开发者都知道，之前的混乱拖了自己的后腿。但开发者们背负期限的压力，只好制造混乱。简言之，他们没花时间让自己做得更快！

真正的专业人士明白，这道谜题的第二部分说错了。制造混乱无助于赶上期限。混乱只会立刻拖慢你，叫你错过期限。赶上期限的唯一方法——做得快的唯一方法——就是始终尽可能保持代码整洁。

1.3.4 整洁代码的艺术

假设你相信混乱的代码是祸首，假设你接受做得快的唯一方法是保持代码整洁的说法，你一定会自问："我怎么才能写出整洁的代码？"不过，如果你不明白整洁对代码有何意义，尝试去写整洁代码就毫无所益！

坏消息是，写整洁代码很像是绘画。多数人都知道一幅画是好还是坏，但能分辨优劣并

---

[1] 著名 IT 讽刺漫画。——译者注
[2] 原文为 But the fault, dear Dilbert, is not in our stars, but in ourselves.脱胎自莎士比亚戏剧《裘力斯·凯撒》第一幕第二场凯厄斯的台词 The fault, dear Brutus, is not in our stars, but in ourselves, that we are underlings.（若我们受人所制，亲爱的勃鲁托斯，那错也在我们身上，不能怪罪命运。）——译者注
[3] 1847 年 Ignaz Semmelweis（伊纳兹·塞麦尔维斯）提出医生应洗手的建议时，遭到了反对，人们认为医生太忙，接诊时无暇洗手。

不表示懂得绘画。能分辨整洁代码和肮脏代码，也不意味着会写整洁代码！

写整洁代码，需要遵循大量的小技巧，贯彻刻苦习得的"整洁感"。这种"代码感"就是关键所在。有些人生而有之，有些人费点儿劲才能得到。它不仅让我们看到代码的优劣，还予我们以借戒规之力化劣为优的攻略。

缺乏"代码感"的程序员，看混乱只是混乱，无处着手。有"代码感"的程序员能从混乱中看出其他的可能与变化。"代码感"帮助程序员选出最好的方案，并指导程序员制订修改行动计划，按图索骥。

简言之，编写整洁代码的程序员就像是艺术家，他能用一系列变换把一块白板变作由优雅代码构成的系统。

1.3.5 什么是整洁代码

有多少程序员，就有多少对整洁代码的定义。所以我只询问了一些非常知名且经验丰富的程序员。

Bjarne Stroustrup，C++语言发明者，《C++程序设计语言》（*C++ Programming Language*）一书作者。

> 我喜欢优雅和高效的代码。代码逻辑应当直截了当，令缺陷难以隐藏；尽量减少依赖关系，使之便于维护；依据某种分层战略完善错误处理代码；性能调至最优，省得引诱别人做没规矩的优化，搞出一堆混乱来。整洁的代码只做好一件事。

Bjarne 用了"优雅"一词。说得好！我 MacBook 上的词典提供了如下定义：外表或举止上令人愉悦的优美和雅观；令人愉悦的精致和简单。注意对"愉悦"一词的强调。Bjarne 显然认为整洁的代码读起来令人愉悦。读这种代码，就像见到手工精美的音乐盒或者设计精良的汽车一般，让你会心一笑。

Bjarne 也提到效率——而且两次提及。这话出自 C++发明者之口，或许并不出奇；不过我认为并非是在单纯追求速度。被浪费掉的运算周期并不雅观，也不令人愉悦。留意 Bjarne 怎么描述那种不雅观的结果，他用了"引诱"这个词。诚哉斯言。糟糕的代码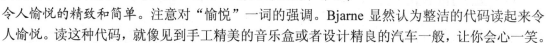引发混乱！别人修改糟糕的代码时，往往会越改越烂。

务实的 Dave Thomas 和 Andy Hunt 从另一角度阐述了这种情况，他们提到破窗理论：窗户破损了的建筑让人觉得似乎无人照管，于是无人再去关心。他们放任窗户继续破损，最终自己也参加破坏活动，在外墙上涂鸦，任由垃圾堆积。一扇破损的窗户开辟了大厦走向倾颓的道路。

Bjarne 也提到完善错误处理代码，往深处说就是在细节上花心思。敷衍了事的错误处理代码只是程序员忽视细节的一种表现。此外还有内存泄漏，还有静态条件代码，还有前后不一致的命名方式。结果就凸显出整洁代码对细节的重视。

　　Bjarne 以"整洁的代码只做好一件事"结束论断。毋庸置疑，软件设计的许多原则最终都会归结为这句警语。有那么多人发表过类似的言论。糟糕的代码想做太多事，它意图混乱、目的含混。而整洁的代码力求集中，每个函数、每个类和每个模块都全神贯注于一事，完全不受四周细节的干扰和污染。

Grady Booch，《面向对象分析与设计》（*Object Oriented Analysis and Design with Applications*）一书作者

　　整洁的代码简单直接。整洁的代码如同优美的散文。整洁的代码从不隐藏设计者的意图，充满了干净利落的抽象和直截了当的控制语句。

　　Grady 的观点与 Bjarne 的观点有类似之处，但他从可读性的角度来定义。我特别喜欢"整洁的代码如同优美的散文"这种看法。想想你读过的某本好书。回忆一下，那些文字如何在脑中形成影像？就像是看了场电影，对吧？还不止！你还看到那些人物，听到那些声音，体验到那些喜怒哀乐。

　　阅读整洁的代码和阅读《指环王》（*Lord of the Rings*）自然不同。不过，仍有可类比之处。如一本好的小说一般，整洁的代码应当明确地展现出要解决问题的张力，它应当将这种张力推至高潮，以某种显而易见的方案解决问题，使读者发出"啊哈！本当如此！"的感叹。

　　窃以为 Grady 所谓"干净利落的抽象"（crisp abstraction），乃是绝妙的矛盾修辞法。毕竟 crisp 几乎就是"具体"（concrete）的同义词。我 MacBook 上的词典这样定义 crisp 一词：果断决绝，就事论事，没有犹豫或不必要的细节。尽管有两种不同的定义，但是该词还是承载了有力的信息。代码应当讲述事实，不引人猜测。它只该包含必需之物。读者应当感受到我们的果断决绝。

"老大"Dave Thomas，OTI 公司创始人，Eclipse 战略"教父"

　　整洁的代码应可由作者之外的开发者阅读和增补。它应当有单元测试和验收测试。它使用有意义的命名。它只提供一种而非多种做一件事的途径。它只有尽量少的依赖关系，而且要明确地定义和提供清晰的、尽量少的 API。代码应通过其字面表达含义，因为不同的语言导致并非所有必需的信息均可通过代码自身清晰表达。

Dave 老大在可读性上和 Grady 持相同观点,但有一个重要的不同之处。Dave 断言,整洁的代码便于其他人予以增补。这看似显而易见,但亦不可过分强调。毕竟易读的代码和易修改的代码之间还是有区别的。

Dave 将整洁系于测试之上!要在十年之前,这会让人大跌眼镜。但测试驱动开发(Test Driven Development)已在行业中造成了深远影响,成为基础规程之一。Dave 说得对,没有测试的代码不干净。不管它有多优雅,也不管有多可读、多易理解,微乎测试,其不洁亦可知也。

Dave 两次提及"尽量少"。显然,他推崇小块的代码。实际上,从有软件起人们就在反复强调这一点,越小越好。

Dave 也提到,代码应在字面上表达其含义。这一观点源自 Knuth 的"字面编程"(literate programming)①。结论就是应当用人类可读的方式来写代码。

Michael Feathers,《修改代码的艺术》(Working Effectively with Legacy Code)一书作者

> 我可以列出我留意到的整洁代码的所有特点,但其中有一条是根本性的。整洁的代码总是看起来像是某位特别在意它的人写的。几乎没有改进的余地。代码的作者什么都想到了,如果你企图改进它,总会回到原点,赞叹某人留给你的代码——全心投入的某人留下的代码。

一言以蔽之:在意。这就是本书的题旨所在。或许该加个副标题,如何在意代码。

Michael 一针见血。整洁代码就是作者着力照料的代码。有人曾花时间让它保持简单有序。他们适当地关注到了细节,他们在意过。

Ron Jeffries,《极限编程实施》(Extreme Programming Installed)以及《C#极限编程探险》(Extreme Programming Adventures in C#)作者

Ron 初入行就在战略空军司令部(Strategic Air Command)编写 Fortran 程序,此后几乎在每种机器上编写过每种语言的代码。他的言论值得咀嚼。

> 近年来,我开始研究贝克的简单代码规则,差不多也都琢磨透了。简单代码,依其重要顺序:
> - 能通过所有测试;

---

① [Knuth92]。

- 没有重复代码；
- 体现系统中的全部设计理念；
- 包括尽量少的实体，比如类、方法、函数等。

在以上诸项中，我最在意代码重复。如果同一段代码反复出现，就表示某种想法未在代码中得到良好的体现。我尽力去找出那到底是什么，然后再尽力将其更清晰地表达出来。

在我看来，有意义的命名是体现表达力的一种方式，我往往会修改好几次才会定下名字来。借助 Eclipse 这样的现代编码工具，重命名代价极低，所以我无所顾忌。然而，表达力还不只体现在命名上。我也会检查对象或方法是否想做的事太多。如果对象功能太多，最好切分为两个或多个对象。如果方法功能太多，我总是使用抽取手段（Extract Method）重构之，从而得到一个能较为清晰地说明其自身功能的方法，以及另外数个说明如何实现这些功能的方法。

消除重复和提高表达力让我在整洁代码方面获益良多，只要铭记这两点，改进脏代码时就会大为改观。不过，我时常关注的另一规则就不太好解释了。

这么多年下来，我发现所有程序都由极为相似的元素构成。例如"在集合中查找某物"。不管是雇员记录数据库还是键值对哈希表，或者某类条目的数组，我们都会发现自己想要从集合中找到某一特定条目。一旦出现这种情况，我通常会把实现手段封装到更抽象的方法或类中。这样做好处多多。

可以先用某种简单的手段，比如哈希表来实现这一功能，由于对搜索功能的引用指向了我那个小小的抽象，因此能随需应变，修改实现手段。这样既能快速前进，又能为未来的修改预留余地。

另外，该集合抽象常常提醒我留意"真正"在发生的事，避免随意实现集合行为，因为我真正需要的不过是某种简单的查找手段。

减少重复代码，提高表达力，提早构建简单抽象。这就是我写整洁代码的方法。

Ron 以寥寥数段文字概括了本书的全部内容。不要重复代码，只做一件事——表达力，小规模抽象。该有的都有了。

Ward Cunningham，Wiki 发明者，极限编程（eXtreme Programming）的创始人之一，Smalltalk 语言和面向对象的思想领袖。所有在意代码者的"教父"。

如果每个例程都让你感到深合己意，那就是整洁代码。如果代码让编程语言看起来像是专为解决那个问题而存在的，就可以称之为漂亮的代码。

这种说法很 Ward。它使你听了之后就点头，然后继续听下去。如此在理，如此浅显，绝不故作高深。你大概以为此言深合己意吧。再走近点看看。

"……深合己意"。你最近一次看到深合己意的模块是什么时候？模块多半都繁复难解吧？难道没有触犯规则吗？你不是也曾挣扎着想抓住一些从整个系统中散落而出的线索，编织进你在读的那个模块吗？你最近一次读到某段代码，并且如同对 Ward 的说法点头一般对这段代码点头，是什么时候的事了？

Ward 期望你不会为整洁代码所震惊。你无须花太多力气，那么代码就是深合你意的。它明确、简单、有力。每个模块都为下一个模块做好准备。每个模块都告诉你下一个模块会是怎样的。整洁的程序好到你根本不会注意到它。设计者把它做得像一切其他设计般简单。

那 Ward 有关"美"的说法又如何呢？我们都曾面临语言不是为要解决的问题而设计的困境，但 Ward 的说法又把球踢回我们这边。他说，漂亮的代码让编程语言像是专为解决那个问题而存在！所以，让语言变得简单的责任就在我们身上了！当心，语言是冥顽不化的！是程序员让语言显得简单。

1.4 思想流派

我（"鲍勃大叔"）又是怎么想的呢？在我眼中整洁代码是什么样的？本书将以详细到吓死人的程度告诉你，我和我的同道对整洁代码的看法。我们会告诉你关于整洁变量名的想法，关于整洁函数的想法，关于整洁类的想法，等等。我们视这些观点为当然，且不为其逆耳而致歉。对我们而言，在职业生涯的这个阶段，这些观点确属当然，也是我们整洁代码派的圭旨。

武术家从不认同所谓最好的武术，也不认同所谓绝招。武术大师们常常创建自己的流派，聚徒而授。因此我们才看到格雷西家族在巴西开创并传授的格雷西柔术（Gracie Jiu Jistu），看到奥山龙峰（Okuyama Ryuho）在东京开创并传授的八光流柔术（Hakkoryu Jiu Jistu），看到李小龙（Bruce Lee）在美国开创并传授的截拳道（Jeet Kune Do）。

弟子们沉浸于创始人的授业。他们全心师从某位师傅，排斥其他师傅。弟子有所成就后，可以转投另一位师傅，扩展自己的知识与技能。有些弟子最终百炼成钢，创出新招数，开宗立派。

任何门派都并非绝对正确。不过，身处某一门派时，我们总以其所传之技为善。归根结底，练习八光流柔术或截拳道，自有其善法，但这并不能否定其他门派所授之法。

可以把本书看作是对象导师（Object Mentor）[1]整洁代码派的说明。书中要传授的就是我们勤操己艺的方法。如果你遵从这些教诲，你就会如我们一般乐受其益，你将学会如何编写整洁而专业的代码。但无论如何也别错以为我们是"正确的"。其他门派和师傅同我们一样专业。你有必要也向他们学习。

实际上，书中很多建议都存在争议。或许你并不完全同意这些建议，甚至可能会强烈反对其中的一些建议。这样挺好的。我们不能要求做最终权威。另外，书中列出的建议，乃是我们长久苦思，从数十年的从业经验和无数尝试与错误中得来的。无论你同意与否，如果你没看到或是不尊重我们的观点，就真该自己害臊。

1.5 我们是作者

Javadoc 中的@author 字段告诉我们自己是什么人。我们是作者。作者都有读者。实际上，作者有责任与读者做良好沟通。下次你写代码的时候，记得自己是作者，要为评判你工作的读者写代码。

你或许会问：代码真正"读"的成分有多少呢？难道主要力量不是应该用在"写"上吗？

你是否玩过"编辑器回放"？20 世纪 80、90 年代，Emac 之类的编辑器可以记录每次的击键动作。你可以在工作一小时之后，回放击键过程，就像是看一部高速电影。我这么做过，结果很有趣。

回放过程显示，多数时间都是在滚动屏幕，浏览其他模块！

> 鲍勃进入模块。
> 他向下滚动到要修改的函数。
> 他停下来考虑可以做什么。
> 哦，他滚动到模块顶端，检查变量初始化。
> 现在他回到修改处，开始键入。
> 喔，他删掉了键入的内容。
> 他重新键入。
> 他又删除了！
> 他键入了一半的东西，又被删除掉。
> 他滚动到调用要修改函数的另一函数，看看是怎么调用的。
> 他回到修改处，重新键入刚才删掉的代码。
> 他停下来。
> 他再一次删掉代码！

---

[1] 本书主要作者 Robert C.Martin 开办的技术咨询和培训公司。——译者注

他打开另一个窗口，查看别的子类。那是个复载函数吗？

……

你该明白了。花费读与写时间的比例超过 10∶1。写新代码时，我们一直在读旧代码。

既然比例如此之高，我们就想让读的过程变得轻松，即便那会使编写过程更难。不可能光写不读，所以使之易读实际也就是使之易写。

这事概无例外。不读周边代码的话就没法写代码。编写代码的难度，取决于读周边代码的难度。要想干得快，要想早点做完，要想轻松写代码，先让代码易读吧。

1.6 童子军军规

光把代码写好可不够。必须时时保持代码整洁。我们都见过代码随时间流逝而腐坏。我们应当更积极地阻止腐坏的发生。

借用美国童子军一条简单的军规，应用到我们的专业领域：

让营地比你来时更干净。①

如果每次签入时，代码都比签出时干净，那么代码就不会腐坏。清理并不一定要花多少功夫，也许只是改好一个变量名，拆分一个有点过长的函数，消除一点点重复代码，清理一个嵌套 if 语句。

你想要为一个代码随时间流逝而越变越好的项目工作吗？你还能相信有其他更专业的做法吗？难道持续改进不是专业性的内在组成部分吗？

1.7 前传与原则

从许多角度看，本书都是我 2002 年写的那本《敏捷软件开发：原则、模式与实践》（*Agile Software Development: Principles, Patterns, and Practices*，简称 PPP）的"前传"。PPP 关注面向对象设计的原则，以及专业开发者采用的许多实践方法。假如你没读过 PPP，你会发现它像本书的延续。如果你读过，你会发现那本书的主张在代码层面于本书中回响。

在本书中，你会发现对不同设计原则的引用，包括单一权责原则（Single Responsibility Principle，SRP）、开放闭合原则（Open Closed Principle，OCP）和依赖倒置原则（Dependency Inversion Principle，DIP）等。

---

① 摘自 Robert Stephenson Smyth Baden-Powell（英国人，童子军创始者）对童子军的遗言："努力，让世界比你来时干净些……"。

1.8 小结

艺术书并不保证你读过之后能成为艺术家,只能告诉你其他艺术家用过的工具、技术和思维过程。本书同样也不担保让你成为好程序员。它不担保能给你"代码感"。它所能做的,只是展示好程序员的思维过程,还有他们使用的技巧、技术和工具。

和艺术书一样,本书也充满了细节。代码会很多,你会看到好代码,也会看到糟糕的代码。你会看到糟糕的代码如何转化为好代码。你会看到启发、规条和技巧的列表。你会看到一个又一个例子。但最终结果取决于你自己。

还记得那个关于小提琴家在去表演的路上迷路的老笑话吗?他在街角拦住一位长者,问他怎么才能去卡耐基音乐厅(Carnegie Hall)。长者看了看小提琴家,又看了看他手中的琴,说道:"你还得练,孩子,还得练!"

1.9 文献

[Beck07]:*Implementation Patterns*, Kent Beck, Addison-Wesley, 2007.

[Knuth92]:*Literate Programming*, Donald E. Knuth, Center for the Study of Language and Information, Leland Stanford Junior University, 1992.

第 2 章

有意义的命名

Tim Ottinger

2.1 介绍

软件中随处可见命名。我们给变量、函数、参数、类和封包命名。我们给源代码及源代码所在目录命名。我们给 jar 文件、war 文件和 ear 文件命名。我们命名、命名，不断命名。既然有这么多命名要做，不妨做好它。本章列出了起个好名字应遵从的几条简单规则。

2.2 名副其实

名副其实说起来简单。我们想要强调，这事很严肃。选个好名字要花时间，但省下来的时间比花掉的多。注意命名，而且一旦发现有更好的名称，就换掉旧的。这么做，读你代码的人（包括你自己）都会更开心。

变量、函数或类的名称应该已经答复了所有的大问题。它该告诉你，它为什么会存在，它做什么事，应该怎么用。如果名称需要注释来补充，那就不算是名副其实。

```
int d;  // elapsed time in days
```

名称 d 什么也没说明。它没有引起读者对时间消逝的感觉，更别说以日计了。我们应该选择指明了计量对象和计量单位的名称：

```
int elapsedTimeInDays;
int daysSinceCreation;
int daysSinceModification;
int fileAgeInDays;
```

选择体现本意的名称能让人更容易理解和修改代码。下列代码的目的何在？

```
public List<int[]> getThem() {
  List<int[]> list1 = new ArrayList<int[]>();
  for (int[] x : theList)
    if (x[0] == 4)
      list1.add(x);
  return list1;
}
```

为什么难以说明上述代码要做什么事？里面并没有复杂的表达式，空格和缩进中规中矩，只用到三个变量和两个常量，甚至没有涉及任何其他类或多态方法，只是（或者看起来是）一个数组的列表而已。

问题不在于代码的简洁度，而在于代码的模糊度：即上下文在代码中未被明确体现的程度。上述代码要求我们了解类似以下问题的答案：

（1）theList 中是什么类型的东西？
（2）theList 零下标条目的意义是什么？
（3）值 4 的意义是什么？
（4）我怎么使用返回的列表？

问题的答案没体现在代码段中，可代码段就是它们该在的地方。比方说，我们在开发一种扫雷游戏，我们发现，盘面是名为 `theList` 的单元格列表，那就将其名称改为 `gameBoard`。

盘面上每个单元格都用一个简单数组表示。我们还发现，零下标条目是一种状态值，而该种状态值为 4 表示"已标记"。只要改为有意义的名称，代码就会得到相当程度的改进：

```java
public List<int[]> getFlaggedCells() {
  List<int[]> flaggedCells = new ArrayList<int[]>();
  for (int[] cell : gameBoard)
    if (cell[STATUS_VALUE] == FLAGGED)
      flaggedCells.add(cell);
  return flaggedCells;
}
```

注意，代码的简洁性并未被触及。运算符和常量的数量全然保持不变，嵌套数量也全然保持不变，但代码变得明确多了。

还可以更进一步，不用 `int` 数组表示单元格，而是另写一个类。该类包括一个名副其实的函数（称为 `isFlagged`），从而掩盖住那个魔术数[①]。于是得到函数的新版本：

```java
public List<Cell> getFlaggedCells() {
  List<Cell> flaggedCells = new ArrayList<Cell>();
  for (Cell cell : gameBoard)
    if (cell.isFlagged())
      flaggedCells.add(cell);
  return flaggedCells;
}
```

只要简单改一下名称，就能轻易知道发生了什么。这就是选用好名称的力量。

2.3 避免误导

程序员必须避免留下掩藏代码本意的错误线索。应当避免使用与本意相悖的词，例如，`hp`、`aix` 和 `sco` 都不该用作变量名，因为它们都是 Unix 平台或类 Unix 平台的专有名称。即便你是在编写三角计算程序，`hp` 看起来是一个不错的缩写[②]，但那也可能会提供错误信息。

别用 `accountList` 来指称一组账号，除非它真的是 `List` 类型。`List` 一词对程序员有特殊意义。如果包纳账号的容器并非真是一个 `List`，就会引起错误的判断[③]。所以，用 `accountGroup` 或 `bunchOfAccounts`，甚至直接用 `accounts` 都会好一些。

提防使用外形相似度较高的名称。例如，想区分模块中某处的 `XYZControllerFor-EfficientHandlingOfStrings` 和另一处的 `XYZControllerForEfficientStorage-OfStrings`，会花多长时间呢？这两个词的外形实在太相似了。

[①] 即表示已标记的 4。——译者注
[②] 即 hypotenuse 的缩写。——译者注
[③] 如后文提到的，即便容器就是一个 `List`，最好也别在名称中写出容器类型名。

以同样的方式拼写出同样的概念才是信息。拼写前后不一致就是**误导**。我们很享受现代Java编程环境的自动代码完成特性。键入某个名称的前几个字母，按一下某个热键组合（如果有的话），就能得到一列该名称的可能形式。假如相似的名称依字母顺序放在一起，且差异很明显，那就会相当有助益，因为程序员多半会压根不看你的详细注释，甚至不看该类的方法列表就直接看名字挑一个对象。

误导性名称真正可怕的例子，是用小写字母 l 和大写字母 O 作为变量名，尤其是在组合使用的时候。当然，问题在于它们看起来完全像是常量"壹"和"零"。

```
int a = l;
if (O == l)
  a = Ol;
else
  l = Ol;
```

读者可能会认为这纯属虚构，但我们确曾见过充斥这类名称的代码。有一次，代码作者建议用不同字体写变量名，好显得更清楚些，但前提是这种方案得要通过口头和书面传递给未来所有的开发者才行。后来，只是做了简单的重命名操作，就解决了问题，而且也没引起别的问题。

2.4 做有意义的区分

如果程序员只是为满足编译器或解释器的需要而写代码，就会制造麻烦。例如，因为同一作用范围内两样不同的东西不能重名，你可能会随手改掉其中一个的名称，有时干脆以错误的拼写充数，结果就会出现在更正拼写错误后导致编译器出错的情况。①

光是添加数字系列或是废话远远不够，即便这足以让编译器满意。如果名称必须相异，那么其意思也应该不同才对。

以数字系列命名（a1、a2…aN）是依义命名的对立面。这样的名称纯属误导——完全没有提供正确信息，没有提供导向作者意图的线索。试看：

```
public static void copyChars(char a1[], char a2[]) {
  for (int i = 0; i < a1.length; i++) {
    a2[i] = a1[i];
  }
}
```

如果参数名改为 source 和 destination，这个函数就会像样许多。

---

① 例如，就因为 class 已有他用，就给一个变量命名为 klass，这真是可怕的做法。

废话是另一种没意义的区分。假设你有一个 `Product` 类，如果还有一个名为 `ProductInfo` 或 `ProductData` 的类，那它们的名称虽然不同，意思却无区别。`Info` 和 `Data` 就像 `a`、`an` 和 `the` 一样，是意义含混的废话。

注意，只要体现出有意义的区分，使用 `a` 和 `the` 这样的前缀就没错。例如，你可能把 `a` 用在域内变量，而把 `the` 用于函数参数[①]。但如果你已经有一个名为 `zork` 的变量，又想调用一个名为 `theZork` 的变量，麻烦就来了。

废话都是冗余。`variable` 一词永远不应当出现在变量名中。`table` 一词永远不应当出现在表名中。`NameString` 会比 `Name` 好吗？难道 `Name` 会是一个浮点数？如果是这样，就违反了关于误导的规则。设想有一个名为 `Customer` 的类，还有一个名为 `CustomerObject` 的类，它们的区别何在呢？哪一个是表示客户历史支付情况的最佳方式？

有一个应用反映了这种状况。为当事者讳，我们改了一下，不过犯错的代码的确就是这个样子：

```
getActiveAccount();
getActiveAccounts();
getActiveAccountInfo();
```

程序员怎么知道该调用哪个函数呢？

如果缺少明确约定，那么变量 `moneyAmount` 与 `money` 就没区别，`customerInfo` 与 `customer` 没区别，`accountData` 与 `account` 没区别，`theMessage` 也与 `message` 没区别。要区分名称，就要以读者能鉴别不同之处的方式来区分。

2.5 使用读得出来的名称

人类长于记忆和使用单词。大脑的相当一部分就是用来容纳和处理单词的。单词能读得出来。人类的大脑中有那么大的一块地方用来处理言语，若不善加利用，实在是种耻辱。

如果名称读不出来，讨论的时候就会像个傻鸟。"哎，这儿，鼻涕阿三喜摁踢（bee cee arr three cee enn tee）[②]上头，有个皮挨死极翘（pee ess zee kyew）[③]整数，看见没？"这不是小事，因为编程本就是一种社会活动。

有一家公司，程序里面写了一个 `genymdhms`（生成日期，年、月、日、时、分、秒），他们一般读作"gen why emm dee aich emm ess"[④]。我有见字照拼读的恶习，于是开口就念"gen-yah-mudda-hims"。后来好些设计师和分析师都有样学样，听起来傻乎乎的。我们知道

[①] "鲍勃大叔"惯于在C++中这样做，但后来放弃了，因为现代 IDE 使这种做法变得没必要了。
[②] BCR3CNT 的逐个字母读音。——译者注
[③] PSZQ 的逐个字母读音。——译者注
[④] YMDHMS 的逐个字母读音。——译者注

典故，所以会觉得很搞笑。搞笑归搞笑，实际是在强忍糟糕的命名。在给新开发者解释变量名的意义时，他们总是读出傻乎乎的自造词，而非恰当的英语词。比较

```
class DtaRcrd102 {
  private Date genymdhms;
  private Date modymdhms;
  private final String pszqint = "102";
  /* ... */
};
```

和

```
class Customer {
  private Date generationTimestamp;
  private Date modificationTimestamp;
  private final String recordId = "102";
  /* ... */
};
```

现在读起来就像人话了："喂，Mikey，看看这条记录！生成时间戳（generation timestamp）[①]被设置为明天了！不能这样吧？"

2.6　使用可搜索的名称

对于单字母名称和数字常量，有一个问题，就是很难在一大篇文字中找出来。

找 `MAX_CLASSES_PER_STUDENT` 很容易，但想找数字 7 就麻烦了，它可能是某些文件名或其他常量定义的一部分，出现在因不同意图而采用的各种表达式中。如果该常量是个长数字，又被人错改过，就会逃过搜索，从而造成错误。

同样，e 也不是一个便于搜索的好变量名，它是英文中最常用的字母，在每个程序、每段代码中都有可能出现。由此而见，长名称胜于短名称，搜得到的名称胜于用自造编码代写就的名称。

窃以为单字母名称仅用于短方法中的本地变量。名称长短应与其作用域大小相对应 [N5]。若变量或常量可能在代码中多处使用，则应赋予其便于搜索的名称。再比较

```
for (int j=0; j<34; j++) {
  s += (t[j]*4)/5;
}
```

和

```
int realDaysPerIdealDay = 4;
const int WORK_DAYS_PER_WEEK = 5;
int sum = 0;
```

---

① 读到 generation timestamp 时，立刻就能与代码中的 generationTimestamp 变量对应上。——译者注

```
for (int j=0; j < NUMBER_OF_TASKS; j++) {
  int realTaskDays = taskEstimate[j] * realDaysPerIdealDay;
  int realTaskWeeks = (realTaskdays / WORK_DAYS_PER_WEEK);
  sum += realTaskWeeks;
}
```

注意，上面代码中的 sum 并非特别有用的名称，不过至少搜得到它。采用能表达意图的名称，貌似拉长了函数代码，但要想想看，WORK_DAYS_PER_WEEK 比数字 5 好找得多，而列表中也只剩下了体现作者意图的名称。

2.7 避免使用编码

编码已经太多，无谓再自找麻烦。把类型或作用域编进名称里面，徒然增加了解码的负担。没理由要求每位新人都在弄清要应付的代码之外（那算是正常的），还要再搞懂另一种编码"语言"。这对解决问题而言，纯属多余的负担。带编码的名称通常也不便发音，容易打错。

2.7.1 匈牙利语标记法

在往昔名称长短很重要的时代，我们毫无必要地破坏了不编码的规矩，如今后悔不迭。Fortran 语言要求首字母体现出类型，导致了编码的产生。BASIC 语言的早期版本只允许使用一个字母再加上一位数字。匈牙利语标记法[①]（Hungarian Notation，HN）将这种态势愈演愈烈。

在 Windows 的 C 语言 API 的时代，HN 相当重要，那时所有名称要么是一个整数句柄，要么是一个长指针或者 void 指针，要不然就是 string 的几种实现（有不同的用途和属性）之一。那时候编译器并不做类型检查，程序员需要匈牙利语标记法来帮助自己记住类型。

现代编程语言具有更丰富的类型系统，编译器也记得并强制使用类型。而且，程序员趋向于使用更小的类、更短的方法，好让每个变量的定义都在视野范围之内。

Java 程序员不需要类型编码，因为对象是强类型的，代码编辑环境已经先进到在编译开始前就能监测到类型错误的程度！所以，如今 HN 和其他类型编码形式都纯属多余。它们增加了修改变量、函数或类的名称或类型的难度，它们增加了阅读代码的难度，它们制造了让编码系统误导读者的可能性。

```
PhoneNumber phoneString;
// name not changed when type changed!
```

[①] Charles Simonyi 在微软任首席架构师时推广了这种命名法。Simonyi 是匈牙利人，在匈牙利语中，姓置于名前。匈牙利语命名法明确写出类型，并将类型置于实际名称前，以大写字母间隔，就像匈牙利语姓名的排列一般。例如，bBusy 代表一个类型为 boolean、名称为 Busy 的变量。——译者注

2.7.2 成员前缀

也不必用 m_ 前缀来标明成员变量。应当把类和函数做得足够小，以消除对成员前缀的需要。你应当使用某种可以高亮或用颜色标出成员的编辑环境。

```
public class Part {
  private String m_dsc; // The textual description
  void setName(String name) {
    m_dsc = name;
  }
}
_____
public class Part {
  String description;
  void setDescription(String description) {
    this.description = description;
  }
}
```

此外，人们会很快学会无视前缀（或后缀），而只看到名称中有意义的部分。代码读得越多，眼中就越没有前缀。最终，前缀变作了不入法眼的废料，变作了旧代码的标志物。

2.7.3 接口和实现

有时也会出现采用编码的特殊情形。比如，你在做一个创建形状用的抽象工厂（Abstract Factory），该工厂是一个接口，要用具体类来实现。你怎么来命名工厂和具体类呢？`IShapeFactory` 和 `ShapeFactory` 吗？我喜欢不加修饰的接口。前导字母 `I` 被滥用到了说好听点儿是干扰，说难听点儿根本就是废话的程度。我不想让用户知道我给他们的是接口，而就想让他们知道那是一个 `ShapeFactory`。如果在接口和实现中必须选其一来编码的话，我宁肯选择实现。`ShapeFactoryImp`，甚至是丑陋的 `CShapeFactory`，都比对接口名称编码好。

2.8 避免思维映射

不应当让读者在脑中把你的名称翻译为他们熟知的名称，这种问题经常出现在选择是使用问题领域术语还是解决方案领域术语时。

单字母变量名就是个问题。在作用域较小、没有名称冲突时，循环计数器自然有可能被命名为 i、j 或 k。（但千万别用字母 l！）这是因为传统上惯用单字母名称做循环计数器。然而，在多数其他情况下，单字母名称不是个好选择；读者必须在脑中将它映射为真实概念。仅仅是因为有了 a 和 b，就要起名为 c，这实在不是像样的理由。

程序员通常都是聪明人。聪明人有时会借脑筋急转弯炫耀其聪明。总而言之，假使你记得 r 代表不包含主机名和模式（scheme）的小写字母版 url 的话，那你真是太聪明了。

聪明程序员和专业程序员之间的区别在于，专业程序员了解，明确是王道。专业程序员善用其能，编写其他人能理解的代码。

2.9 类名

类名和对象名应该是名词或名词短语，例如 Customer、WikiPage、Account 或 AddressParser。应避免使用 Manager、Processor、Data 或 Info 这样的类名。类名不应当是动词。

2.10 方法名

方法名应当是动词或动词短语，如 postPayment、deletePage 或 save。属性访问器（accessor）、修改器（mutator）和断言（predicate）应该根据其值命名，并依 Javabean 标准加上前缀 get、set 和 is。

```
string name = employee.getName();
customer.setName("mike");
if (paycheck.isPosted())...
```

重载构造器时，使用描述了参数的静态工厂方法名。例如，

```
Complex fulcrumPoint = Complex.FromRealNumber(23.0);
```

通常好于

```
Complex fulcrumPoint = new Complex(23.0);
```

可以考虑将相应的构造器设置为 private，强制使用这种命名手段。

2.11 别抖机灵

如果名称太耍宝，那就只有同作者一般有幽默感的人才能记得住，而且还是在他们记得那

个笑话的时候才行。谁会知道名为 `HolyHand-Grenade`①的函数是用来做什么的呢？没错，这名字挺有趣，不过 `DeleteItems`②或许是更好的名称。宁可明确，毋为好玩。

抖机灵在代码中经常体现为使用俗话或俚语。例如，别用 `whack()`③来表示 `kill()`。别用 `eatMyShorts()`④这类与文化紧密相关的笑话来表示 `abort()`。

言到意到。意到言到。

2.12 每个概念对应一个词

给每个抽象概念选一个词，并且一以贯之。例如，使用 `fetch`、`retrieve` 和 `get` 来给在多个类中的同种方法命名。你怎么记得住是哪个类中的哪个方法呢？很悲哀，你总得记住编写库或类的公司、机构或个人，才能想得起来用的是哪个术语。否则，就得耗费大把时间浏览各个文件头及前面的代码。

Eclipse 和 IntelliJ 之类现代编程环境提供了与环境相关的线索，比如某个对象能调用的方法列表。不过要注意，列表中通常不会给出你为函数名和参数列表编写的注释。如果参数名称来自函数声明，你就太幸运了。函数名称应当独一无二，而且要保持一致，这样你才能不借助多余的浏览就找到正确的方法。

同样，如果在同一堆代码中有 `controller`，有 `manager`，还有 `driver`，就会令人困惑。`DeviceManager` 和 `Protocol Controller` 之间有何根本区别？为什么不全用 `controller` 或 `manager` 呢？它们都是 `Drivers` 吗？这种名称，让人觉得这两个对象是不同类型的，也分属不同的类。

对于那些会用到你代码的程序员，一以贯之的命名法简直就是天降福音。

2.13 别用双关语

避免将同一单词用于不同目的。同一术语用于不同概念，基本上就是双关语了。如果遵

---

① 即"圣手雷"。在英国喜剧团体 Monty Python 的电影《巨蟒与圣杯》（*Monty Python and the Holy Grail*）中亚瑟王使用"圣手雷"（影射象征英国王权的权珠）炸死了一只凶恶的兔子。——译者注
② 意为"删除条目"。——译者注
③ 美国俚语，劈砍。——译者注
④ 美国俚语，去死吧。——译者注

循"一词一义"规则，可能在好多个类里面都会有 add 方法。只要这些 add 方法的参数列表和返回值在语义上等价，就一切顺利。

但是，可能会有人决定为"保持一致"而使用 add 这个词来命名，即便并非真的想表示这种意思。比如，在多个类中都有 add 方法，该方法通过增加或连接两个现存值来获得新值。假设要写一个新类，该类中有一个方法，把单个参数放到群集（collection）中。该把这个方法叫作 add 吗？这样做貌似和其他 add 方法保持了一致，但实际上语义却不同，应该用 insert 或 append 之类的词来命名才对。把该方法命名为 add，就是双关语了。

代码作者应尽力写出易于理解的代码。我们想把代码写得让别人能一目了然，而不必殚精竭虑地研究。我们想要那种大众化的作者尽责写清楚的平装书模式，而不想要那种学者挖地三尺才能明白个中意义的学院派模式。

2.14 使用解决方案领域名称

记住，只有程序员才会读你的代码。所以，尽管用那些计算机科学（Computer Science，CS）术语、算法名、模式名、数学术语吧。依据问题所涉领域来命名可不算是聪明的做法，因为不该让协作者老是跑去问客户每个名称的含义，其实他们早该通过另一名称了解这个概念了。

对于熟悉访问者（VISITOR）模式的程序员来说，名称 AccountVisitor 富有意义。哪个程序员会不知道 JobQueue 的意思呢？程序员要做太多技术性工作，给这些事起个技术性的名称，通常是最靠谱的做法。

2.15 使用源自所涉问题领域的名称

如果不能用程序员熟悉的术语来给手头的工作命名，就采用从所涉问题领域而来的名称吧。至少，负责维护代码的程序员就能去请教领域专家了。

优秀的程序员和设计师，其工作之一就是分离解决方案领域和问题领域的概念。与所涉问题领域更为贴近的代码，应当采用源自问题领域的名称。

2.16 添加有意义的语境

很少有名称是能自我说明的——多数都不能。反之，你需要用命名良好的类、函数或名称空间来放置名称，给读者提供语境。如果没这么做，给名称添加前缀就是最后一招了。

2.16 添加有意义的语境

设想你有名为 firstName、lastName、street、houseNumber、city、state 和 zipcode 的变量。把它们放一块儿的时候，很明显构成了一个地址。但是，假使只是在某个方法中看见孤零零的一个 state 变量呢？你会理所当然地推断那是某个地址的一部分吗？

可以添加前缀 addrFirstName、addrLastName、addrState 等，以此提供语境。至少，读者会明白这些变量是某个更大结构的一部分。当然，更好的方案是创建名为 Address 的类。这样，即便是编译器也会知道这些变量隶属某个更大的概念了。

看看代码清单 2-1 中的方法，以下变量是否需要更有意义的语境呢？函数名仅给出了部分语境；算法提供了剩下的部分。遍览函数后，你会知道 number、verb 和 pluralModifier 这 3 个变量是"测估"信息的一部分。不幸的是这样的语境得靠读者推断出来。第一眼看到这个方法时，这些变量的含义完全不清楚。

代码清单 2-1　语境不明确的变量

```
private void printGuessStatistics(char candidate, int count) {
  String number;
  String verb;
  String pluralModifier;
  if (count == 0) {
    number = "no";
    verb = "are";
    pluralModifier = "s";
  } else if (count == 1) {
    number = "1";
    verb = "is";
    pluralModifier = "";
  } else {
    number = Integer.toString(count);
    verb = "are";
    pluralModifier = "s";
  }
  String guessMessage = String.format(
    "There %s %s %s%s", verb, number, candidate, pluralModifier
  );
  print(guessMessage);
}
```

上述函数有点儿过长，变量的使用贯穿始终。要分解这个函数，需要创建一个名为 GuessStatisticsMessage 的类，把 3 个变量做成该类的成员字段，这样它们就在定义上变作了 GuessStatisticsMessage 的一部分。语境的增强也让算法能够通过分解为更小的函数而变得更为干净利落（如代码清单 2-2 所示）。

代码清单 2-2　有语境的变量

```
public class GuessStatisticsMessage {
  private String number;
  private String verb;
  private String pluralModifier;
```

```java
public String make(char candidate, int count) {
  createPluralDependentMessageParts(count);
  return String.format(
    "There %s %s %s%s",
     verb, number, candidate, pluralModifier );
}

private void createPluralDependentMessageParts(int count) {
  if (count == 0) {
    thereAreNoLetters();
  } else if (count == 1) {
    thereIsOneLetter();
  } else {
    thereAreManyLetters(count);
  }
}

private void thereAreManyLetters(int count) {
  number = Integer.toString(count);
  verb = "are";
  pluralModifier = "s";
}

private void thereIsOneLetter() {
  number = "1";
  verb = "is";
  pluralModifier = "";
}

private void thereAreNoLetters() {
  number = "no";
  verb = "are";
  pluralModifier = "s";
}
```

2.17 不要添加没用的语境

假若有一个名为"加油站豪华版"（Gas Station Deluxe）的应用，在其中给每个类添加 GSD 前缀就不是什么好点子。说白了，你是在和自己在用的工具过不去。输入 G，按下自动完成键，结果会得到系统中全部类的列表，列表恨不得有一英里那么长。这样做聪明吗？为什么要搞得 IDE 没法帮助你？

再比如，你在 GSD 应用程序中的记账模块创建了一个表示邮件地址的类，然后给该类命名为 GSDAccountAddress。稍后，你的客户联络应用中需要用到邮件地址，你会用 GSDAccountAddress 吗？这名字听起来没问题吗？在这 17 个字母里面，有 10 个字母纯属多余，和当前语境毫无关联。

只要短名称足够清楚，就比长名称好。别给名称添加不必要的语境。

对 `Address` 类的实体来说，`accountAddress` 和 `customerAddress` 都是不错的名称，不过用在类名上就不太好了。`Address` 是一个好类名。如果需要与邮政编码 MAC 地址和 Web 地址相区分，我会考虑使用 `PostalAddress`、`MAC` 和 `URI`。这样的名称更为精确，而精确正是命名的要点。

2.18 最后的话

起好名字最难的地方在于需要良好的描述技巧和共有文化背景。与其说这是一种技术、商业或管理问题，还不如说是一种教学问题。其结果是，这个领域内的许多人都没能学会做得很好。

我们有时会担心其他开发者反对重命名。讨论一下就会知道，如果名称改得更好，那大家真的会感激你。多数时候我们并不记忆类名和方法名，而使用现代工具应对这些细节，好让自己集中精力把代码写得像词句篇章，至少像是表和数据结构（词句并非总是呈现数据的最佳手段）。改名可能会让某些人吃惊，就像你做到其他代码改善工作一样。别让这种事阻碍你前进的步伐。

不妨试试上面这些规则，看看你的代码可读性是否有所提升。如果你是在维护别人写的代码，使用重构工具来解决问题，效果会立竿见影，而且会持续下去。

第 3 章

函 数

在编程的早期岁月,系统由程序和子程序组成。后来,到 Fortran 和 PL/1 的年代,系统由程序、子程序和函数组成。如今,只有函数存活下来。函数是所有程序中的第一组代码。本章将讨论如何写好函数。

请看代码清单 3-1。在 FitNesse[①]中，很难找到长函数，不过我还是搜寻到一个。它不光长，而且很复杂，有大量字符串、怪异且不显见的数据类型和 API。花 3 分钟时间，你能读懂多少？

代码清单 3-1　HtmlUtil.java（FitNesse 20070619）

```java
public static String testableHtml(
  PageData pageData,
  boolean includeSuiteSetup
) throws Exception {
  WikiPage wikiPage = pageData.getWikiPage();
  StringBuffer buffer = new StringBuffer();
  if (pageData.hasAttribute("Test")) {
    if (includeSuiteSetup) {
      WikiPage suiteSetup =
        PageCrawlerImpl.getInheritedPage(
                SuiteResponder.SUITE_SETUP_NAME, wikiPage
        );
      if (suiteSetup != null) {
        WikiPagePath pagePath =
          suiteSetup.getPageCrawler().getFullPath(suiteSetup);
        String pagePathName = PathParser.render(pagePath);
        buffer.append("!include -setup .")
              .append(pagePathName)
              .append("\n");
      }
    }
    WikiPage setup =
      PageCrawlerImpl.getInheritedPage("SetUp", wikiPage);
    if (setup != null) {
      WikiPagePath setupPath =
        wikiPage.getPageCrawler().getFullPath(setup);
      String setupPathName = PathParser.render(setupPath);
      buffer.append("!include -setup .")
            .append(setupPathName)
            .append("\n");
    }
  }
  buffer.append(pageData.getContent());
  if (pageData.hasAttribute("Test")) {
    WikiPage teardown =
      PageCrawlerImpl.getInheritedPage("TearDown", wikiPage);
    if (teardown != null) {
      WikiPagePath tearDownPath =
        wikiPage.getPageCrawler().getFullPath(teardown);
      String tearDownPathName = PathParser.render(tearDownPath);
      buffer.append("\n")
            .append("!include -teardown .")
            .append(tearDownPathName)
            .append("\n");
    }
    if (includeSuiteSetup) {
```

[①] 一种开源测试工具。

```
      WikiPage suiteTeardown =
        PageCrawlerImpl.getInheritedPage(
                SuiteResponder.SUITE_TEARDOWN_NAME,
                wikiPage
        );
      if (suiteTeardown != null) {
        WikiPagePath pagePath =
          suiteTeardown.getPageCrawler().getFullPath (suiteTeardown);
        String pagePathName = PathParser.render(pagePath);
        buffer.append("!include -teardown .")
              .append(pagePathName)
              .append("\n");
      }
    }
  }
  pageData.setContent(buffer.toString());
  return pageData.getHtml();
}
```

读懂这个函数了吗？大概没有。有太多事发生，有太多不同层级的抽象。奇怪的字符串和函数调用，混以双重嵌套、用标识来控制的 `if` 语句等，不一而足。

不过，只要做几个简单的方法抽离和重命名操作，加上一点点重构，就能在 9 行代码之内解决问题（如代码清单 3-2 所示）。花 3 分钟阅读代码清单 3-2，看你能理解吗？

代码清单 3-2　HtmlUtil.java（重构之后）

```
public static String renderPageWithSetupsAndTeardowns(
  PageData pageData, boolean isSuite
) throws Exception {
  boolean isTestPage = pageData.hasAttribute("Test");
  if (isTestPage) {
    WikiPage testPage = pageData.getWikiPage();
    StringBuffer newPageContent = new StringBuffer();
    includeSetupPages(testPage, newPageContent, isSuite);
    newPageContent.append(pageData.getContent());
    includeTeardownPages(testPage, newPageContent, isSuite);
    pageData.setContent(newPageContent.toString());
  }

  return pageData.getHtml();
}
```

除非你正在研究 FitNesse，否则就理解不了所有细节。不过，你大概能明白，该函数包含把一些设置和拆解页放入一个测试页面，再渲染为 HTML 的操作。如果你熟悉 JUnit[①]，或许会想到，该函数归属某个基于 Web 的测试框架，而且，这当然没错。从代码清单 3-2 中获得信息很容易，而代码清单 3-1 则晦涩难明。

是什么让代码清单 3-2 易于阅读和理解？怎么才能让函数表达其意图？该给函数赋予哪

---

① 一种开源 Java 单元测试工具。

些属性，好让读者一看就明白函数是属于怎样的程序呢？

3.1 短小

函数的第一条规则是要短小。第二条规则是还要更短小。我没办法证明这个断言。我给不出任何研究结果来证实小函数更佳。我能说的是，几十年来，我写过各种大小不同的函数。我写过令人憎恶的长达 3000 行的函数，也写过许多 100～300 行的函数，还写过 20～30 行的函数。经过漫长的试错过程，经验告诉我，函数就应该短小。

在 20 世纪 80 年代，我们常说函数不该长于一屏。当然，说这话的时候，VT100 屏幕只有 24 行、80 列，而编辑器就先占去 4 行空间放菜单。如今，用上了精致的字体和宽大的显示器，一屏里面可以显示 100 行，每行能容纳 150 个字符。每行都不应该有 150 个字符那么长。函数也不该有 100 行那么长，20 行封顶最佳。

函数到底该有多短？1991 年，我去 Kent Beck 位于俄勒冈州的家中拜访。我们坐到一起写了一些代码。他给我看一个叫作 Sparkle（火花闪耀）的有趣的 Java/Swing 小程序。程序在屏幕上描画电影《灰姑娘》(*Cinderella*) 中仙女用魔棒造出的视觉效果。只要移动鼠标，光标所在处就会爆发出一团令人欣喜的火花，沿着模拟重力场滑落到窗口底部。Kent 给我看代码的时候，我惊讶于其中那些函数的尺寸如此之小。我看惯了 Swing 程序中长度以英里计的函数，而这个程序中的每个函数都只有两行、三行或四行长。每个函数都一目了然。每个函数都只说一件事。而且，每个函数都依序把你带到下一个函数。这就是函数应该达到的短小程度！[①]

函数应该有多短小？通常来说，应该短于代码清单 3-2 中的函数！代码清单 3-2 实在应该缩短成代码清单 3-3。

代码清单 3-3　HtmlUtil.java（再次重构之后）

```
public static String renderPageWithSetupsAndTeardowns(
  PageData pageData, boolean isSuite) throws Exception {
  if (isTestPage(pageData))
    includeSetupAndTeardownPages(pageData, isSuite);
  return pageData.getHtml();
}
```

代码块和缩进

if 语句、else 语句、while 语句等，其中的代码块应该只占一行，该行大抵应该是一个函数调用语句。这样不但能保持函数短小，而且，因为块内调用的函数拥有较具说明性的

[①] 我问 Kent 是否留存了这段程序，他说找不到了。我搜遍自己的计算机也没找到。现在只有在记忆中的这段往事了。

名称,所以增加了文档上的价值。

这也意味着函数不应该大到足以容纳嵌套结构。所以,函数的缩进层级不该多于一层或两层。当然,这样的函数易于阅读和理解。

3.2 只做一件事

代码清单 3-1 显然想做好几件事,它创建缓冲区、获取页面、搜索继承下来的页面、渲染路径、添加神秘的字符串、生成 HTML,如此等等。代码清单 3-1 手忙脚乱。而代码清单 3-3 则只做一件简单的事,即将设置和拆解功能包纳到测试页面中。

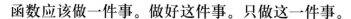

过去几十年以来,以下建议以不同形式一再出现:

函数应该做一件事。做好这件事。只做这一件事。

问题在于很难知道那件该做的事是什么。代码清单 3-3 只做了一件事,对吧?其实也很容易看作以下是 3 件事:

(1) 判断是否为测试页面;
(2) 如果是,则容纳进设置和分拆步骤;
(3) 渲染成 HTML。

那件事是什么?函数是做了一件事,还是做了 3 件事呢?注意,这 3 个步骤均在该函数名下的同一抽象层上。可以用简洁的 TO[①] 起头段落来描述这个函数:

> TO RenderPageWithSetupsAndTeardowns, we check to see whether the page is a test page and if so, we include the setups and teardowns. In either case we render the page in HTML.
>
> (要 RenderPageWithSetupsAndTeardowns,检查页面是否为测试页,如果是测试页,就容纳进设置和分拆步骤。无论是否是测试页,都渲染成 HTML。)

如果函数只是做了该函数名下同一抽象层上的步骤,则函数还是只做了一件事。编写函数毕竟是为了把较大的概念(换言之,函数的名称)拆分为另一抽象层上的一系列步骤。

代码清单 3-1 明显包括了多个处于不同抽象层级的步骤。显然,它所做的不止一件事。即便是代码清单 3-2 也有两个抽象层,这已被我们可将其缩短的能力所证明。但再将代码清单 3-3 做有意义的缩短却很难。可以将 `if` 语句拆出来做一个名为 `includeSetupsAnd-TeardownsIfTestpage` 的函数,但那只是重新诠释代码,并未改变抽象层级。

---

① LOGO 语言中的 `TO` 关键字,与 Ruby 和 Python 中 `def` 关键字的用法一致。所以,每个函数都以 `TO` 起头。这对函数的设计产生了有趣的影响。

所以，要判断函数是否不止做了一件事，还有一个方法，就是看它是否能再拆出一个函数，该函数不仅只是单纯地重新诠释其实现[G34]。

函数中的区段

请看代码清单 4-7。注意，generatePrimes 函数被切分为 declarations、initializations 和 sieve 等区段。这就是函数做事太多的明显征兆。只做一件事的函数无法被合理地切分为多个区段。

3.3 每个函数一个抽象层级

要确保函数只做一件事，函数中的语句就要在同一抽象层级上。一眼就能看出，代码清单 3-1 违反了这条规则。那里面有 getHtml() 等位于较高抽象层的概念，也有 String pagePathName = PathParser.render(pagePath) 等位于中间抽象层的概念，还有 .append("\n") 等位于相当低的抽象层的概念。

函数中混杂不同抽象层级，往往会让人迷惑，读者可能无法判断某个表达式是基础概念还是细节。更恶劣的是，就像破损的窗户，一旦细节与基础概念混杂，更多的细节就会在函数中纠结起来。

自顶向下读代码：向下规则

我们想要让代码拥有自顶向下的阅读顺序[①]。我们想要让每个函数后面都跟着位于下一抽象层级的函数，这样一来，在查看函数列表时，就能循抽象层级向下阅读了。我把这叫作向下规则。

换一种说法。我们想要这样读程序：程序就像是一系列 TO 起头的段落，每一段都描述当前抽象层级，并引用位于下一抽象层级的后续 TO 起头段落。

> To include the setups and teardowns, we include setups, then we include the test page content, and then we include the teardowns. （要容纳设置和分拆步骤，就先容纳设置步骤，然后纳入测试页面内容，再纳入分拆步骤。）

> To include the setups, we include the suite setup if this is a suite, then we include the regular setup. （要容纳设置步骤，如果是套件，就纳入套件设置步骤，然后再纳入普通设置步骤。）

---

① [KP78]。

To include the suite setup, we search the parent hierarchy for the "SuiteSetUp" page and add an include statement with the path of that page.（要容纳套件设置步骤，先搜索 "SuiteSetUp" 页面的上级继承关系，再添加一个包括该页面路径的语句。）

To search the parent. . . （要搜索……）

程序员往往很难学会遵循这条规则，写出只停留于一个抽象层级上的函数。尽管如此，学习这个技巧还是很重要。这是保持函数短小、确保只做一件事的要诀。让代码读起来像是一系列自顶向下的 *TO* 起头段落是保持抽象层级协调一致的有效技巧。

请看本章末尾的代码清单 3-7，它展示了遵循这条原则重构的完整 `testableHtml` 函数。留意每个函数是如何引出下一个函数，并且如何保持在同一抽象层上的。

3.4 switch 语句

写出短小的 `switch` 语句很难[①]。即便是只有两种条件的 `switch` 语句也比我想要的单个代码块或函数大得多。写出只做一件事的 `switch` 语句也很难。`switch` 天生要做 *N* 件事。不幸的是，我们总无法避开 `switch` 语句，不过还是能够确保每个 `switch` 都埋藏在较低的抽象层级，而且永远不重复。当然，我们可以利用多态来实现这一点。

请看代码清单 3-4，它呈现了可能依赖雇员类型的仅仅一种操作。

代码清单 3-4 Payroll.java

```java
public Money calculatePay(Employee e)
throws InvalidEmployeeType {
  switch (e.type) {
    case COMMISSIONED:
      return calculateCommissionedPay(e);
    case HOURLY:
      return calculateHourlyPay(e);
    case SALARIED:
      return calculateSalariedPay(e);
    default:
      throw new InvalidEmployeeType(e.type);
  }
}
```

该函数有好几个问题。首先，它太长，当出现新的雇员类型时，还会变得更长。其次，它明显做了不止一件事。第三，它违反了单一权责原则（Single Responsibility Principle，SRP），因为有好几个修改它的理由。第四，它违反了开放闭合原则（Open Closed Principle，OCP），因为每当添加新类型时，就必须修改该函数。不过，该函数最麻烦的可能是到处皆有类似结构的函数。例如，可能会有

---

① 当然，这也包括 if/else 语句在内。

```
isPayday(Employee e, Date date),
```

或

```
deliverPay(Employee e, Money pay),
```

如此等等。它们的结构都有同样的问题。

该问题的解决方案（如代码清单 3-5 所示）是将 switch 语句埋藏到抽象工厂[①]底下，不让任何人看到。该工厂使用 switch 语句为 Employee 的派生物创建适当的实体，而不同的函数，如 calculatePay、isPayday 和 deliverPay 等，则借由 Employee 接口多态地接受派遣。

对于 switch 语句，我的规则是，如果只出现一次用于创建多态对象，而且隐藏在某个继承关系中，在系统其他部分看不到，就还能容忍[G23]。当然也要就事论事，有时我也会部分或全部违反这条规则。

代码清单 3-5　Employee 与工厂

```java
public abstract class Employee {
  public abstract boolean isPayday();
  public abstract Money calculatePay();
  public abstract void deliverPay(Money pay);
}
-----------------
public interface EmployeeFactory {
  public Employee makeEmployee(EmployeeRecord r) throws InvalidEmployeeType;
}
-----------------
public class EmployeeFactoryImpl implements EmployeeFactory {
  public Employee makeEmployee(EmployeeRecord r) throws InvalidEmployeeType {
    switch (r.type) {
      case COMMISSIONED:
        return new CommissionedEmployee(r) ;
      case HOURLY:
        return new HourlyEmployee(r);
      case SALARIED:
        return new SalariedEmploye(r);
      default:
        throw new InvalidEmployeeType(r.type);
    }
  }
}
```

3.5　使用具有描述性的名称

在代码清单 3-7 中，我把示例函数的名称从 testableHtml 改为了 SetupTeardown-Includer.render。这个名称好得多，因为它较好地描述了函数做的事。我也给每个私有方

---

① [GOF]。

法起个同样具有描述性的名称，如 `isTestable` 或 `includeSetupAndTeardownPages`。好名称的价值怎么好评都不为过。记住 Ward 原则："如果每个例程都让你感到深合己意，那就是整洁代码。"要遵循这一原则，泰半工作都在于为只做一件事的小函数起个好名字。函数越短小、功能越集中，就越便于起个好名字。

别害怕长名称。长而具有描述性的名称，要比短而令人费解的名称好。长而具有描述性的名称，要比描述性的长注释好。使用某种命名约定，让函数名称中的多个单词容易阅读，然后使用这些单词给函数起个能说清其功用的名称。

别害怕花时间起名字。你当尝试不同的名称，实测其阅读效果。在 Eclipse 或 IntelliJ 等现代 IDE 中改名称易如反掌。使用这些 IDE 测试不同名称，直至找到最具有描述性的那一个为止。

选择描述性的名称能理清你关于模块的设计思路，并帮你改进之。追索好名称，往往导致对代码的改善重构。

命名方式要保持一致。使用与模块名一脉相承的短语、名词和动词给函数命名。例如，`includeSetupAndTeardownPages`、`includeSetupPages`、`includeSuiteSetupPage` 和 `includeSetupPage` 等。这些名称使用了类似的措辞，依序讲出一个故事。实际上，假使我只给你看上述函数序列，你就会自问："`includeTeardownPages`、`includeSuite-TeardownPage` 和 `includeTeardownPage` 又会如何？"这就是所谓"深合己意"了。

3.6 函数参数

最理想的参数数量是 0（零参数函数），其次是 1（单参数函数），再次是 2（双参数函数），应尽量避免 3（三参数函数）。有足够特殊的理由才能用 3 个以上参数（多参数函数）——所以无论如何也不要这么做。

参数不易对付。它们带有太多概念性。所以我在代码范例中几乎不加参数。比如，以 `StringBuffer` 为例，我们可能不把它作为实体变量，而是当作参数来传递，那样的话，读者每次看到它都得要翻译一遍。阅读模块所讲述的故事时，`includeSetupPage()` 要比 `includeSetupPageInto(newPageContent)` 易于理解。参数与函数名处在不同的抽象层级，它要求你了解目前并不特别重要的细节（即那个 `StringBuffer`）。

从测试的角度看，参数甚至更叫人为难。想想看，要编写能确保参数的各种组合运行正常的测试用例，是多么困难的事。如果没有参数，就是小菜一碟。如果只有一个参数，也不太困难。有两个参数，问题就麻烦多了。如果参数多于两个，测试覆盖所有可能值的组合简直令人生畏。

输出参数比输入参数还要难以理解。读函数时，我们惯于认为信息通过参数输入函数，通过返回值从函数中输出。我们不太期望信息通过参数输出。所以，输出参数往往让人苦思之后才恍然大悟。

与没有参数相比，只有一个输入参数算是第二好的做法。`SetupTeardownIncluder.render(pageData)` 也相当易于理解。很明显，我们将渲染 `pageData` 对象中的数据。

3.6.1 单参数函数的普遍形式

向函数传入单个参数有两种极普遍的理由。你也许会问关于那个参数的问题，就像在 `boolean fileExists("MyFile")` 中那样；也可能是操作该参数，将其转换为其他的东西，再输出之。例如，`InputStream fileOpen("MyFile")` 把 `String` 类型的文件名转换为 `InputStream` 类型的返回值。这就是读者看到函数时所期待的东西。你应当选用较能区分这两种理由的名称，而且总在一致的上下文中使用这两种形式。

还有一种虽不那么普遍但仍极有用的单参数函数形式，那就是事件（event）。在这种形式中，有输入参数而无输出参数。程序将函数看作是一个事件，使用该参数修改系统状态，例如 `void passwordAttemptFailedNtimes(int attempts)`。请小心使用这种形式。应该让读者很清楚地了解它是一个事件，谨慎地选用名称和上下文语境。

尽量避免编写不遵循这些形式的单参数函数，例如，`void includeSetupPageInto(StringBuffer pageText)`。对于转换，使用输出参数而非返回值会令人迷惑。如果函数要对输入参数进行转换操作，转换结果就该体现为返回值。实际上，`StringBuffertransform(StringBuffer in)` 比 `void transform(StringBuffer out)` 强，即便第一种形式只简单地返回输入参数也是这样，至少，它遵循了转换的形式。

3.6.2 标识参数

标识参数丑陋不堪。向函数传入布尔值简直就是骇人听闻的做法。这样做，方法签名会立刻变得复杂起来，这相当于大声宣布本函数不止做一件事，即如果标识为 `true` 将会这样做，则标识为 `false` 会那样做！

在代码清单 3-7 中，我们别无选择，因为调用者已经传入了那个标识，而我想把重构范围限制在该函数及该函数以下的范围之内。方法调用 `render(true)` 对于可怜的读者来说仍然摸不着头脑。滚动屏幕，看到 `render(Boolean isSuite)`，稍许有点帮助，不过仍然不够。应该把该函数一分为二：`reanderForSuite()` 和 `renderForSingleTest()`。

3.6.3 双参数函数

有两个参数的函数要比单参数函数难懂。例如，writeField(name)比 writeField(outputStream, name)①好懂。

尽管两种情况下意义都很清楚，但第一个只要扫一眼就能明白，更好地表达了其意义。第二个就得暂停一下才能明白，除非我们学会忽略第一个参数，而且如果这样，最终也会导致问题，因为我们根本就不该忽略任何代码，忽略掉的部分就是缺陷藏身之地。

当然，有些时候两个参数正好。例如，Point p = new Point(0, 0);就相当合理，因为笛卡儿点天生拥有两个参数。如果看到new Point(0)，我们会倍感惊讶。然而，本例中的两个参数却只是单个值的有序组成部分！而 outputStream 和 name 则既非自然的组合，也不是自然的排序。

即便是如 assertEquals(expected, actual)这样的双参数函数也有其问题。你有多少次会搞错 actual 和 expected 的位置呢？这两个参数没有自然的顺序。expected 在前，actual 在后，只是一种需要学习的约定罢了。

双参数函数不算恶劣，而且你当然也会编写双参数函数。不过，你得小心，使用双参数函数要付出代价。你应该尽量利用一些机制将其转换成单参数函数。例如，可以把 writeField 方法写成 outputStream 的成员之一，从而能这样用：outputStream.writeField(name)。或者，也可以把 outputStream 写成当前类的成员变量，从而无须再传递它。还可以分离出类似于 FieldWriter 的新类，在其构造器中采用 outputStream，并且包含一个 write 方法。

3.6.4 三参数函数

有3个参数的函数要比双参数函数难懂得多。排序、琢磨、忽略的问题都会加倍体现。建议你在写三参数函数前一定要想清楚。

例如，设想 assertEquals 有3个参数，该函数形式为 assertEquals(message, expected, actual)。有多少次，你读到 message 时，会错以为它是 expected 呢？我就常栽在这个三参数函数上。实际上，每次我看到这里，总会绕半天圈子，最后学会了忽略 message 参数。

另外，这里有一个并不那么险恶的三参数函数：assertEquals(1.0, amount, 0.001)。虽然也要费点儿神，但还是值得的。得到"浮点值的等值是相对而言"的提示总

---

① 我刚重构了一个使用了二元形式的模块。现在就能把 outputStream 做成该类的一个字段，并把所有对 writeField 的调用都变作一元形式。结果就干净多了。

是好的。

3.6.5 参数对象

如果函数看起来需要 2 个、3 个或 3 个以上参数，就说明其中一些参数应该封装为类了。例如，下面两个声明的差别：

```
Circle makeCircle(double x, double y, double radius);
Circle makeCircle(Point center, double radius);
```

从参数创建对象，从而减少参数数量，看起来像是在作弊，但实则并非如此。当一组参数被共同传递，就像上例中的 x 和 y 那样，往往就是该有自己名称的某个概念的一部分。

3.6.6 参数列表

有时，我们想要向函数传入数量可变的参数。例如，String.format 方法：

```
String.format("%s worked %.2f hours.", name, hours);
```

如果可变参数像上例中那样被同等对待，就和类型为 List 的单个参数没什么两样。这样一来，String.format 实则是双参数函数。下列 String.format 的声明也很明显是二元的：

```
public String format(String format, Object... args)
```

同理，有可变参数的函数可能是单参数、双参数甚至三参数的。超过这个数量就可能要犯错了。

```
void monad(Integer... args);
void dyad(String name, Integer... args);
void triad(String name, int count, Integer... args);
```

3.6.7 动词与关键字

给函数起个好名字，能较好地解释函数的意图，以及参数的顺序和意图。对于单参数函数，函数和参数应当形成一种非常良好的动词/名词对形式。例如，write(name) 就相当令人认可。不管这个"name"是什么，都要被"write"。更好的名称大概是 writeField(name)，它告诉我们，"name" 是一个 "field"。

最后那个例子展示了函数名称的关键字（keyword）形式。使用这种形式，我们把参数的名称编码成了函数名。例如，把 assertEquals 改成 assertExpectedEqualsActual(expected, actual) 可能会好些，这大大减轻了记忆参数顺序的负担。

3.7 无副作用

副作用是一种谎言。函数承诺只做一件事，但还是会做其他被藏起来的事。有时，它会对自己类中的变量做出未能预期的改动。有时，它会把变量搞成向函数传递的参数或是系统全局变量。无论哪种情况，都是具有破坏性的，会导致古怪的时序性耦合及顺序依赖。

以代码清单 3-6 中看似无伤大雅的函数为例。该函数使用标准算法来匹配 `userName` 和 `password`。如果匹配成功，则返回 `true`，如果匹配失败，则返回 `false`，但它会有副作用。你知道问题所在吗？

代码清单 3-6　UserValidator.java

```java
public class UserValidator {
  private Cryptographer cryptographer;

  public boolean checkPassword(String userName, String password) {
    User user = UserGateway.findByName(userName);
    if (user != User.NULL) {
      String codedPhrase = user.getPhraseEncodedByPassword();
      String phrase = cryptographer.decrypt(codedPhrase, password);
      if ("Valid Password".equals(phrase)) {
        Session.initialize();
        return true;
      }
    }
    return false;
  }
}
```

显然，副作用就在于对 `Session.initialize()` 的调用。`checkPassword` 函数，顾名思义，就是用来检查密码的，它的名称并未暗示它会初始化该次会话。所以，当某个误信了函数名的调用者想要检查用户有效性时，就得冒抹除现有会话数据的风险。

这一副作用造成了一次时序性耦合。也就是说，`checkPassword` 只能在特定时刻调用（换言之，在初始化会话是安全的时候调用）。如果在不合适的时候调用，会话数据就有可能沉默地丢失。时序性耦合令人迷惑，特别是当它躲在副作用后面时。如果一定要时序性耦合，就应该在函数名称中说明。在本例中，可以将函数重命名为 `checkPasswordAnd-InitializeSession`，虽然那还是违反了"只做一件事"的规则。

输出参数

参数多数会被自然而然地看作是函数的输入项。如果你编过好多年程序，我担保你一定被用作输出而非输入的参数迷惑过。例如：

```
appendFooter(s);
```

这个函数是把 s 添加到什么东西后面吗？或者它把什么东西添加到了 s 后面？s 是输入参数还是输出参数？稍许花点时间看看函数签名：

```
public void appendFooter(StringBuffer report)
```

事情弄清楚了，但付出了检查函数声明的代价。如果你被迫检查函数签名，就得花上一点儿时间。应该避免这种中断思路的事。

在面向对象编程出现之前的岁月里，有时的确需要输出参数。然而，面向对象语言中对输出参数的大部分需求已经消失了，因为 this 也有输出函数的意味。换言之，最好是这样调用 appendFooter：

```
report.appendFooter();
```

普遍而言，应避免使用输出参数。如果函数必须要修改某种状态，就修改所属对象的状态吧。

3.8　分隔指令与询问

函数要么做什么事，要么回答什么事，但二者不可得兼。函数应该修改某对象的状态，或是返回该对象的有关信息。如果两样都干，常会导致混乱。看看下面的例子：

```
public boolean set(String attribute, String value);
```

该函数设置某个指定属性，如果成功，就返回 true，如果不存在那个属性，就返回 false。这样就导致了以下语句：

```
if (set("username", "unclebob"))...
```

从读者的角度考虑一下吧。这是什么意思呢？它是在问 username 属性值是否之前已设置为 unclebob，还是在问 username 属性值是否成功设置为 unclebob 呢？从该行调用语句很难判断其含义，因为 set 是动词还是形容词并不清楚。

作者本意是，set 是一个动词，但在 if 语句的上下文中，感觉它像是一个形容词。该语句读起来像是在说"如果 username 属性值之前已被设置为 unclebob"，而不是"设置 username 属性值为 unclebob，看看是否可行，然后……"。要解决这个问题，可以将 set 函数重命名为 setAndCheckIfExists，但这对提高 if 语句的可读性帮助不大。真正的解决方案是把指令与询问分隔开来，防止混淆的产生：

```
if (attributeExists("username")) {
  setAttribute("username", "unclebob");
  ...
}
```

3.9 使用异常替代返回错误码

从指令式函数返回错误码略微违反了指令与询问分隔的规则。它鼓励了在 `if` 语句判断中把指令当作表达式使用。

```
if (deletePage(page) == E_OK)
```

这不会引起动词/形容词混淆，但会导致更深层次的嵌套结构。当返回错误码时，就是在要求调用者立刻处理错误。

```
if (deletePage(page) == E_OK) {
  if (registry.deleteReference(page.name) == E_OK) {
    if (configKeys.deleteKey(page.name.makeKey()) == E_OK){
      logger.log("page deleted");
    } else {
      logger.log("configKey not deleted");
    }
  } else {
    logger.log("deleteReference from registry failed");
  }
} else {
  logger.log("delete failed");
  return E_ERROR;
}
```

另外，如果使用异常替代返回错误码，错误处理代码就能从主路径代码中分离出来，从而得到简化：

```
try {
  deletePage(page);
  registry.deleteReference(page.name);
  configKeys.deleteKey(page.name.makeKey());
}
catch (Exception e) {
  logger.log(e.getMessage());
}
```

3.9.1 抽离 try/catch 代码块

try/catch 代码块丑陋不堪。它们搞乱了代码结构，把错误处理与正常流程混为一谈。最好把 try 和 catch 代码块的主体部分抽离出来，另外形成函数。

```
public void delete(Page page) {
  try {
    deletePageAndAllReferences(page);
  }
```

```
    catch (Exception e) {
      logError(e);
    }
  }

  private void deletePageAndAllReferences(Page page) throws Exception {
    deletePage(page);
    registry.deleteReference(page.name);
    configKeys.deleteKey(page.name.makeKey());
  }

  private void logError(Exception e) {
    logger.log(e.getMessage());
  }
```

在上例中，`delete` 函数只与错误处理有关，很容易理解然后可以忽略。`deletePage-AndAllReferences` 函数只与完全删除一个 `page` 有关，错误处理可以忽略。有了这样美妙的区隔，代码就更易于理解和修改了。

3.9.2 错误处理就是一件事

函数应该只做一件事。错误处理就是一件事。因此，处理错误的函数不该做其他事。这意味着（如上例所示）如果关键字 `try` 在某个函数中存在，它就应该是这个函数的第一个单词，而且在 `catch/finally` 代码块后面也不该有其他内容。

3.9.3 Error.java 依赖磁铁

返回错误码通常暗示某处有个类或是枚举，其定义了所有错误码。

```
public enum Error {
  OK,
  INVALID,
  NO_SUCH,
  LOCKED,
  OUT_OF_RESOURCES,
  WAITING_FOR_EVENT;
}
```

这样的类就是一块依赖磁铁（dependency magnet），其他许多类都得导入和使用它。当 `Error` 枚举修改时，其他所有类都需要重新编译和部署[1]。这对 `Error` 类造成了负面压力。程序员不愿增加新的错误代码，因为如果这样他们就得重新构建和部署所有东西。于是他们就复用旧的错误码，而不添加新的。

使用异常替代错误码，新异常就可以从异常类派生出来，而无须重新编译或重新部署[2]。

---

[1] 那些以为可以不重新编译和部署就扬长而去的家伙最终都将自尝恶果。
[2] 这也是开放闭合原则（OCP）的一个范例[PPP02]。

3.10 别重复自己[①]

回头仔细看看代码清单 3-1，你会注意到，有个算法在 SetUp、SuiteSetUp、TearDown 和 SuiteTearDown 中总共被重复了 4 次。识别重复不太容易，因为这 4 次重复与其他代码混在一起，而且也不完全一样。这样的重复还会导致问题，因为代码因此而臃肿，且当算法改变时需要修改 4 处地方，就会增加 4 次放过错误的可能性。

使用代码清单 3-7 中的 include 方法修正了这些重复。再读一遍那段代码，你会注意到，整个模块的可读性因为重复的消除而得到了提升。

重复可能是软件中一切邪恶的根源。许多原则与实践规则都是为控制与消除重复而创建的。例如，全部 Codd[②] 数据库范式就是为消除数据重复而服务的。再想想看，面向对象编程如何将代码集中到基类，从而避免了冗余。面向方面编程（Aspect Oriented Programming）、面向组件编程（Component Oriented Programming）多少也都是消除重复的一种策略。看来，自子程序发明以来，软件开发领域的所有创新都是在不断尝试从源代码中消灭重复。

3.11 结构化编程

有些程序员遵循 Edsger Dijkstra 的结构化编程规则[③]。Dijkstra 认为，每个函数、函数中的每个代码块都应该有一个入口、一个出口。遵循这些规则，意味着在每个函数中只该有一个 return 语句，循环中不能有 break 或 continue 语句，而且永远不能有任何 goto 语句。

我们赞成结构化编程的目标和规范，但对于小函数，这些规则助益不大。只有在大函数中，这些规则才会有明显的好处。

所以，只要函数保持短小，偶尔出现的 return、break 或 continue 语句没有坏处，甚至比单入单出原则更具有表达力。另外，goto 只在大函数中才有道理，所以应该尽量避免使用。

[①] DRY 原则。[PRAG]。
[②] Edgar F. Codd，关系数据库之父。
[③] [SP72]。

3.12 如何写出这样的函数

写代码和写别的东西很像。在写论文或文章时,你先想什么就写什么,然后再打磨它。初稿也许粗陋无序,你可以对其斟酌推敲,直至达到你心目中的样子。

我写函数时,一开始都冗长而复杂。有太多缩进和嵌套循环,有过长的参数列表。名称是随意起的,也会有重复的代码。不过我会配上一套单元测试,覆盖每行丑陋的代码。

然后我打磨这些代码,分解函数、修改名称、消除重复。我缩短和重新安置方法。有时我还拆解类,同时保持测试通过。

最后,遵循本章列出的规则,我组装好这些函数。

我并不从一开始就按照规则写函数。我想没人做得到。

3.13 小结

每个系统都是使用某种领域特定语言搭建的,而这种语言是程序员设计来描述那个系统的。函数是语言的动词,类是名词。这并非是要退回到最初设想的那种认为需求文档中的名词和动词就是系统中类和函数的可怕的旧观念。其实这是个历史更久的真理。编程艺术是且一直是语言设计的艺术。

大师级程序员把系统当作故事来讲,而不是当作程序来写。他们使用选定编程语言提供的工具构建一种更为丰富且更具表达力的语言,用来讲那个故事。那种领域特定语言的一个部分,就是描述在系统中发生的各种行为的函数层级。在一种狡猾的递归操作中,这些行为使用它们定义的与领域紧密相关的语言讲述自己那个小故事。

本章所讲述的是有关编写良好函数的机制。如果你遵循这些规则,函数就会短小、有个好名字,而且被很好地归置。不过永远别忘记,真正的目标在于讲述系统的故事,而你编写的函数必须干净利落地拼装到一起,形成一种精确而清晰的语言,帮助你讲故事。

3.14 SetupTeardownIncluder 程序

SetupTeardownIncluder 程序如代码清单 3-7 所示。

代码清单 3-7 SetupTeardownIncluder.java

```
package fitnesse.html;
```

```java
import fitnesse.responders.run.SuiteResponder;
import fitnesse.wiki.*;

public class SetupTeardownIncluder {
  private PageData pageData;
  private boolean isSuite;
  private WikiPage testPage;
  private StringBuffer newPageContent;
  private PageCrawler pageCrawler;


  public static String render(PageData pageData) throws Exception {
    return render(pageData, false);
  }

  public static String render(PageData pageData, boolean isSuite)
    throws Exception {
    return new SetupTeardownIncluder(pageData).render(isSuite);
  }

  private SetupTeardownIncluder(PageData pageData) {
    this.pageData = pageData;
    testPage = pageData.getWikiPage();
    pageCrawler = testPage.getPageCrawler();
    newPageContent = new StringBuffer();
  }

  private String render(boolean isSuite) throws Exception {
    this.isSuite = isSuite;
    if (isTestPage())
      includeSetupAndTeardownPages();
    return pageData.getHtml();
  }

  private boolean isTestPage() throws Exception {
    return pageData.hasAttribute("Test");
  }

  private void includeSetupAndTeardownPages() throws Exception {
    includeSetupPages();
    includePageContent();
    includeTeardownPages();
    updatePageContent();
  }

  private void includeSetupPages() throws Exception {
    if (isSuite)
      includeSuiteSetupPage();
    includeSetupPage();
  }

  private void includeSuiteSetupPage() throws Exception {
    include(SuiteResponder.SUITE_SETUP_NAME, "-setup");
  }
```

```java
  private void includeSetupPage() throws Exception {
    include("SetUp", "-setup");
  }

  private void includePageContent() throws Exception {
    newPageContent.append(pageData.getContent());
  }

  private void includeTeardownPages() throws Exception {
    includeTeardownPage();
    if (isSuite)
      includeSuiteTeardownPage();
  }

  private void includeTeardownPage() throws Exception {
    include("TearDown", "-teardown");
  }

  private void includeSuiteTeardownPage() throws Exception {
    include(SuiteResponder.SUITE_TEARDOWN_NAME, "-teardown");
  }

  private void updatePageContent() throws Exception {
    pageData.setContent(newPageContent.toString());
  }

  private void include(String pageName, String arg) throws Exception {
    WikiPage inheritedPage = findInheritedPage(pageName);
    if (inheritedPage != null) {
      String pagePathName = getPathNameForPage(inheritedPage);
      buildIncludeDirective(pagePathName, arg);
    }
  }

  private WikiPage findInheritedPage(String pageName) throws Exception {
    return PageCrawlerImpl.getInheritedPage(pageName, testPage);
  }

  private String getPathNameForPage(WikiPage page) throws Exception {
    WikiPagePath pagePath = pageCrawler.getFullPath(page);
    return PathParser.render(pagePath);
  }

  private void buildIncludeDirective(String pagePathName, String arg) {
    newPageContent
      .append("\n!include ")
      .append(arg)
      .append(" .")
      .append(pagePathName)
      .append("\n");
  }
}
```

3.15 文献

[KP78]：Kernighan and Plaugher, *The Elements of Programming Style*, 2d. ed., McGraw-Hill, 1978.

[PPP02]：Robert C. Martin, *Agile Software Development: Principles, Patterns, and Practices*, Prentice Hall, 2002.

[GOF]：*Design Patterns: Elements of Reusable Object Oriented Software*, Gamma et al., Addison-Wesley, 1996.

[PRAG]：*The Pragmatic Programmer*, Andrew Hunt, Dave Thomas, Addison-Wesley, 2000.

[SP72]：*Structured Programming*, O.-J. Dahl, E. W. Dijkstra, C. A. R. Hoare, Academic Press, London, 1972.

第 4 章

注 释

> "别给糟糕的代码加注释——重新写吧。"
> ——Brian W. Kernighan 与 P. J. Plaugher[1]

什么也比不上放置良好的注释来得有用。什么也不会比乱七八糟的注释更有本事搞乱一个模块。什么也不会比陈旧、提供错误信息的注释更有破坏性。

注释并不像辛德勒的名单。它们并不"纯然地好"。实际上，注释最多也就是一种必需的恶。若编程语言足够有表达力，或者我们长于用这些语言来表达意图，就不那么需要

---
[1] [KP78], p. 144。

注释——也许根本不需要。

注释的恰当用法是弥补我们在用代码表达意图时遭遇的失败。注意，我用了"失败"一词，我是说真的，注释总是一种失败。我们总无法找到不用注释就能表达自我的方法，所以总要有注释，这并不值得庆贺。

如果你发现自己需要写注释，就再想想看是否有办法翻盘，用代码来表达。每次用代码表达，你都该夸奖一下自己。每次写注释，你都该做个鬼脸，感受自己在表达能力上的失败。

我为什么要极力贬低注释？因为注释会撒谎，也不是说总是如此或有意如此，但出现得实在太频繁。注释存在的时间越久，离其所描述的代码就越远，就越来越变得全然错误。原因很简单：程序员不能坚持维护注释。

代码在变动，在演化，从这里移到那里，彼此分离、重造又合到一处。很不幸，注释并不总是随之变动——不能总是跟着代码走。注释常常会与其所描述的代码分隔开来，孑然飘零，越来越不准确。例如，看看以下注释以及它本来要描述的代码行变成了什么样子：

```
MockRequest request;
private final String HTTP_DATE_REGEXP =
  "[SMTWF][a-z]{2}\\,\\s[0-9]{2}\\s[JFMASOND][a-z]{2}\\s"+
  "[0-9]{4}\\s[0-9]{2}\\:[0-9]{2}\\:[0-9]{2}\\sGMT";
private Response response;
private FitNesseContext context;
private FileResponder responder;
private Locale saveLocale;
//   Example: "Tue, 02 Apr 2003 22:18:49 GMT"
```

在 HTTP_DATE_REGEXP 常量及其注释之间，有可能插入其他实体变量。

程序员应当负责将注释保持在可维护、有关联、精确的高度。我同意这种说法，但我更主张把力气用在写清楚代码上，这样可以直接保证无须编写注释。

不准确的注释要比没注释糟糕得多。它们满口胡言，它们预期的东西永不能实现，它们设定了无须也不应再遵循的旧规则。

真实只在一处地方有：代码。只有代码能忠实地告诉你它做的事。那是唯一真正准确的信息来源。所以，尽管有时也需要注释，但是我们也该多花心思尽量减少注释量。

4.1 注释不能美化糟糕的代码

写注释的常见动机之一是糟糕的代码的存在。我们编写一个模块，发现它令人困扰、乱七八糟。我们知道，它烂透了。我们告诉自己："喔，最好写点注释！"不！最好是把代码弄干净！

带有少量注释的整洁而有表达力的代码，要比带有大量注释的零碎而复杂的代码像样得多。与其花时间编写解释你写出的糟糕的代码的注释，不如花时间清理那堆糟糕的代码。

4.2 用代码来阐述

有时，代码本身不足以解释其行为。不幸的是，许多程序员据此以为代码很少——如果有的话——能做好解释工作。这种观点纯属错误。你愿意看到这个：

```
// Check to see if the employee is eligible for full benefits
if ((employee.flags & HOURLY_FLAG) &&
    (employee.age > 65))
```

还是下面这个？

```
if (employee.isEligibleForFullBenefits())
```

只要想上几秒，就能用代码解释你大部分的意图。很多时候，简单到只需要创建一个描述了与注释所言同一事物的函数即可。

4.3 好注释

有些注释是必需的，也是有利的。来看看一些我认为值得写的注释。不过要记住，唯一真正好的做法是你想办法不去写注释。

4.3.1 法律信息

有时，公司代码规范要求编写与法律有关的注释。例如，版权及著作权声明就是必须和有理由在每个源文件开头注释处放置的内容。

下例是我们在 FitNesse 项目每个源文件开头放置的标准注释。我可以很开心地说，IDE 自动卷起这些注释，这样就不会显得凌乱了。

```
// Copyright (C) 2003,2004,2005 by Object Mentor, Inc. All rights reserved.
// Released under the terms of the GNU General Public License version 2 or later.
```

这类注释不应是合同或法典。只要有可能，就指向一份标准许可或其他外部文档，而不要把所有条款放到注释中。

4.3.2 提供信息的注释

有时，用注释来提供基本信息也有其用处。例如，以下注释解释了某个抽象方法的返回值：

```
// Returns an instance of the Responder being tested.
protected abstract Responder responderInstance();
```

这类注释有时管用,但更好的方式是尽量利用函数名称传达信息。比如,在本例中,只要把函数重新命名为 `responderBeingTested`,注释就是多余的了。

下例稍好一些:

```
// format matched kk:mm:ss EEE, MMM dd, yyyy
Pattern timeMatcher = Pattern.compile(
  "\\d*:\\d*:\\d* \\w*, \\w* \\d*, \\d*");
```

在本例中,注释说明,该正则表达式意在匹配一个经由 `SimpleDateFormat.format` 函数利用特定格式字符串格式化的时间和日期。同样,如果把这段代码移到某个转换日期和时间格式的类中,就会更好、更清晰,而注释也就变得多此一举了。

4.3.3 对意图的解释

有时,注释不仅提供了有关实现的有用信息,而且还提供了某个决定后面的意图。在下例中,我们看到注释反映出来的一个有趣决定。在对比两个对象时,作者决定将他的类放置在比其他东西更高的位置。

```
public int compareTo(Object o)
{
  if(o instanceof WikiPagePath)
  {
    WikiPagePath p = (WikiPagePath) o;
    String compressedName = StringUtil.join(names, "");
    String compressedArgumentName = StringUtil.join(p.names, "");
    return compressedName.compareTo(compressedArgumentName);
  }
  return 1; // we are greater because we are the right type.
}
```

下面的例子更好。你也许不同意程序员给这个问题提供的解决方案,但至少你知道他想干什么。

```
public void testConcurrentAddWidgets() throws Exception {
  WidgetBuilder widgetBuilder =
    new WidgetBuilder(new Class[]{BoldWidget.class});
  String text = "'''bold text'''";
  ParentWidget parent =
    new BoldWidget(new MockWidgetRoot(), "'''bold text'''");
  AtomicBoolean failFlag = new AtomicBoolean();
  failFlag.set(false);

  //This is our best attempt to get a race condition
  //by creating large number of threads.
  for (int i = 0; i < 25000; i++) {
    WidgetBuilderThread widgetBuilderThread =
```

```
      new WidgetBuilderThread(widgetBuilder, text, parent, failFlag);
    Thread thread = new Thread(widgetBuilderThread);
    thread.start();
  }
  assertEquals(false, failFlag.get());
}
```

4.3.4 阐释

有时，注释把某些晦涩难明的参数或返回值的意义翻译为某种可读形式，也是有用的。通常，更好的方法是尽量让参数或返回值本身就足够清楚；但如果参数或返回值是某个标准库的一部分，或是你不能修改的代码，帮助阐释其含义的代码就会有用。

```
public void testCompareTo() throws Exception
{
  WikiPagePath a = PathParser.parse("PageA");
  WikiPagePath ab = PathParser.parse("PageA.PageB");
  WikiPagePath b = PathParser.parse("PageB");
  WikiPagePath aa = PathParser.parse("PageA.PageA");
  WikiPagePath bb = PathParser.parse("PageB.PageB");
  WikiPagePath ba = PathParser.parse("PageB.PageA");

  assertTrue(a.compareTo(a) == 0);      // a == a
  assertTrue(a.compareTo(b) != 0);      // a != b
  assertTrue(ab.compareTo(ab) == 0);    // ab == ab
  assertTrue(a.compareTo(b) == -1);     // a < b
  assertTrue(aa.compareTo(ab) == -1);   // aa < ab
  assertTrue(ba.compareTo(bb) == -1);   // ba < bb
  assertTrue(b.compareTo(a) == 1);      // b > a
  assertTrue(ab.compareTo(aa) == 1);    // ab > aa
  assertTrue(bb.compareTo(ba) == 1);    // bb > ba
}
```

当然，这也会冒阐释性注释本身就不正确的风险。回头看看上例，你会发现想要确认注释的正确性有多难。这一方面说明了阐释有多必要，另一方面也说明了它有风险。所以，在写这类注释之前，考虑一下是否还有更好的办法，然后再加倍小心地确认注释的正确性。

4.3.5 警示

有时，用于警示其他程序员可能会出现某种后果的注释也是有用的。例如，下面的注释解释了为什么要关闭某个特定的测试用例：

```
// Don't run unless you
// have some time to kill.
public void _testWithReallyBigFile()
{
  writeLinesToFile(10000000);
```

```
        response.setBody(testFile);
        response.readyToSend(this);
        String responseString = output.toString();
        assertSubString("Content-Length: 1000000000", responseString);
        assertTrue(bytesSent > 1000000000);
    }
```

当然，如今我们在多数情况下会利用附上恰当解释性字符串的@Ignore属性来关闭测试用例。比如@Ignore("Takes too long to run"[①])。但在JUnit4之前的日子里，惯常的做法是在方法名前面加上下划线。如果注释足够有说服力，就会很有用了。

这里有一个更麻烦的例子：

```
public static SimpleDateFormat makeStandardHttpDateFormat()
{
    //SimpleDateFormat is not thread safe,
    //so we need to create each instance independently.
    SimpleDateFormat df = new SimpleDateFormat("EEE, dd MMM  yyyy HH:mm:ss z");
    df.setTimeZone(TimeZone.getTimeZone("GMT"));
    return df;
}
```

你也许会抱怨说，还会有更好的解决方法。我大概会同意。不过上面的注释绝对有道理，它能阻止某位急切的程序员以效率之名使用静态初始器。

4.3.6 TODO注释

有时，有理由用//TODO形式在源代码中放置要做的工作列表。在下例中，TODO注释解释了为什么该函数的实现部分无所作为，将来应该是怎样。

```
//TODO-MdM these are not needed
//  We expect this to go away when we do the checkout model
protected VersionInfo makeVersion() throws Exception
{
  return null;
}
```

TODO是一种程序员认为应该做，但由于某些原因目前还没做的工作。它可能是要提醒删除某个不必要的特性，或者要求他人注意某个问题。它可能是恳请别人起个好名字，或者提示对依赖某个计划的事件的修改。无论TODO的目的如何，它都不是在系统中留下糟糕的代码的借口。

如今，大多数好IDE都提供了特别的手段来定位所有TODO注释，这些注释看来丢不了。你不会愿意代码因为TODO的存在而变成一堆垃圾，所以要定期查看，删除不再需要的TODO注释。

---

① 意为"运行时间过长"。——译者注

4.3.7 放大

注释可以用来放大某种看来不合理之物的重要性。

```
String listItemContent = match.group(3).trim();
// the trim is real important. It removes the starting
// spaces that could cause the item to be recognized
// as another list.
new ListItemWidget(this, listItemContent, this.level + 1);
return buildList(text.substring(match.end()));
```

4.3.8 公共 API 中的 Javadoc

没有什么比描述良好的公共 API 更有用和令人满意的了。标准 Java 库中的 Javadoc 就是一例。没有它们，写 Java 程序就会变得很难。

如果你在编写公共 API，就该为它编写良好的 Javadoc。不过要记住本章中的其他建议。就像其他注释一样，Javadoc 也可能误导、不适用或者提供错误信息。

4.4 坏注释

大多数注释都属此类。通常，坏注释都是糟糕的代码的支撑或借口，或者是对错误决策的修正，基本上等于程序员自说自话。

4.4.1 喃喃自语

如果只是因为你觉得应该或者因为过程需要就添加注释，那就是无谓之举。如果你决定写注释，就要花必要的时间确保写出最好的注释。

例如，我在 FitNesse 中找到的这个例子，例子中的注释大概确实有用。不过，作者太着急，或者没太花心思。他的喃喃自语变成了一个谜团。

```
public void loadProperties()
{
  try
  {
    String propertiesPath = propertiesLocation + "/" + PROPERTIES_FILE;
    FileInputStream propertiesStream = new FileInputStream(propertiesPath);
    loadedProperties.load(propertiesStream);
  }
  catch(IOException e)
```

```
        {
            // No properties files means all defaults are loaded
        }
    }
```

catch 代码块中的注释是什么意思呢？显然对于作者有其意义，不过并没有好到足够的程度。很明显，如果出现 IOException，就表示没有属性文件，在那种情况下，载入默认设置。但谁来装载默认设置呢？会在对 loadProperties.load 调用之前装载吗？抑或 loadProperties.load 捕获异常、装载默认设置、再向上传递异常以忽略它？再或 loadProperties.load 在尝试载入文件前就装载所有默认设置？还是作者只是在安慰自己别在意 catch 代码块的留空？或者——这种可能最可怕——作者是想告诉自己，将来再回过头来写装载默认设置的代码？

我们唯有检视系统其他部分的代码，弄清事情原委。任何迫使读者查看其他模块的注释，都没能与读者沟通好，不值所费。

4.4.2 多余的注释

代码清单 4-1 展示的简单函数，其头部位置的注释全属多余。读这段注释花的时间没准比读代码花的时间还要长。

代码清单 4-1　waitForClose

```
// Utility method that returns when this.closed is true. Throws an exception
// if the timeout is reached.
public synchronized void waitForClose(final long timeoutMillis)
throws Exception
{
    if(!closed)
    {
        wait(timeoutMillis);
        if(!closed)
            throw new Exception("MockResponseSender could not be closed");
    }
}
```

这段注释起了什么作用呢？它并不能比代码本身提供更多的信息。它没有证明代码的意义，也没有给出代码的意图或逻辑。读它并不比读代码更容易。事实上，它不如代码精确，误导读者接受不精确的信息，而不是正确地理解代码。它就像个自来熟的二手车贩子，满口保证你不用打开发动机盖查验。

来看看代码清单 4-2 中摘自 Tomcat 项目的无用而多余的 Javadoc 吧。这些注释只是一味将代码搞得含混不清，完全没有文档的价值。下面只列出了靠前面的一些代码，后续模块中还有许多类似情况。

代码清单 4-2　ContainerBase.java（Tomcat）

```java
public abstract class ContainerBase
  implements Container, Lifecycle, Pipeline,
  MBeanRegistration, Serializable {

  /**
   * The processor delay for this component.
   */
  protected int backgroundProcessorDelay = -1;


  /**
   * The lifecycle event support for this component.
   */
  protected LifecycleSupport lifecycle =
    new LifecycleSupport(this);


  /**
   * The container event listeners for this Container.
   */
  protected ArrayList listeners = new ArrayList();


  /**
   * The Loader implementation with which this Container is
   * associated.
   */
  protected Loader loader = null;


  /**
   * The Logger implementation with which this Container is
   * associated.
   */
  protected Log logger = null;


  /**
   * Associated logger name.
   */
  protected String logName = null;


  /**
   * The Manager implementation with which this Container is
   * associated.
   */
  protected Manager manager = null;


  /**
   * The cluster with which this Container is associated.
   */
```

```
    protected Cluster cluster = null;


    /**
     * The human-readable name of this Container.
     */
    protected String name = null;


    /**
     * The parent Container to which this Container is a child.
     */
    protected Container parent = null;


    /**
     * The parent class loader to be configured when we install a
     * Loader.
     */
    protected ClassLoader parentClassLoader = null;


    /**
     * The Pipeline object with which this Container is
     * associated.
     */
    protected Pipeline pipeline = new StandardPipeline(this);


    /**
     * The Realm with which this Container is associated.
     */
    protected Realm realm = null;


    /**
     * The resources DirContext object with which this Container
     * is associated.
     */
    protected DirContext resources = null;
```

4.4.3 误导性注释

有时，尽管初衷可嘉，但是程序员还是会写出不够精确的注释。想想代码清单 4-1 中那些多余而又有误导嫌疑的注释吧。

你有没有发现那样的注释是如何误导读者的？在 this.closed 变为 true 的时候，方法并没有返回。方法只在判断到 this.closed 为 true 的时候才返回，否则，就只是等待遥遥无期的超时，然后如果判断 this.closed 还是非 true，就抛出一个异常。

这一细微的误导信息，放在比代码本身更难阅读的注释里面，有可能导致其他程序员快活地调用这个函数,并期望在 `this.closed` 变为 `true` 时立即返回。那位可怜的程序员将会发现自己陷于调试困境之中,拼命想找出代码执行得如此之慢的原因。

4.4.4 循规式注释

所谓每个函数都要有 Javadoc 或每个变量都要有注释的规矩全然是愚蠢可笑的。这类注释徒然让代码变得散乱,满口胡言,令人迷惑不解。

例如,要求每个函数都要有 Javadoc,就会得到类似代码清单 4-3 那样面目可憎的代码。这类废话只会搞乱代码,有可能会误导读者。

代码清单 4-3

```
/**
 *
 * @param title The title of the CD
 * @param author The author of the CD
 * @param tracks The number of tracks on the CD
 * @param durationInMinutes The duration of the CD in minutes
 */
public void addCD(String title, String author,
                  int tracks, int durationInMinutes) {
    CD cd = new CD();
    cd.title = title;
    cd.author = author;
    cd.tracks = tracks;
    cd.duration = durationInMinutes;
    cdList.add(cd);
}
```

4.4.5 日志式注释

有人会在每次编辑代码时,在模块开始处添加一条注释。这类注释就像是一种记录每次修改的日志。我见过满篇尽是这类日志的代码模块。

```
 * Changes (from 11-Oct-2001)
 * --------------------------
 * 11-Oct-2001 : Re-organised the class and moved it to new package
 *               com.jrefinery.date (DG);
 * 05-Nov-2001 : Added a getDescription() method, and eliminated NotableDate
 *               class (DG);
 * 12-Nov-2001 : IBD requires setDescription() method, now that NotableDate
 *               class is gone (DG);  Changed getPreviousDayOfWeek(),
 *               getFollowingDayOfWeek() and getNearestDayOfWeek() to correct
 *               bugs (DG);
 * 05-Dec-2001 : Fixed bug in SpreadsheetDate class (DG);
 * 29-May-2002 : Moved the month constants into a separate interface
 *               (MonthConstants) (DG);
```

第 4 章 注释

```
 * 27-Aug-2002 : Fixed bug in addMonths() method, thanks to N???levka Petr (DG);
 * 03-Oct-2002 : Fixed errors reported by Checkstyle (DG);
 * 13-Mar-2003 : Implemented Serializable (DG);
 * 29-May-2003 : Fixed bug in addMonths method (DG);
 * 04-Sep-2003 : Implemented Comparable.  Updated the isInRange javadocs (DG);
 * 05-Jan-2005 : Fixed bug in addYears() method (1096282) (DG);
```

如果在很久以前，在模块开始处创建并维护这些记录还算有道理，因为那时我们还没有源代码控制系统可用，但是如今，这种冗长的记录只会让模块变得凌乱不堪，应当全部删除。

4.4.6 废话注释

有时，你会看到纯然是废话的注释。它们对于显然之事喋喋不休，毫无新意。

```
/**
 *  Default constructor.
 */
protected AnnualDateRule() {
}
```

对吧？再看看这个：

```
/** The day of the month. */
    private int dayOfMonth;
```

还有这样的废话模范：

```
/**
 * Returns the day of the month.
 *
 * @return the day of the month.
 */
public int getDayOfMonth() {
  return dayOfMonth;
}
```

这类注释废话连篇，我们都学会了视而不见。读代码时，眼光不会停留在它们上面。最终，当代码修改之后，这类注释就变作了谎言一堆。

代码清单 4-4 中的第一条注释貌似还行[①]。它解释了 catch 代码块为何被忽略。不过第二条注释就纯是废话了。显然，该程序员对编写函数中那些 try/catch 代码块感到沮丧。

代码清单 4-4　startSending

```
private void startSending()
{
  try
  {
    doSending();
```

---

① IDE 对注释中拼写检查的支持对我们这些看大量代码的人实在是一种妙事。

```
    }
    catch(SocketException e)
    {
      // normal. someone stopped the request.
    }
    catch(Exception e)
    {
      try
      {
        response.add(ErrorResponder.makeExceptionString(e));
        response.closeAll();
      }
      catch(Exception e1)
      {
        // Give me a break!
      }
    }
}
```

程序员与其纠缠毫无价值的废话注释，不如意识到，他的挫败感可以由改进代码结构而消除。他应该把力气花在将最末一个 `try/catch` 代码块拆解到单独的函数中，如代码清单 4-5 所示。

代码清单 4-5　startSending（重构之后）

```
private void startSending()
{
  try
  {
    doSending();
  }
  catch(SocketException e)
  {
    //normal. someone stopped the request.
  }
  catch(Exception e)
  {
    addExceptionAndCloseResponse(e);
  }
}

private void addExceptionAndCloseResponse(Exception e)
{
  try
  {
    response.add(ErrorResponder.makeExceptionString(e));
    response.closeAll();
  }
  catch(Exception e1)
  {
  }
}
```

用整理代码的决心替代创造废话的冲动吧，这样你就会发现自己将成为更优秀、更快乐的程序员。

4.4.7 可怕的废话

Javadoc 也可能是废话。下列 Javadoc（来自某知名开源库）的目的是什么？答案：无。它们只是源自某种提供文档的不当愿望的废话注释。

```
/** The name. */
private String name;

/** The version. */
private String version;

/** The licenceName. */
private String licenceName;

/** The version. */
private String info;
```

再仔细读读这些注释。你是否发现了剪切-粘贴错误？如果作者在写（或粘贴）注释时都没花心思，又怎么能指望读者从中获益呢？

4.4.8 能用函数或变量时就别用注释

看看以下代码概要：

```
// does the module from the global list <mod> depend on the
// subsystem we are part of?
if (smodule.getDependSubsystems().contains(subSysMod.getSubSystem()))
```

可以改成以下没有注释的版本：

```
ArrayList moduleDependees = smodule.getDependSubsystems();
String ourSubSystem = subSysMod.getSubSystem();
if (moduleDependees.contains(ourSubSystem))
```

代码原作者可能（不太像）是先写注释再编写代码的。不过，作者应该重构代码，如我所做的那样，从而删掉注释。

4.4.9 位置标记

有时，程序员喜欢在源代码中标记某个特别位置。例如，最近我在程序中看到这样一行：

```
// Actions //////////////////////////////////
```

把特定函数茔放在这种标记栏下面，多数时候实属无理。鸡零狗碎，理当删除——特别是尾部那一长串无用的斜杠。

这么说吧，如果标记栏不多，它就会显而易见。所以，尽量少用标记栏，只在特别有价值的时候用。如果滥用标记栏，就会沉没在背景噪音中而被忽略。

4.4.10 括号后面的注释

有时，程序员会在括号后面放置特殊的注释，如代码清单 4-6 所示。尽管这对于含有深度嵌套结构的长函数可能有意义，但只会给我们更愿意编写的短小、封装的函数带来混乱。如果你发现自己想标记右括号，其实应该做的是缩短函数。

代码清单 4-6　wc.java

```java
public class wc {
  public static void main(String[] args) {
    BufferedReader in = new BufferedReader(new InputStreamReader(System.in));
    String line;
    int lineCount = 0;
    int charCount = 0;
    int wordCount = 0;
    try {
      while ((line = in.readLine()) != null) {
        lineCount++;
        charCount += line.length();
        String words[] = line.split("\\W");
        wordCount += words.length;
      } //while
      System.out.println("wordCount = " + wordCount);
      System.out.println("lineCount = " + lineCount);
      System.out.println("charCount = " + charCount);
    } // try
    catch (IOException e) {
      System.err.println("Error:" + e.getMessage());
    } //catch
  } //main
}
```

4.4.11 归属与署名

```
/* Added by Rick   */
```

源代码控制系统非常善于记住是谁在何时添加了什么。没必要用那些小小的签名搞脏代码。你也许会认为，这种注释大概有助于他人了解应该和谁讨论这段代码。不过，事实却是注释在那儿放了一年又一年，越来越不准确，越来越和原作者没关系。

重申一下，源代码控制系统是这类信息最好的归属地。

4.4.12 注释掉的代码

直接把代码注释掉是讨厌的做法。别这么干！

```
        InputStreamResponse response = new InputStreamResponse();
        response.setBody(formatter.getResultStream(), formatter.getByteCount());
//      InputStream resultsStream = formatter.getResultStream();
//      StreamReader reader = new StreamReader(resultsStream);
//      response.setContent(reader.read(formatter.getByteCount()));
```

其他人不敢删除注释掉的代码。他们会想，代码依然放在那儿，一定有其原因，而且这段代码很重要，不能删除。注释掉的代码堆积在一起，就像破酒瓶底的渣滓一般。

看看以下来自 Apache 公共库的代码：

```
this.bytePos = writeBytes(pngIdBytes, 0);
//hdrPos = bytePos;
writeHeader();
writeResolution();
//dataPos = bytePos;
if (writeImageData()) {
  writeEnd();
  this.pngBytes = resizeByteArray(this.pngBytes, this.maxPos);
}
else{
  this.pngBytes=null;
}
return this.pngBytes;
```

这两行代码为什么要注释掉？它们重要吗？它们搁在那儿，是为了给未来的修改做提示吗？或者，只是某人在多年以前注释掉、懒得清理的过时玩意？

20 世纪 60 年代，曾经有一段时间，注释掉的代码可能有用。但我们已经拥有优良的源代码控制系统如此之久，这些系统可以为我们记住不要的代码。我们无须再用注释来标记，删掉即可，它们丢不了。我担保。

4.4.13 HTML 注释

源代码注释中的 HTML 标记是一种厌物，如你在下面代码中所见。编辑器/IDE 中的代码本来易于阅读，却因为 HTML 注释的存在而变得难以卒读。如果注释将由某种工具（例如 Javadoc）抽取出来，呈现到网页，那么该是工具而非程序员来负责给注释加上合适的 HTML 标签。

```
/**
 * Task to run fit tests.
 * This task runs fitnesse tests and publishes the results.
 * <p/>
 * <pre>
```

```
 * Usage:
 * &lt;taskdef name=&quot;execute-fitnesse-tests&quot;
 *     classname=&quot;fitnesse.ant.ExecuteFitnesseTestsTask&quot;
 *     classpathref=&quot;classpath&quot; /&gt;
 * OR
 * &lt;taskdef classpathref=&quot;classpath&quot;
 *           resource=&quot;tasks.properties&quot; /&gt;
 * <p/>
 * &lt;execute-fitnesse-tests
 *     suitepage=&quot;FitNesse.SuiteAcceptanceTests&quot;
 *     fitnesseport=&quot;8082&quot;
 *     resultsdir=&quot;${results.dir}&quot;
 *     resultshtmlpage=&quot;fit-results.html&quot;
 *     classpathref=&quot;classpath&quot; /&gt;
 * </pre>
 */
```

4.4.14 非本地信息

假如你一定要写注释，请确保它描述了离它最近的代码。别在本地注释的上下文环境中给出系统级的信息。以下面的 Javadoc 注释为例，除了可怕的冗余，它还给出了有关默认端口的信息。不过该函数完全没控制那个所谓的默认值。这个注释并未描述该函数，而是在描述系统中远在其他地方的其他函数。当然，也无法确保在包含那个默认值的代码修改之后，这里的注释也会随之修改。

```
/**
 * Port on which fitnesse would run. Defaults to <b>8082</b>.
 *
 * @param fitnessePort
 */
public void setFitnessePort(int fitnessePort)
{
  this.fitnessePort = fitnessePort;
}
```

4.4.15 信息过多

别在注释中添加有趣的历史性话题或者无关的细节描述。下列注释来自某个用来测试 base64 编解码函数的模块。除 RFC 文档编号之外，注释中的其他细节信息对于读者完全没有必要。

```
/*
    RFC 2045 - Multipurpose Internet Mail Extensions (MIME)
    Part One: Format of Internet Message Bodies
    section 6.8.  Base64 Content-Transfer-Encoding
    The encoding process represents 24-bit groups of input bits as output
    strings of 4 encoded characters. Proceeding from left to right, a
```

```
    24-bit input group is formed by concatenating 3 8-bit input groups.
    These 24 bits are then treated as 4 concatenated 6-bit groups, each
    of which is translated into a single digit in the base64 alphabet.
    When encoding a bit stream via the base64 encoding, the bit stream
    must be presumed to be ordered with the most-significant-bit first.
    That is, the first bit in the stream will be the high-order bit in
    the first 8-bit byte, and the eighth bit will be the low-order bit in
    the first 8-bit byte, and so on.
 */
```

4.4.16 不明显的联系

注释及其描述的代码之间的联系应该显而易见。如果你不嫌麻烦要写注释，至少让读者能看到注释和代码，并且理解注释所谈何物。

以来自 Apache 公共库的这段注释为例：

```
/*
 * start with an array that is big enough to hold all the pixels
 * (plus filter bytes), and an extra 200 bytes for header info
 */
this.pngBytes = new byte[((this.width + 1) * this.height * 3) + 200];
```

过滤器字节是什么？与那个+1 有关系吗？或与*3 有关？还是与两者皆有关？为什么用 200？注释的作用是解释未能自行解释的代码。如果注释本身还需要解释，就太遗憾了。

4.4.17 函数头

短函数不需要太多描述。为只做一件事的短函数选个好名字，通常要比写函数头注释好。

4.4.18 非公共代码中的 Javadoc

虽然 Javadoc 对于公共 API 非常有用，但对于不打算作公共用途的代码就令人厌恶了。为系统中的类和函数生成 Javadoc 页并非总有用，而对 Javadoc 注释额外的形式要求几乎等同于八股文章。

4.4.19 范例

我曾为首个 XP Immersion[①] 课程编写了代码清单 4-7 列出的模块。这个模块几乎是糟糕的代码和坏注释风格的典范。后来 Kent Beck 当着几十位满腔热情的学生的面重构了这些代码，将其变得令人愉悦。后来，我在拙著《敏捷软件开发：原则、模式与实践》(*Agile Software*

[①] Object Mentor 公司开办的极限程深入课程。——译者注

Development, Principles, Patterns, and Practices)和 *Software Development* 杂志的"技艺"专栏的第一篇文章中引用了这个例子。

代码清单 4-7　GeneratePrimes.java

```java
/**
 * This class Generates prime numbers up to a user specified
 * maximum.  The algorithm used is the Sieve of Eratosthenes.
 * <p>
 * Eratosthenes of Cyrene, b. c. 276 BC, Cyrene, Libya --
 * d. c. 194, Alexandria.  The first man to calculate the
 * circumference of the Earth.  Also known for working on
 * calendars with leap years and ran the library at Alexandria.
 * <p>
 * The algorithm is quite simple.  Given an array of integers
 * starting at 2.  Cross out all multiples of 2.  Find the next
 * uncrossed integer, and cross out all of its multiples.
 * Repeat untilyou have passed the square root of the maximum
 * value.
 *
 * @author Alphonse
 * @version 13 Feb 2002 atp
 */
import java.util.*;

public class GeneratePrimes
{
  /**
   * @param maxValue is the generation limit.
   */
  public static int[] generatePrimes(int maxValue)
  {
    if (maxValue >= 2) // the only valid case
    {
      // declarations
      int s = maxValue + 1; // size of array
      boolean[] f = new boolean[s];
      int i;
      // initialize array to true.
      for (i = 0; i < s; i++)
        f[i] = true;
      // get rid of known non-primes
      f[0] = f[1] = false;

      // sieve
      int j;
      for (i = 2; i < Math.sqrt(s) + 1; i++)
      {
        if (f[i]) // if i is uncrossed, cross its multiples.
        {
          for (j = 2 * i; j < s; j += i)
            f[j] = false; // multiple is not prime
        }
      }
```

```
      // how many primes are there?
      int count = 0;
      for (i = 0; i < s; i++)
      {
        if (f[i])
          count++; // bump count.
      }

      int[] primes = new int[count];

      // move the primes into the result
      for (i = 0, j = 0; i < s; i++)
      {
        if (f[i])             // if prime
          primes[j++] = i;
      }

      return primes;  // return the primes
    }
    else // maxValue < 2
      return new int[0]; // return null array if bad input.
  }
}
```

这个模块最迷人的地方是，有那么一阵，我们中的许多人都认为它"文档做得很好"。如今，我们认为它是一小团乱麻。看看你能发现多少个不同的注释问题吧。

在代码清单 4-8 中，你可以看到该模块重构后的版本。注意，注释的使用被明显地限制了。在整个模块中只有两个注释。每个注释都足具说明意义。

代码清单 4-8　PrimeGenerator.java（重构后）

```
/**
 * This class Generates prime numbers up to a user specified
 * maximum.  The algorithm used is the Sieve of Eratosthenes.
 * Given an array of integers starting at 2:
 * Find the first uncrossed integer, and cross out all its
 * multiples.  Repeat until there are no more multiples
 * in the array.
 */

public class PrimeGenerator
{
  private static boolean[] crossedOut;
  private static int[] result;

  public static int[] generatePrimes(int maxValue)
  {
    if (maxValue < 2)
      return new int[0];
    else
    {
      uncrossIntegersUpTo(maxValue);
```

```java
    crossOutMultiples();
    putUncrossedIntegersIntoResult();
    return result;
  }
}

private static void uncrossIntegersUpTo(int maxValue)
{
  crossedOut = new boolean[maxValue + 1];
  for (int i = 2; i < crossedOut.length; i++)
    crossedOut[i] = false;
}

private static void crossOutMultiples()
{
  int limit = determineIterationLimit();
  for (int i = 2; i <= limit; i++)
    if (notCrossed(i))
      crossOutMultiplesOf(i);
}

private static int determineIterationLimit()
{
  // Every multiple in the array has a prime factor that
  // is less than or equal to the root of the array size,
  // so we don't have to cross out multiples of numbers
  // larger than that root.
  double iterationLimit = Math.sqrt(crossedOut.length);
  return (int) iterationLimit;
}

private static void crossOutMultiplesOf(int i)
{
  for (int multiple = 2*i;
       multiple < crossedOut.length;
       multiple += i)
    crossedOut[multiple] = true;
}

private static boolean notCrossed(int i)
{
  return crossedOut[i] == false;
}

private static void putUncrossedIntegersIntoResult()
{
  result = new int[numberOfUncrossedIntegers()];
  for (int j = 0, i = 2; i < crossedOut.length; i++)
    if (notCrossed(i))
      result[j++] = i;
}

private static int numberOfUncrossedIntegers()
{
```

```
    int count = 0;
    for (int i = 2; i < crossedOut.length; i++)
      if (notCrossed(i))
        count++;

    return count;
  }
}
```

很容易说明，第一个注释完全是多余的，因为它读起来非常像是 `generatePrimes` 函数自身。不过，我认为这段注释还是省了读者去读具体算法的精力，所以我倾向于保留它。

第二个注释显然很有必要。它解释了平方根作为循环限制的理由。我找不到能说明白这个问题的简单变量名或者其他编程结构。另外，对平方根的使用可能也有点武断。通过限制平方根循环，我是否真节省了许多时间？计算平方根所花的时间会不会比省下的时间还要多？这些都值得考虑。使用平方根作为循环限制，满足了我这种旧式 C 语言和汇编语言黑客，不过我可不敢说抵得上其他人为理解它而花的时间和精力。

4.5 文献

[KP78]：Kernighan and Plaugher, *The Elements of Programming Style*, 2d. ed., McGraw-Hill, 1978.

第 5 章

格 式

　　当有人查看底层代码实现时，我们希望他们为代码的整洁、一致及所感知到的对细节的关注而震惊。我们希望他们高高扬起眉毛，一路看下去。我们希望他们感受到专业人士们的劳作。但若他们看到的只是一堆像是由酒醉的水手写出的鬼画符，那他们多半会得出结论，认为项目其他任何部分也同样对细节漠不关心。

　　你应该保持良好的代码格式。你应该选用一套管理代码格式的简单规则，然后贯彻这些规则。如果你在团队中工作，则团队应该一致同意采用一套简单的格式规则，所有成员都要遵从这套规则。使用能帮你应用这套格式规则的自动化工具会很有帮助。

5.1 格式的目的

先明确一下，代码格式很重要。代码格式不可忽略，必须严肃对待。代码格式关乎沟通，而沟通是专业开发者的头等大事。

或许你认为"让代码能工作"才是专业开发者的头等大事。然而，我希望本书能让你抛开那种想法。你今天编写的功能代码，极有可能在下一版本中被修改，但代码的可读性却会对以后可能发生的修改行为产生深远影响。原始代码修改之后很久，其代码风格和可读性仍会影响代码的可维护性和可扩展性。即便代码已不复存在，你的风格和律条也会存活下来。

那么，代码格式的哪些相关方面能帮我们更好地沟通呢？

5.2 垂直格式

从垂直尺寸开始吧。源代码文件该有多大？在 Java 中，文件尺寸与类尺寸极其相关。讨论类时再说类的尺寸。现在先考虑文件尺寸。

多数 Java 源代码文件有多大呢？事实表明，尺寸各有不同，长度殊异，如图 5-1 所示。

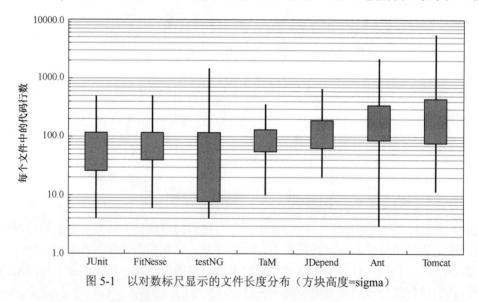

图 5-1　以对数标尺显示的文件长度分布（方块高度=sigma）

图 5-1 中涉及 7 个不同项目：JUnit、FitNesse、testNG、Time and Money、JDepend、Ant 和 Tomcat。贯穿方块的直线两端显示这些项目中最小和最大的文件长度。方块表示在平均值

以上或以下的大约三分之一文件（一个标准偏差[①]）的长度。方块中间位置就是平均数。所以 FitNesse 项目的文件平均长度是 65 行，而平均长度上面三分之一在 40～100 行及 100 行以上之间。FitNesse 中的文件最大大约有 400 行，最小有 6 行。这是个对数标尺，所以较小的垂直位置差异意味着文件绝对长度的较大差异。

JUnit、FitNesse 和 Time and Money 由相对较小的文件组成，没有一个文件长度超过 500 行，多数都短于 200 行。Tomcat 和 Ant 则有些文件达到数千行，将近一半文件长度长于 200 行。

对我们来说，这意味着什么？意味着有可能用大多数为 200 行、最长 500 行的单个文件构造出色的系统（FitNesse 总长约 50000 行）。尽管这并非不可违背的原则，但也应该乐于接受。短文件通常比长文件易于理解。

5.2.1 向报纸学习

想想写得很好的报纸文章。你从上到下阅读，在顶部，你期望有个头条，告诉你故事主题，好让你决定是否读下去。第一段是整个故事的大纲，给出粗线条概述，但隐藏了故事细节。接着读下去，细节渐次增加，直至你了解所有的日期、名字、引语、说法及其他细节。

源文件也要像报纸文章那样。名称应当简单且一目了然。名称本身应该足以告诉我们是否在正确的模块中。源文件最顶部应该给出高层次的概念和算法。细节应该往下渐次展开，直至找到源文件中最底层的函数和细节。

报纸由许多篇文章组成，多数短小精悍，有些稍微长点儿，很少有占满一整页的。这样做，报纸才可用。假若一份报纸只登载一篇长故事，其中充斥毫无组织的事实、日期、名字等，就没人会去读它。

5.2.2 概念间垂直方向上的区隔

几乎所有的代码都是从上往下读，从左往右读。每行展现一个表达式或一个子句，每组代码行展示一条完整的思路。这些思路用空白行区隔开来。

以代码清单 5-1 为例。在封包声明、导入声明和每个函数之间，都有空白行隔开。这条极其简单的规则极大地影响代码的视觉外观。每个空白行都是一条线索，标识出新的独立概念。往下读代码时，你的目光总会停留于空白行之后的那一行。

代码清单 5-1   BoldWidget.java

```
package fitnesse.wikitext.widgets;

import java.util.regex.*;
```

[①] 方块显示平均数的 sigma/2 以上及以下长度。没错，我知道文件长度分布不太寻常，所以标准偏差也并非那么精确。不过在此并不寻求精确，只是找个感觉罢了。

```
public class BoldWidget extends ParentWidget {
  public static final String REGEXP = "'''.+?'''";
  private static final Pattern pattern = Pattern.compile("'''(.+?)'''",
    Pattern.MULTILINE + Pattern.DOTALL
  );

  public BoldWidget(ParentWidget parent, String text) throws Exception {
    super(parent);
    Matcher match = pattern.matcher(text);
    match.find();
    addChildWidgets(match.group(1));
  }

  public String render() throws Exception {
    StringBuffer html = new StringBuffer("<b>");
    html.append(childHtml()).append("</b>");
    return html.toString();
  }
}
```

如代码清单 5-2 所示，抽掉这些空白行，代码可读性就减弱了不少。

代码清单 5-2　BoldWidget.java

```
package fitnesse.wikitext.widgets;
import java.util.regex.*;
public class BoldWidget extends ParentWidget {
  public static final String REGEXP = "'''.+?'''";
  private static final Pattern pattern = Pattern.compile("'''(.+?)'''",
    Pattern.MULTILINE + Pattern.DOTALL);
  public BoldWidget(ParentWidget parent, String text) throws Exception {
    super(parent);
    Matcher match = pattern.matcher(text);
    match.find();
    addChildWidgets(match.group(1));}
  public String render() throws Exception {
    StringBuffer html = new StringBuffer("<b>");
    html.append(childHtml()).append("</b>");
    return html.toString();
  }
}
```

在你不特意注视时，后果就更严重了。在第一个例子中，代码组会跳到你眼前，而第二个例子就像一团乱麻。两段代码的区别，展示了垂直方向上区隔的作用。

5.2.3　垂直方向上的靠近

如果说空白行隔开了概念，靠近的代码行则暗示了它们之间的紧密关系。所以，紧密相关的代码应该互相靠近。注意，代码清单 5-3 中的注释是如何割断两个实体变量间的联系的。

代码清单 5-3

```java
public class ReporterConfig {

    /**
     * The class name of the reporter listener
     */
    private String m_className;

    /**
     * The properties of the reporter listener
     */
    private List<Property> m_properties = new ArrayList<Property>();

    public void addProperty(Property property) {
        m_properties.add(property);
    }
}
```

代码清单 5-4 更易于阅读。它刚好"一览无余",至少对我来说是这样。我一眼就能看到,这是一个有两个变量和一个方法的类。看上面的代码时,我不得不更多地移动头部和眼球,才能获得相同的理解度。

代码清单 5-4

```java
public class ReporterConfig {
    private String m_className;
    private List<Property> m_properties = new ArrayList<Property>();

    public void addProperty(Property property) {
        m_properties.add(property);
    }
}
```

5.2.4 垂直距离

你是否曾经在某个类中摸索,从一个函数跳到另一个函数,上下求索,想要弄清楚这些函数如何操作、如何互相相关,最后却被搞糊涂了?你是否曾经苦苦追索某个变量或函数的继承链条?这让人沮丧,因为你的目的是想要理解系统做什么,但是花的时间和精力却用于找到和记住那些代码碎片在哪里。

关系密切的概念应该互相靠近[G10]。显然,这条规则并不适用于分布在不同文件中的概念。除非有很好的理由,否则就不要把关系密切的概念放到不同的文件中。实际上,这也是避免使用 protected 变量的理由之一。

对于那些关系密切、放置于同一源文件中的概念,它们之间的区隔应该成为对彼此的易懂度影响有多重要的衡量标准。应避免迫使读者在源文件和类中跳来跳去。

变量声明。变量声明应尽可能靠近其使用位置。因为函数很短,本地变量应该在函数的顶部出现，就像 JUnit 4.3.1 中这个稍长的函数中那样。

```
private static void readPreferences() {
  InputStream is= null;
  try {
    is= new FileInputStream(getPreferencesFile());
    setPreferences(new Properties(getPreferences()));
    getPreferences().load(is);
  } catch (IOException e) {
    try {
      if (is != null)
        is.close();
    } catch (IOException e1) {
    }
  }
}
```

循环中的控制变量应该总是在循环语句中声明，如下列来自同一项目的绝妙小函数所示。

```
public int countTestCases() {
  int count= 0;
  for (Test each : tests)
    count += each.countTestCases();
  return count;
}
```

偶尔，在较长的函数中，变量也可能在某个代码块顶部，或在循环之前声明。你可以在以下摘自 TestNG 中一个长函数的代码片段中找到类似的变量。

```
...
for (XmlTest test : m_suite.getTests()) {
  TestRunner tr = m_runnerFactory.newTestRunner(this, test);
  tr.addListener(m_textReporter);
  m_testRunners.add(tr);

  invoker = tr.getInvoker();

  for (ITestNGMethod m : tr.getBeforeSuiteMethods()) {
    beforeSuiteMethods.put(m.getMethod(), m);
  }

  for (ITestNGMethod m : tr.getAfterSuiteMethods()) {
    afterSuiteMethods.put(m.getMethod(), m);
  }
}
...
```

实体变量。实体变量应该在类的顶部声明。这应该不会增加变量的垂直距离，因为在设计良好的类中，它们如果不是被该类的所有方法所用，也会被大多数方法所用。

关于实体变量应该放在哪里，争论不断。在 C++ 中，通常会采用所谓剪刀原则（scissors rule），即所有实体变量都放在底部。而在 Java 中，惯例是放在类的顶部。没理由一定要遵循其他惯例，而重点是在谁都知道的地方声明实体变量。大家都应该知道在哪儿能看到这些声明。

例如，JUnit 4.3.1 中的这个奇怪情形。我极力删减了这个类，以说明问题。如果你看到代码清单大致一半的位置，会看到在那里声明了两个实体变量。如果放在更好的位置，它们就会更明显。而现在，读代码者只能在无意中看到这些声明（就像我一样）。

```java
public class TestSuite implements Test {
  static public Test createTest(Class<? extends TestCase> theClass,
                                String name) {
    ...
  }

  public static Constructor<? extends TestCase>
  getTestConstructor(Class<? extends TestCase> theClass)
  throws NoSuchMethodException {
    ...
  }

  public static Test warning(final String message) {
    ...
  }

  private static String exceptionToString(Throwable t) {
    ...
  }

  private String fName;

  private Vector<Test> fTests= new Vector<Test>(10);

  public TestSuite() {
  }

   public TestSuite(final Class<? extends TestCase> theClass) {
    ...
  }

  public TestSuite(Class<? extends TestCase>  theClass, String name) {
    ...
  }
  ... ... ... ... ...
}
```

相关函数。若某个函数调用了另外一个，就应该把它们放到一起，而且调用者应该尽可能放在被调用者上面。这样，程序就会有自然的顺序。若坚定地遵循这条约定，读者将能够确信函数声明总会在其调用后很快出现。以源自 FitNesse 的代码清单 5-5 为例，注意，顶部的函数如何调用其下的函数，而这些被调用的函数又如何调用更下面的函数。这样就能容易找到被调用的函数，从而极大地增强整个模块的可读性。

代码清单 5-5　WikiPageResponder.java

```java
public class WikiPageResponder implements SecureResponder {
  protected WikiPage page;
  protected PageData pageData;
  protected String pageTitle;
  protected Request request;
  protected PageCrawler crawler;

  public Response makeResponse(FitNesseContext context, Request request)
    throws Exception {
    String pageName = getPageNameOrDefault(request, "FrontPage");
    loadPage(pageName, context);
    if (page == null)
      return notFoundResponse(context, request);
    else
      return makePageResponse(context);
  }

  private String getPageNameOrDefault(Request request, String defaultPageName)
  {
    String pageName = request.getResource();
    if (StringUtil.isBlank(pageName))
      pageName = defaultPageName;
    return pageName;
  }

  protected void loadPage(String resource, FitNesseContext context)
    throws Exception {
    WikiPagePath path = PathParser.parse(resource);
    crawler = context.root.getPageCrawler();
    crawler.setDeadEndStrategy(new VirtualEnabledPageCrawler());
    page = crawler.getPage(context.root, path);
    if (page != null)
      pageData = page.getData();
  }

  private Response notFoundResponse(FitNesseContext context, Request request)
    throws Exception {
    return new NotFoundResponder().makeResponse(context, request);
  }

  private SimpleResponse makePageResponse(FitNesseContext context)
    throws Exception {
    pageTitle = PathParser.render(crawler.getFullPath(page));
    String html = makeHtml(context);

    SimpleResponse response = new SimpleResponse();
    response.setMaxAge(0);
    response.setContent(html);
    return response;
  }
  ...
```

说句题外话，以上代码片段也是把常量保持在恰当级别的好例子 [G35]。"FrontPage" 常量可以埋在 getPageNameOrDefault 函数中，但那样就会把一个众人皆知的常量埋藏到不太合适的底层函数中。更好的做法是把它放在易于找到的位置，然后再传递到真实使用的位置。

概念相关。概念相关的代码应该放到一起。代码的相关性越强，彼此之间的距离就该越短。

如上所述，相关性应建立在直接依赖的基础上，如函数间调用，或函数使用某个变量，但也有其他相关性的可能。相关性可能来自执行相似操作的一组函数。请看以下来自 JUnit 4.3.1 的代码片段：

```
public class Assert {
  static public void assertTrue(String message, boolean condition) {
    if (!condition)
      fail(message);
  }

  static public void assertTrue(boolean condition) {
    assertTrue(null, condition);
  }

  static public void assertFalse(String message, boolean condition) {
    assertTrue(message, !condition);
  }

  static public void assertFalse(boolean condition) {
    assertFalse(null, condition);
  }
...
```

这些函数有着极强的概念相关性，因为它们拥有共同的命名模式，属于执行同一基础任务的不同变种。互相调用是第二位的。即便没有互相调用，也应该放在一起。

5.2.5 垂直顺序

一般而言，我们想自上向下展示函数调用依赖顺序。也就是说，被调用的函数应该放在执行调用的函数下面①。这样就建立了一种自顶向下贯穿源代码模块的良好信息流。

像报纸文章一般，我们期望最重要的概念先出现，并期望以包括最少细节的方式表述

---

① 在 Pascal、C 和 C++等语言中则完全不同，在这些语言中，函数应该在被调用之前定义，至少是声明。

它们，而期望底层细节最后出现。这样，我们就能扫过源代码文件，自最前面的几个函数获知要旨，而不至于沉溺于细节。代码清单 5-5 就是如此组织的。或许，更好的例子是代码清单 15-5 及代码清单 3-7。

5.3 横向格式

　　一行代码应该有多宽？要回答这个问题，来看看典型的程序中代码行的宽度。我们再一次检验 7 个不同项目。图 5-2 展示了这 7 个项目的代码行宽度分布情况。其中展现的规律性令人印象深刻，45 个字符左右的宽度分布尤为如此。其实，20~60 个字符的每个宽度，都代表全部代码行数的 1%。也就是总共 40%！或许其余 30% 的代码行短于 10 个字符。记住，这是个对数标尺，所以图中长于 80 个字符部分的线性下降在实际情况中会极其可观。程序员们显然更喜爱短代码行。

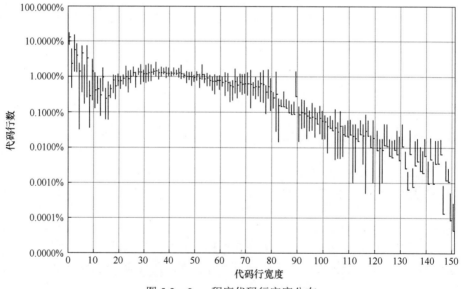

图 5-2　Java 程序代码行宽度分布

　　这说明，应该尽力保持代码行短小。死守 80 个字符的上限有点僵化，而且我也不反对代码行宽度达到 100 个字符或 120 个字符。再多的话，大抵就是肆意妄为了。

　　我一向遵循无须拖动滚动条即可看到最右边代码的原则，但近年来显示器越来越宽，而年轻程序员又能将显示字符缩小到如此程度，屏幕上甚至能容纳 200 个字符的宽度。别那么做！我个人的上限是 120 个字符。

5.3.1 水平方向上的区隔与靠近

我们使用空格字符将彼此紧密相关的事物连接到一起,也用空格字符把相关性较弱的事物区隔开。请看以下函数:

```
private void measureLine(String line) {
  lineCount++;
  int lineSize = line.length();
  totalChars += lineSize;
  lineWidthHistogram.addLine(lineSize, lineCount);
  recordWidestLine(lineSize);
}
```

我在赋值操作符周围加上空格字符,以达到强调的目的。赋值语句有两个确定而重要的要素:左边和右边。空格字符加强了区隔效果。

另一方面,我不在函数名和左圆括号之间加空格。这是因为函数与其参数密切相关,如果隔开,就会显得互无关系。我把函数调用括号中的参数一一隔开,强调逗号,表示参数是互相分离的。

空格字符的另一种用法是强调其前面的运算符。

```
public class Quadratic {
  public static double root1(double a, double b, double c) {
    double determinant = determinant(a, b, c);
    return (-b + Math.sqrt(determinant)) / (2*a);
  }

  public static double root2(int a, int b, int c) {
    double determinant = determinant(a, b, c);
    return (-b - Math.sqrt(determinant)) / (2*a);
  }

  private static double determinant(double a, double b, double c) {
    return b*b - 4*a*c;
  }
}
```

看看这些等式读起来多舒服。乘法因子之间没加空格,因为乘法具有较高优先级。加减法运算项之间用空格隔开,因为加法和减法优先级较低。

不幸的是,多数代码格式化工具都会漠视运算符优先级,从头到尾采用同样的空格方式。在重新格式化代码后,以上这些微妙的空格用法就消失殆尽了。

5.3.2 水平对齐

当我还是个汇编语言程序员时[①]，使用水平对齐来强调某些程序结构。开始用 C、C++ 编码，最终转向 Java 后，我继续尽力对齐一组声明中的变量名，或一组赋值语句中的右值。我的代码看起来大概是这样：

```java
public class FitNesseExpediter implements ResponseSender
{
  private     Socket          socket;
  private     InputStream     input;
  private     OutputStream    output;
  private     Request         request;
  private     Response        response;
  private     FitNesseContext context;
  protected   long            requestParsingTimeLimit;
  private     long            requestProgress;
  private     long            requestParsingDeadline;
  private     boolean         hasError;

  public FitNesseExpediter(Socket          s,
                           FitNesseContext context) throws Exception
  {
    this.context =          context;
    socket =                s;
    input =                 s.getInputStream();
    output =                s.getOutputStream();
    requestParsingTimeLimit = 10000;
  }
```

我发现这种对齐方式没什么用。对齐，像是在强调不重要的东西，把我的目光从真正的意义上拉开。例如，在上面的声明列表中，你会从上到下阅读变量名，而忽视了它们的类型。同样，在赋值语句代码清单中，你也会从上到下阅读右值，而对赋值运算符视而不见。更麻烦的是，代码自动格式化工具通常会把这类对齐消除掉。

所以，我最终放弃了这种做法。如今，我更喜欢用不对齐的声明和赋值，如下所示，因为它们指出了重点。如果有较长的列表需要做对齐处理，那问题就是在列表的长度上而不是对齐上。下面 FitNesseExpediter 类中声明列表的长度说明该类应该被拆分了。

```java
public class FitNesseExpediter implements ResponseSender
{
  private Socket socket;
  private InputStream input;
  private OutputStream output;
  private Request request;
  private Response response;
  private FitNesseContext context;
```

[①] 开什么玩笑！到现在我仍是一个汇编语言程序员。把男孩从铁旁边赶走容易，从男孩身边把铁拿走可难！

```
  protected long requestParsingTimeLimit;
  private long requestProgress;
  private long requestParsingDeadline;
  private boolean hasError;

  public FitNesseExpediter(Socket s, FitNesseContext context) throws Exception
  {
    this.context = context;
    socket = s;
    input = s.getInputStream();
    output = s.getOutputStream();
    requestParsingTimeLimit = 10000;
  }
```

5.3.3 缩进

　　源文件是一种继承结构，而不是一种大纲结构。其中的信息涉及整个文件、文件中每个类、类中的方法、方法中的代码块，也涉及代码块中的代码块。这种继承结构中的每一层级都圈出一个范围，名称可以在其中声明，而声明和执行语句也可以在其中解释。

　　要让这种范围式继承结构可见，我们依源代码行在继承结构中的位置对源代码行做缩进处理。在文件顶层的语句，例如大多数的类声明，根本不缩进。类中的方法相对该类缩进一个层级。方法的实现相对方法声明缩进一个层级。代码块的实现相对于其容器代码块缩进一个层级，以此类推。

　　程序员相当依赖这种缩进模式。他们从代码行左边查看自己在什么范围内工作，这让他们能快速跳过与当前关注的情形无关的范围，例如 if 或 while 语句的实现之类。他们的眼光扫过左边，查找新的方法声明、新变量，甚至新类。没有缩进的话，程序就会变得无法阅读。

　　试看以下在语法和语义上等价的两个程序：

```
public class FitNesseServer implements SocketServer { private FitNesseContext
context; public FitNesseServer(FitNesseContext context) { this.context =
context; } public void serve(Socket s) { serve(s, 10000); } public void
serve(Socket s, long requestTimeout) { try { FitNesseExpediter sender = new
FitNesseExpediter(s, context);
sender.setRequestParsingTimeLimit(requestTimeout); sender.start(); }
catch(Exception e) { e.printStackTrace(); } } }

-----

public class FitNesseServer implements SocketServer {
  private FitNesseContext context;
  public FitNesseServer(FitNesseContext context) {
    this.context = context;
  }

  public void serve(Socket s) {
    serve(s, 10000);
```

```
    }

    public void serve(Socket s, long requestTimeout) {
      try {
        FitNesseExpediter sender = new FitNesseExpediter(s, context);
        sender.setRequestParsingTimeLimit(requestTimeout);
        sender.start();
      }
      catch (Exception e) {
        e.printStackTrace();
      }
    }
  }
```

你能很快洞悉有缩进的那个文件的结构。你几乎能立即辨识出那些变量、构造器、存取器和方法。只需要几秒就能了解这是一个套接字的简单前端，其中包括了超时设定。而对于未缩进的版本，则不经过一番折腾就无法明白。

违反缩进规则。 有时，我会忍不住想要在短小的 `if` 语句、`while` 循环或小函数中违反缩进规则。一旦这么做了，我多数时候还是会回过头去加上缩进。这样就避免了出现以下这种范围层级坍塌到一行的情况：

```
public class CommentWidget extends TextWidget
{
  public static final String REGEXP = "^#[^\r\n]*(?:(?:\r\n)|\n|\r)?";

  public CommentWidget(ParentWidget parent, String text){super(parent, text);}
  public String render() throws Exception {return ""; }
}
```

我更喜欢扩展和缩进范围，就像这样：

```
public class CommentWidget extends TextWidget {
  public static final String REGEXP = "^#[^\r\n]*(?:(?:\r\n)|\n|\r)?";

  public CommentWidget(ParentWidget parent, String text) {
    super(parent, text);
  }

  public String render() throws Exception {
    return "";
  }
}
```

5.3.4 空范围

有时，`while` 或 `for` 语句的语句体为空，如下所示。我不喜欢这种结构，所以尽量不使用它。如果无法避免，就确保空范围体的缩进，并用括号包围起来。我无法告诉你，我曾经多少次被静静安坐在与 `while` 循环语句同一行末尾的分号所欺骗，除非把那个分号放到另一行再加以缩进，否则很难看到它。

```
while (dis.read(buf, 0, readBufferSize) != -1)
  ;
```

5.4 团队规则

每个程序员都有自己喜欢的格式规则，但如果在团队中工作，就是团队说了算[①]。

一组开发者应当认同一种格式风格，每个成员都应该采用大家都认同的那种风格。我们想要让软件拥有一以贯之的风格，而不想让它显得是由一大票意见相左的个人所写成的。

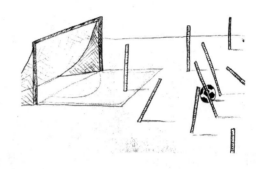

2002 年启动 FitNesse 项目时，我和开发团队一起制订了一套编码规则。这只花了我们 10 分钟时间。我们决定了在什么地方放置括号，缩进几个字符，如何命名类、变量和方法，如此等等。然后，我们把这些规则编写进 IDE 的代码格式功能，接着就一直沿用。这些规则并非全是我喜爱的，但它们是团队共同确定的规则。作为团队一员，在为 FitNesse 项目编写代码时，我遵循这些规则。

记住，好的软件系统是由一系列读起来不错的代码文件组成的。它们需要拥有一致和顺畅的风格。读者要确信，他们在一个源文件中看到的格式风格在其他文件中也是适用的。绝对不要用各种不同的风格来编写源代码，这样会增加代码的复杂度。

5.5 "鲍勃大叔"的格式规则

我个人使用的规则相当简单，如代码清单 5-6 所示。可以把这段代码看作是展示如何把代码写成最好的编码标准文档的范例。

代码清单 5-6　CodeAnalyzer.java

```
public class CodeAnalyzer implements JavaFileAnalysis {
  private int lineCount;
  private int maxLineWidth;
  private int widestLineNumber;
  private LineWidthHistogram lineWidthHistogram;
  private int totalChars;
```

[①] 团队规则，原文 team rules。单词 rule 在这里有两个意思，一个是名词"规则"，另一个是动词"管辖"，所以本节标题玩了个文字游戏。中文不易译出，故采取意译加注。——译者注

```java
public CodeAnalyzer() {
  lineWidthHistogram = new LineWidthHistogram();
}

public static List<File> findJavaFiles(File parentDirectory) {
  List<File> files = new ArrayList<File>();
  findJavaFiles(parentDirectory, files);
  return files;
}

private static void findJavaFiles(File parentDirectory, List<File> files) {
  for (File file : parentDirectory.listFiles()) {
    if (file.getName().endsWith(".java"))
      files.add(file);
    else if (file.isDirectory())
      findJavaFiles(file, files);
  }
}

public void analyzeFile(File javaFile) throws Exception {
  BufferedReader br = new BufferedReader(new FileReader(javaFile));
  String line;
  while ((line = br.readLine()) != null)
    measureLine(line);
}

private void measureLine(String line) {
  lineCount++;
  int lineSize = line.length();
  totalChars += lineSize;
  lineWidthHistogram.addLine(lineSize, lineCount);
  recordWidestLine(lineSize);
}

private void recordWidestLine(int lineSize) {
  if (lineSize > maxLineWidth) {
    maxLineWidth = lineSize;
    widestLineNumber = lineCount;
  }
}

public int getLineCount() {
  return lineCount;
}

public int getMaxLineWidth() {
  return maxLineWidth;
}

public int getWidestLineNumber() {
  return widestLineNumber;
}

public LineWidthHistogram getLineWidthHistogram() {
```

```
      return lineWidthHistogram;
    }

    public double getMeanLineWidth() {
      return (double) totalChars / lineCount;
    }

    public int getMedianLineWidth() {
      Integer[] sortedWidths = getSortedWidths();
      int cumulativeLineCount = 0;
      for (int width : sortedWidths) {
        cumulativeLineCount += lineCountForWidth(width);
        if (cumulativeLineCount > lineCount / 2)
          return width;
      }
      throw new Error("Cannot get here");
    }

    private int lineCountForWidth(int width) {
      return lineWidthHistogram.getLinesforWidth(width).size();
    }
    private Integer[] getSortedWidths() {
      Set<Integer> widths = lineWidthHistogram.getWidths();
      Integer[] sortedWidths = (widths.toArray(new Integer[0]));
      Arrays.sort(sortedWidths);
      return sortedWidths;
    }
  }
```

第 6 章

对象和数据结构

我们将变量设置为私有（private）有一个理由：不想让其他人依赖这些变量。我们还想在心血来潮时能自由修改其类型或实现。那么，为什么还是有那么多程序员不假思索就给对象添加赋值器（setter）和取值器（getter），将私有变量公之于众，如同它们是公共变量一般呢？

6.1 数据抽象

看看代码清单 6-1 和代码清单 6-2 之间的区别。每段代码都表示笛卡儿平面上的一个点。

不过，其中一个曝露了其实现，而另一个则完全隐藏了其实现。

代码清单 6-1　具象点

```
public class Point {
  public double x;
  public double y;
}
```

代码清单 6-2　抽象点

```
public interface Point {
  double getX();
  double getY();
  void setCartesian(double x, double y);
  double getR();
  double getTheta();
  void setPolar(double r, double theta);
}
```

代码清单 6-2 的漂亮之处在于，你不知道该实现会是在矩形坐标系中还是在极坐标系中。可能两个都不是！然而，该接口还是明白无误地呈现了一种数据结构。

不过它呈现的还不止是一个数据结构。那些方法固化了一套存取策略，你可以单独读取某个坐标，但必须通过一次原子操作设定所有坐标。

而代码清单 6-1 则非常清楚地表明是在矩形坐标系中实现的，并要求我们单个操作那些坐标。这就曝露了实现。实际上，即便变量都是私有的，而且我们也通过变量取值器和赋值器使用变量，其实现也被曝露了。

隐藏实现并非只是在变量之间放上一个函数层那么简单。隐藏实现关乎抽象！类并不简单地用取值器和赋值器将其变量推向外界，而是曝露抽象接口，以便用户无须了解数据的实现就能操作数据本体（essence）。

看看代码清单 6-3 和代码清单 6-4。前者采用具象手段与机动车的燃料层通信，而后者则采用百分比抽象。你能确定前者里面都是一些变量存取器，而却无法得知后者中的数据形态。

代码清单 6-3　具象机动车

```
public interface Vehicle {
  double getFuelTankCapacityInGallons();
  double getGallonsOfGasoline();
}
```

代码清单 6-4　抽象机动车

```
public interface Vehicle {
  double getPercentFuelRemaining();
}
```

以上两段代码以后者为佳。我们不愿意曝露数据细节，而更愿意以抽象形态表述数据。这并不意味着只是用接口和/或赋值器、取值器就万事大吉。要以最好的方式呈现某个对象包含的数据，需要进行严肃的思考。随意乱加取值器和赋值器是最坏的选择。

6.2 数据、对象的反对称性

这两个例子展示了对象与数据结构之间的差异。对象把数据隐藏于抽象之后，曝露操作数据的函数；而数据结构曝露其数据，没有提供有意义的函数。回过头再读一遍，注意这两种定义的本质，其实它们是对立的。这种差异貌似微小，但却有深远的含义。

例如，代码清单 6-5 中的过程式形状代码范例。Geometry 类操作 3 个形状类。形状类都是简单的数据结构，没有任何行为。所有行为都在 Geometry 类中。

代码清单 6-5　过程式形状代码

```java
public class Square {
  public Point topLeft;
  public double side;
}

public class Rectangle {
  public Point topLeft;
  public double height;
  public double width;
}

public class Circle {
  public Point center;
  public double radius;
}

public class Geometry {
  public final double PI = 3.141592653589793;

  public double area(Object shape) throws NoSuchShapeException
  {
    if (shape instanceof Square) {
      Square s = (Square)shape;
      return s.side * s.side;
    }
    else if (shape instanceof Rectangle) {
      Rectangle r = (Rectangle)shape;
      return r.height * r.width;
    }
    else if (shape instanceof Circle) {
      Circle c = (Circle)shape;
      return PI * c.radius * c.radius;
    }
    throw new NoSuchShapeException();
  }
}
```

面向对象编程的程序员可能会对此嗤之以鼻，抱怨说这是过程式代码——他们大概是对的，不过这种嘲笑并不完全正确。想想看，如果给 Geometry 类添加一个 primeter() 函数会怎样？那些形状类根本不会因此而受影响！另一方面，如果要添加一个新形状，就得修改 Geometry 中的所有函数来处理它。再读一遍代码。注意，这两种情形也是直接对立的。

现在来看看代码清单 6-6 中的面向对象方案。这里，area() 方法是多态的，不需要有 Geometry 类。所以，如果要添加一个新形状，现有的函数中没有一个会受到影响，而当添加新函数时所有的形状都得做修改[①]！

代码清单 6-6　多态式形状

```java
public class Square implements Shape {
  private Point topLeft;
  private double side;

  public double area() {
    return side*side;
  }
}

public class Rectangle implements Shape {
  private Point topLeft;
  private double height;
  private double width;

  public double area() {
    return height * width;
  }
}

public class Circle implements Shape {
  private Point center;
  private double radius;
  public final double PI = 3.141592653589793;

  public double area() {
    return PI * radius * radius;
  }
}
```

我们再次看到这两种定义的本质：它们是截然对立的。这说明了对象与数据结构之间的二分原理：

> 过程式代码（使用数据结构的代码）便于在不改动既有数据结构的前提下添加新函数；面向对象代码便于在不改动既有函数的前提下添加新类。

反过来讲也说得通：

[①] 经验丰富的面向对象编程的设计人员都知道一些方法，例如，VISITOR 模式或双向分派，但这些方法也有成本，而且通常返回一种过程式程序的结构。

过程式代码难以添加新数据结构，因为必须修改所有函数；面向对象代码难以添加新函数，因为必须修改所有类。

所以，对于面向对象较难完成的事，对于过程式代码却较容易，反之亦然！

在任何一个复杂系统中，都会有需要添加新数据类型而不是新函数的时候。这时，对象和面向对象就比较适合。另一方面，也会有想要添加新函数而不是数据类型的时候。在这种情况下，过程式代码和数据结构就更适合。

老练的程序员知道，一切都是对象的说法只是一个传说。有时候你真的想要在简单数据结构上做一些过程式的操作。

6.3 得墨忒耳律

著名的得墨忒耳律（The Law of Demeter）[1]认为，模块不应了解它所操作对象的内部情形。如上节所见，对象隐藏数据，曝露操作。这意味着对象不应通过存取器曝露其内部结构，因为这样更像是曝露而非隐藏其内部结构。

更准确地说，得墨忒耳律认为，类 C 的方法 f 只应该调用以下对象的方法：

- C；
- 由 f 创建的对象；
- 作为参数传递给 f 的对象；
- 由 C 的实体变量持有的对象。

方法不应调用由任何函数返回的对象的方法。换言之，只跟朋友谈话，不与陌生人谈话。

下列代码[2]违反了得墨忒耳律（除违反其他规则之外），因为它调用了 `getOptions()` 返回值的 `getScratchDir()` 函数，又调用了 `getScratchDir()` 返回值的 `getAbsolutePath()` 方法。

```
final String outputDir = ctxt.getOptions().getScratchDir().getAbsolutePath();
```

6.3.1 火车失事

这类代码常被称作火车失事，因为它看起来就像是一列火车。这类连串的调用通常被认为是肮脏的风格，应该避免[G36]。最好做类似如下的切分：

```
Options opts = ctxt.getOptions();
File scratchDir = opts.getScratchDir();
```

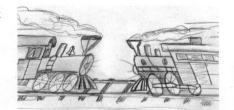

---

[1] 由 Ian Holland 于 1987 年在位于波士顿的美国东北大学提出。——译者注
[2] 来自 Apache 框架中某处。

```
final String outputDir = scratchDir.getAbsolutePath();
```

上述代码是否违反了得墨忒耳律呢？当然，模块知道 ctxt 对象包含多个选项，每个选项中都有一个临时目录，而每个临时目录都有一个绝对路径。对于一个函数，这些知识真够丰富的。调用函数懂得如何在一大堆不同对象间浏览。

这些代码是否违反得墨忒耳律，取决于 ctxt、Options 和 ScratchDir 是对象还是数据结构。如果是对象，则它们的内部结构应当隐藏而不曝露，而有关其内部细节的知识就明显违反了得墨忒耳律。如果 ctxt、Options 和 ScratchDir 只是数据结构，没有任何行为，则它们自然会曝露其内部结构，得墨忒耳律也就不适用了。

属性访问器函数的使用把问题搞复杂了。如果像下面这样写代码，我们大概就不会提及是否违反得墨忒耳律。

```
final String outputDir = ctxt.options.scratchDir.absolutePath;
```

如果数据结构只简单地拥有公共变量，没有函数，而对象则拥有私有变量和公共函数，那么这个问题就没那么复杂。然而，有些框架和标准甚至要求最简单的数据结构也要有访问器和改值器。

6.3.2 混杂

这种混杂有时会不幸地导致混合结构，即一半是对象，另一半是数据结构。这种结构拥有执行操作的函数，也有公共变量或公共访问器及改值器。无论出于怎样的初衷，公共访问器及改值器都把私有变量公开化，诱导外部函数以过程式程序使用数据结构的方式使用这些变量[①]。

此类混杂增加了添加新函数的难度，也增加了添加新数据结构的难度，两头不讨好。应避免创造这种结构。它们的出现，展示了一种乱七八糟的设计，其作者不确定——或者更糟糕，完全无视——他们是否需要函数或类型的保护。

6.3.3 隐藏结构

假使 ctxt、Options 和 ScratchDir 是拥有真实行为的对象又怎样呢？由于对象应隐藏其内部结构，我们就不该看到内部结构。这样一来，如何才能取得临时目录的绝对路径呢？

```
ctxt.getAbsolutePathOfScratchDirectoryOption();
```

[①] 在《重构：改善既有代码的设计》(*Refactoring: Improving the Design of Existing Code*) 一书中，有时把这种情况称作特性依恋 (Feature Envy)。

或者

```
ctx.getScratchDirectoryOption().getAbsolutePath()
```

第一种方案可能导致 ctxt 对象中方法的曝露。第二种方案是在假设 getScratch-DirectoryOption() 返回一个数据结构而非对象。感觉两种方案都不好。

如果 ctxt 是一个对象，就应该要求它做点儿什么，而不该要求它给出内部情形。那我们为何还要得到临时目录的绝对路径呢？我们要它做什么？来看看同一模块（许多行之后）的这段代码：

```
String outFile = outputDir + "/" + className.replace('.', '/') + ".class";
FileOutputStream fout = new FileOutputStream(outFile);
BufferedOutputStream bos = new BufferedOutputStream(fout);
```

这种不同层级细节的混杂（[G34][G6]）有点麻烦。句点、斜杠、文件扩展名和 File 对象不该如此随便地混杂到一起。不过，撇开这些毛病，我们发现，取得临时目录绝对路径的初衷是为了创建指定名称的临时文件。

所以，直接让 ctxt 对象来做这事如何？

```
BufferedOutputStream bos = ctxt.createScratchFileStream(classFileName);
```

这下看起来像是对象做的事了！ctxt 隐藏了其内部结构，防止当前函数因浏览它不该知道的对象而违反得墨忒耳律。

6.4 数据传送对象

最为精练的数据结构，是一个只有公共变量、没有函数的类。这种数据结构有时被称为数据传送对象（Data Transfer Objects，DTO）。DTO 是非常有用的结构，尤其是在与数据库通信或解析套接字传递的消息之类的场景中，在应用程序代码里一系列将原始数据转换为数据库的翻译过程中，它们往往是排头兵。

更常见的是如代码清单 6-7 所示的"bean"结构。"bean"结构拥有由赋值器和取值器操作的私有变量。对"bean"结构的半封装会让某些面向对象纯化论者感觉舒服些，不过通常没有其他好处。

代码清单 6-7　address.java

```
public class Address {
  private String street;
  private String streetExtra;
  private String city;
  private String state;
  private String zip;

  public Address(String street, String streetExtra,
```

```
                String city, String state, String zip) {
    this.street = street;
    this.streetExtra = streetExtra;
    this.city = city;
    this.state = state;
    this.zip = zip;
  }

  public String getStreet() {
    return street;
  }

  public String getStreetExtra() {
    return streetExtra;
  }

  public String getCity() {
    return city;
  }

  public String getState() {
    return state;
  }

  public String getZip() {
    return zip;
  }
}
```

Active Record

Active Record 是一种特殊的 DTO 形式。它们是拥有公共（或可"bean"式访问的）变量的数据结构，但通常也会拥有类似 `save` 和 `find` 这样的可浏览方法。Active Record 一般是对数据库表或其他数据源的直接翻译。

我们不幸经常发现开发者往这类数据结构中塞进业务规则方法，把这类数据结构当成对象来用。这是不智的行为，因为它导致了数据结构和对象的混杂体。

当然，解决方案就是把 Active Record 当作数据结构，并创建包含业务规则、隐藏内部数据（可能就是 Active Record 的实体）的独立对象。

6.5 小结

对象曝露行为，隐藏数据，便于添加新对象类型而无须修改既有行为，同时难以在既有对象中添加新行为；数据结构曝露数据，没有明显的行为，便于向既有数据结构添加新行为，同时难以向既有函数添加新数据结构。

在任何系统中，我们有时会希望能够灵活地添加新数据类型，所以更喜欢在这部分使用对象。另外一些时候，我们希望能灵活地添加新行为，这时我们更喜欢使用数据类型和过程。优秀的软件开发者不带成见地了解这种情形，并依据手边工作的性质选择其中一种适合的手段。

6.6 文献

[Refactoring]：*Refactoring: Improving the Design of Existing Code*, Martin Fowler et al., Addison-Wesley, 1999.

第 7 章

错误处理

Michael Feathers

在一本有关整洁代码的书中,居然有讨论错误处理的章节,看起来有些突兀。错误处理只不过是编程时必须要做的事之一。输入可能出现异常,设备可能失效。简言之,可能会出错,当错误发生时,程序员就有责任确保代码照常工作。

然而,应该弄清楚错误处理与整洁代码的关系。许多程序完全由错误处理所占据。所谓占据,并不是说错误处理就是全部。我的意思是几乎无法看明白代码所做的事,因为到处都是凌乱的错误处理代码。错误处理很重要,但如果它搞乱了代码逻辑,就是错误的做法。

在本章中,我将概要列出编写既整洁又强固的代码——雅致地处理错误代码的一些技巧和思路。

7.1 使用异常而非返回码

在很久以前,许多语言都不支持异常,这些语言汇报和处理错误的手段都有限。你要么设置一个错误标识,要么返回给调用者检查的错误码。代码清单 7-1 中的代码展示了这些手段。

代码清单 7-1 DeviceController.java

```java
public class DeviceController {
  ...
  public void sendShutDown() {
    DeviceHandle handle = getHandle(DEV1);
    // Check the state of the device
    if (handle != DeviceHandle.INVALID) {
      // Save the device status to the record field
      retrieveDeviceRecord(handle);
      // If not suspended, shut down
      if (record.getStatus() != DEVICE_SUSPENDED) {
        pauseDevice(handle);
        clearDeviceWorkQueue(handle);
        closeDevice(handle);
      } else {
        logger.log("Device suspended.  Unable to shut down");
      }
    } else {
      logger.log("Invalid handle for: " + DEV1.toString());
    }
  }
  ...
}
```

这类手段的问题在于,它们搞乱了调用者代码,调用者必须在调用之后即刻检查错误。不幸的是,这个步骤很容易被遗忘。所以,遇到错误时,最好抛出一个异常,这样调用代码会很整洁,其逻辑不会被错误处理搞乱。

代码清单 7-2 展示了在方法中遇到错误时抛出异常的情形。

代码清单 7-2 DeviceController.java(采用异常处理)

```java
public class DeviceController {
  ...
  public void sendShutDown() {
    try {
      tryToShutDown();
    } catch (DeviceShutDownError e) {
      logger.log(e);
    }
  }

  private void tryToShutDown() throws DeviceShutDownError {
```

```
        DeviceHandle handle = getHandle(DEV1);
        DeviceRecord record = retrieveDeviceRecord(handle);

        pauseDevice(handle);
        clearDeviceWorkQueue(handle);
        closeDevice(handle);
    }

    private DeviceHandle getHandle(DeviceID id) {
        ...
        throw new DeviceShutDownError("Invalid handle for: " + id.toString());
        ...
    }

    ...
}
```

注意，这段代码整洁了很多，这不仅关乎美观。这段代码更好一些，因为之前相互纠结的两个主题，设备关闭算法和错误处理，现在被隔离了。你可以查看其中任一主题，分别理解它。

7.2 先写 try-catch-finally 语句

异常的妙处之一是，它们在程序中定义了范围。执行 try-catch-finally 语句中 try 部分的代码时，你是在表明可随时取消执行，并在 catch 语句中接续。

在某种意义上，try 代码块就像是事务。catch 代码块将程序维持在一种持续状态，无论 try 代码块中发生了什么均如此。所以，在编写可能抛出异常的代码时，最好先写出 try-catch-finally 语句。这能帮你定义该代码的用户应该期待什么，无论 try 代码块中执行的代码出什么错都一样。

来看个例子。我们要编写访问某个文件并读出一些序列化对象的代码。
先写一个单元测试，其中显示当文件不存在时将得到一个异常：

```
@Test(expected = StorageException.class)
public void retrieveSectionShouldThrowOnInvalidFileName() {
    sectionStore.retrieveSection("invalid - file");
}
```

该测试令我们创建以下占位代码：

```
public List<RecordedGrip> retrieveSection(String sectionName) {
    // dummy return until we have a real implementation
    return new ArrayList<RecordedGrip>();
}
```

测试失败了，因为以上代码并未抛出异常。下一步，修改实现代码，尝试访问非法文件。该操作抛出一个异常：

```
public List<RecordedGrip> retrieveSection(String sectionName) {
  try {
    FileInputStream stream = new FileInputStream(sectionName)
  } catch (Exception e) {
    throw new StorageException("retrieval error", e);
  }
  return new ArrayList<RecordedGrip>();
}
```

这次测试通过了,因为我们捕获了异常,此时就可以重构了。我们可以缩小异常类型的范围,使之符合 `FileInputStream` 构造器真正抛出的异常,即 `FileNotFoundException`:

```
public List<RecordedGrip> retrieveSection(String sectionName) {
  try {
    FileInputStream stream = new FileInputStream(sectionName);
    stream.close();
  } catch (FileNotFoundException e) {
    throw new StorageException("retrieval error", e);
  }
  return new ArrayList<RecordedGrip>();
}
```

如此一来,我们就用 `try-catch` 结构定义了范围,可以继续用测试驱动开发(TDD)方法构建剩余的代码逻辑。这些代码逻辑将在 `FileInputStream` 和 `close` 之间添加,装作一切正常的样子。

尝试编写强行抛出异常的测试,再往处理器中添加行为,使之满足测试要求。结果就是你要先构造 `try` 代码块的事务范围,而且也会帮助你维护好该范围的事务特征。

7.3 使用未检异常

争辩业已结束。多年来,Java 程序员们一直在争论已检异常(checked exception)的利与弊。Java 的第一个版本中引入已检异常时,已检异常看似是一个极好的点子。每个方法的签名都列出它可能传递给调用者的异常。而且,这些异常就是方法类型的一部分。如果签名与代码实际所做之事不符,代码在字面上就无法编译。

那时,我们认为已检异常是个绝妙的主意,而且,它也有所裨益。然而,现在已经很清楚,对于强固软件的生产,它并非必需的。C#不支持已检异常。尽管做过勇敢的尝试,C++最后也不支持已检异常。Python 和 Ruby 同样如此。不过,用这些语言也有可能写出强固的软件。我们得决定——的确如此——已检异常是否值回票价。

代价是什么?已检异常的代价就是违反开放/闭合原则[1]。如果你在方法中抛出已检异常,而 catch 语句在 3 个层级之上,你就得在 catch 语句和抛出异常处之间的每个方法签名中声明该异常。这意味着对软件中较低层级的修改,都将波及较高层级的签名。修改好的模块

---

[1] [Martin]。

必须重新构建、发布,即便它们自身所关注的任何东西都没改动过。

以某个大型系统的调用层级为例。顶端函数调用它们之下的函数,逐级向下。假设某个位于最低层级的函数被修改为抛出一个异常。如果该异常是已检的,则函数签名就要添加 `throw` 子句。这意味着每个调用该函数的函数都要被修改,捕获新异常,或在其签名中添加合适的 `throw` 子句。以此类推。最终得到的就是一个从软件最底端贯穿到最顶端的修改链!封装被打破了,因为在抛出路径中的每个函数都要去了解下一层级的异常细节。既然异常旨在让你能在较远处处理错误,那么已检异常以这种方式破坏封装简直就是一种耻辱。

如果你在编写一套关键代码库,则已检异常有时也会有用:你必须捕获异常。但对于一般的应用开发,其依赖成本要高于收益。

7.4 给出异常发生的环境说明

你抛出的每个异常,都应当提供足够的环境说明,以便判断错误的来源和位置。在 Java 中,你可以从任何异常里得到栈踪迹(stack trace),然而,栈踪迹却无法告诉你该失败操作的初衷。

应创建信息充分的错误消息,并和异常一起传递出去,在消息中,应包括失败的操作和失败类型。如果你的应用程序有日志系统,可以传递足够的信息给 `catch` 块,并记录下来。

7.5 依调用者需要定义异常类

对异常分类有很多方式。可以依其来源分类:是来自组件还是其他地方?也可以依其类型分类:是设备错误、网络错误还是编程错误?不过,当我们在应用程序中定义异常类时,最重要的考虑应该是它们如何被捕获。

我们来看一个不太好的异常分类的例子。下面的 `try-catch-finally` 语句是对某个第三方代码库的调用。它覆盖了该调用可能抛出的所有异常:

```
ACMEPort port = new ACMEPort(12);

try {
  port.open();
} catch (DeviceResponseException e) {
  reportPortError(e);
  logger.log("Device response exception", e);
} catch (ATM1212UnlockedException e) {
  reportPortError(e);
  logger.log("Unlock exception", e);
} catch (GMXError e) {
```

```
    reportPortError(e);
    logger.log("Device response exception");
  } finally {
    ...
  }
```

语句包含了一大堆重复代码，这并不出奇。在大多数异常处理中，不管真实原因如何，我们总是做相对标准的处理。我们得记录错误，确保能继续工作。

在本例中，既然知道我们所做的事不外如此，就可以通过打包调用 API，确保它返回通用异常类型，从而简化代码。

```
LocalPort port = new LocalPort(12);
try {
  port.open();
} catch (PortDeviceFailure e) {
  reportError(e);
  logger.log(e.getMessage(), e);
} finally {
  ...
}
```

LocalPort 类就是一个简单的打包类，它捕获并翻译由 ACMEPort 类抛出的异常：

```
public class LocalPort {
  private ACMEPort innerPort;

  public LocalPort(int portNumber) {
    innerPort = new ACMEPort(portNumber);
  }

  public void open() {
    try {
      innerPort.open();
    } catch (DeviceResponseException e) {
      throw new PortDeviceFailure(e);
    } catch (ATM1212UnlockedException e) {
      throw new PortDeviceFailure(e);
    } catch (GMXError e) {
      throw new PortDeviceFailure(e);
    }
  }
  ...
}
```

类似于我们为 ACMEPort 定义的这种打包类非常有用。实际上，将第三方 API 打包是个良好的实践手段。当你打包一个第三方 API，你就降低了对它的依赖：未来你可以不太痛苦地改用其他代码库。在你测试自己的代码时，打包也有助于模拟第三方调用。

打包的好处还在于你不必绑死在某个特定厂商的 API 设计上，你可以定义自己感觉舒服的 API。在上例中，我们为 port 设备错误定义了一个异常类型，然后发现这样能写出更整洁的代码。

对于代码的某个特定区域，单一异常类通常可行。伴随异常发送出来的信息能够区分不同错误。如果你想要捕获某个异常，并且放过其他异常，就使用不同的异常类。

7.6 定义常规流程

如果你遵循前文提及的建议，在业务逻辑和错误处理代码之间就会有良好的区隔。大量代码会开始变得像是整洁而简朴的算法。然而，这样做却把错误检测推到了程序的边缘地带。你打包了外部 API 以抛出自己的异常，你在代码的顶端定义了一个处理器来应对任何失败了的运算。在大多数时候，这种手段很棒，不过有时你也许不愿这么做。

我们来看一个例子。下面的笨代码来自某个记账应用的开支总计模块：

```
try {
  MealExpenses expenses = expenseReportDAO.getMeals(employee.getID());
  m_total += expenses.getTotal();
} catch(MealExpensesNotFound e) {
  m_total += getMealPerDiem();
}
```

业务逻辑是：如果消耗了餐食，则计入总额中；如果没有消耗，则员工得到当日餐食补贴。异常打断了业务逻辑。如果不去处理特殊情况会不会好一些？那样的话代码看起来会更简洁，就像这样：

```
MealExpenses expenses = expenseReportDAO.getMeals(employee.getID());
m_total += expenses.getTotal();
```

能把代码写得那样简洁吗？能。可以修改 `ExpenseReportDAO`，使其总是返回 `MealExpense` 对象。如果没有消耗餐食，就返回一个返回餐食补贴的 `MealExpense` 对象。

```
public class PerDiemMealExpenses implements MealExpenses {
  public int getTotal() {
    // return the per diem default
  }
}
```

这种手法叫作特例模式（SPECIAL CASE PATTERN [Fowler]）。创建一个类或配置一个对象，用来处理特例。你来处理特例，客户代码就不用应对异常行为了。异常行为被封装到特例对象中。

7.7 别返回 null 值

我认为，要讨论错误处理，就一定要提及那些容易引发错误的做法。第一项就是返回 null 值。我不想去计算曾经见过多少个几乎每行代码都在检查 null 值的应用程序。下面就是其中的一个例子：

```
public void registerItem(Item item) {
  if (item != null) {
    ItemRegistry registry = peristentStore.getItemRegistry();
    if (registry != null) {
      Item existing = registry.getItem(item.getID());
      if (existing.getBillingPeriod().hasRetailOwner()) {
        existing.register(item);
      }
    }
  }
}
```

这种代码看似不坏，其实糟透了！返回 null 值，基本上是在给自己增加工作量，也是在给调用者添乱。只要有一处没检查 null 值，应用程序就会失控。

你有没有注意到，嵌套 if 语句的第二行没有检查 null 值？如果在运行时 persistentStore 为 null 会发生什么事？我们会在运行时得到一个 NullPointerException 异常，也许有人在代码顶端捕获这个异常，也可能没有捕获。两种情况都很糟糕。对于从应用程序深处抛出的 NullPointerException 异常，你到底该作何反应呢？

可以敷衍说上列代码的问题是少做了一次 null 值检查，其实问题多多。如果你打算在方法中返回 null 值，不如抛出异常，或是返回特例对象。如果你在调用某个第三方 API 中可能返回 null 值的方法，可以考虑用新方法打包这个方法，在新方法中抛出异常或返回特例对象。

在许多情况下，特例对象都是爽口良药。设想有以下一段代码：

```
List<Employee> employees = getEmployees();
if (employees != null) {
  for(Employee e : employees) {
    totalPay += e.getPay();
  }
}
```

现在，getExployees 可能返回 null 值，但是否一定要这么做呢？如果修改 getEmployees，返回空列表，就能使代码整洁起来：

```
List<Employee> employees = getEmployees();
for(Employee e : employees) {
  totalPay += e.getPay();
}
```

所幸 Java 有 `Collections.emptyList()` 方法，该方法返回一个预定义的不可变列表，可用于达到这种目的：

```
public List<Employee> getEmployees() {
  if( .. there are no employees .. )
    return Collections.emptyList();
}
```

这样编码，就能尽量避免 `NullPointerException` 的出现，代码也更整洁了。

7.8 别传递 null 值

在方法中返回 `null` 值是糟糕的做法，将 `null` 值传递给其他方法就更糟糕了。除非 API 要求你向它传递 `null` 值，否则就要尽可能避免传递 `null` 值。

举例说明原因。用下面这个简单的方法计算两点间的一种度量：

```
public class MetricsCalculator
{
  public double xProjection(Point p1, Point p2) {
    return (p2.x - p1.x) * 1.5;
  }
  ...
}
```

如果有人传入 `null` 值会怎样？

```
calculator.xProjection(null, new Point(12, 13));
```

当然，我们会得到一个 `NullPointerException` 异常。

如何修正？可以创建一个新异常类型并抛出：

```
public class MetricsCalculator
{
  public double xProjection(Point p1, Point p2) {
    if (p1 == null || p2 == null) {
      throw InvalidArgumentException(
        "Invalid argument for MetricsCalculator.xProjection");
    }
    return (p2.x - p1.x) * 1.5;
  }
}
```

这样做好些吗？可能比 `null` 指针异常好一些，但要记住，我们还得为 `InvalidArgument-Exception` 异常定义处理器。这个处理器该做什么？还有更好的做法吗？

还有替代方案。可以使用一组断言（assertion）：

```
public class MetricsCalculator
{
  public double xProjection(Point p1, Point p2) {
```

```
    assert p1 != null : "p1 should not be null";
    assert p2 != null : "p2 should not be null";
    return (p2.x - p1.x) * 1.5;
  }
}
```

看上去很美,但仍未解决问题。如果有人传入 null 值,还是会得到运行时错误。

在大多数编程语言中,没有良好的方法能应对由调用者意外传入的 null 值。事已至此,恰当的做法就是禁止传入 null 值。这样,你在编码的时候,就会时时记住参数列表中的 null 值意味着出问题了,从而大量避免这种无心之失。

7.9　小结

整洁代码是可读的,但也要强固。可读与强固并不冲突。如果将错误处理隔离看待,独立于主要逻辑之外,就能写出强固而整洁的代码。做到这一步,我们就能单独处理它,也可以极大地提升代码的可维护性。

7.10　文献

[**Martin**]:*Agile Software Development: Principles, Patterns, and Practices*, Robert C. Martin, Prentice Hall, 2002.

第 8 章

边 界

James Grenning

我们很少能控制系统中的全部软件。有时我们购买第三方程序包或使用开放源代码,有时我们依靠公司中其他团队打造组件或子系统。不管是哪种情况,我们都得将外来代码干净利落地整合进自己的代码中。本章将介绍一些保持软件边界整洁的实践手段和技巧。

8.1 使用第三方代码

在接口提供者和使用者之间，存在与生俱来的矛盾。第三方程序包和框架提供者追求普适性，这样就能在多种环境中工作，从而吸引广泛的用户。而使用者则想要得到集中满足特定需求的接口。这种矛盾会导致系统边界上出现问题。

以 java.util.Map 为例。如图 8-1 中所示，Map 有着广阔的接口和丰富的功能。当然，这种力量和灵活性很有用，但也要付出代价。比如，应用程序可能构造一个 Map 对象并传递它。我们的初衷可能是 Map 对象的所有接收者都不要删除映射图中的任何东西。但图 8-1 的顶端却正好有一个 clear() 方法。Map 的任何使用者都能清除映射图。或许设计惯例是 Map 中只能保存特定的类型，但 Map 并不会可靠地约束存于其中的对象的类型。使用者可随意往 Map 中塞入任何类型的条目。

- clear() void – Map
- containsKey(Object key) boolean – Map
- containsValue(Object value) boolean – Map
- entrySet() Set – Map
- equals(Object o) boolean – Map
- get(Object key) Object – Map
- getClass() Class<? extends Object> – Object
- hashCode() int – Map
- isEmpty() boolean – Map
- keySet() Set – Map
- notify() void – Object
- notifyAll() void – Object
- put(Object key, Object value) Object – Map
- putAll(Map t) void – Map
- remove(Object key) Object – Map
- size() int – Map
- toString() String – Object
- values() Collection – Map
- wait() void – Object
- wait(long timeout) void – Object
- wait(long timeout, int nanos) void – Object

图 8-1 Map 类的方法

如果你的应用程序需要一个包容 Sensor 类对象的 Map 映射图，大概会是这样：

```
Map sensors = new HashMap();
```
当代码的其他部分需要访问这些传感器，就会有这行代码：
```
Sensor s = (Sensor)sensors.get(sensorId);
```
这行代码一再出现。代码的调用端承担了从 Map 中取得对象并将其转换为正确类型的职责。行倒是行，却非整洁代码。而且，这行代码并未说明自己的用途。通过对泛型的使用，这段代码的可读性可以大大提升，如下所示：
```
    Map<String,Sensor> sensors = new HashMap<Sensor>();
...
    Sensor s = sensors.get(sensorId);
```
不过，Map<String, Sensor>提供了超出所需/所愿的功能的问题，仍未得到解决。

在系统中不受限制地传递 Map<String, Sensor>的实体，意味着当 Map 的接口被修改时，有许多地方都要跟着改。你或许会认为这样的改动不太可能发生，不过，当 Java 5 加入对泛型的支持时，的确发生了改动。我们也的确见到一些系统因为要做大量改动才能自由使用 Map 类，而无法使用泛型。

使用 Map 的更整洁的方式大致如下。Sensors 的用户不必关心是否用了泛型，那将是（也该是）实现细节才关心的。
```
public class Sensors {
  private Map sensors = new HashMap();

  public Sensor getById(String id) {
    return (Sensor) sensors.get(id);
  }

  // snip
}
```
边界上的接口（Map）是隐藏的。它能随来自应用程序其他部分的极小的影响而变动。对泛型的使用不再是个大问题，因为转换和类型管理是在 Sensors 类内部处理的。

该接口也经过仔细修整和归置以适应应用程序的需要。结果就是得到易于理解、难以被误用的代码。Sensors 类推动了设计和业务的规则的形成。

我们并不建议总是以这种方式封装 Map 的使用。我们建议不要将 Map（或在边界上的其他接口）在系统中传递。如果你使用类似 Map 这样的边界接口，就把它保留在类或近亲类中。避免从公共 API 中返回边界接口，或将边界接口作为参数传递给公共 API。

8.2　浏览和学习边界

第三方代码帮助我们在更少时间内发布更丰富的功能。在利用第三方程序包时，该从何处入手呢？我们没有测试第三方代码的职责，但为要使用的第三方代码编写测试，可能最符合我们的利益。

设想我们对第三方代码库的使用方法并不清楚。我们可能会花上一两天（或者更多）时间阅读文档，决定如何使用。然后，我们会编写使用第三方代码的代码，看看是否如我们所愿地工作。我们会陷入长时间的调试，找出在我们或他们代码中的缺陷，这可不是什么稀罕事。

学习第三方代码很难，整合第三方代码也很难，同时做这两件事难上加难。如果我们采用不同的做法呢？不要在生产代码中试验新东西，而是编写测试来遍览和理解第三方代码。Jim Newkirk 把这叫作学习性测试（learning test）[①]。

在学习性测试中，我们就像在应用中那样调用第三方代码。我们基本上是在通过核对试验来检测自己对那个 API 的理解程度。测试聚焦于我们想从 API 得到的东西。

8.3 学习 log4j

比如，我们想使用 Apache `log4j` 包来代替自定义的日志代码。我们下载了 `log4j`，打开介绍文档页。无须看太久，就编写了第一个测试用例，希望它能向控制台输出"hello"字样。

```
@Test
public void testLogCreate() {
  Logger logger = Logger.getLogger("MyLogger");
  logger.info("hello");
}
```

运行上述代码，`logger` 发生了一个错误，告诉我们需要用 `Appender`。再多读一点文档，我们发现有个 `ConsoleAppender`。于是我们创建了一个 `ConsoleAppender`，再看是否能解开向控制台输出日志的秘诀。

```
@Test
public void testLogAddAppender() {
  Logger logger = Logger.getLogger("MyLogger");
  ConsoleAppender appender = new ConsoleAppender();
  logger.addAppender(appender);
  logger.info("hello");
}
```

这回，我们发现 `Appender` 没有输出流。奇怪，它该有输出流的。在谷歌上得到一点帮助后，我们写了以下代码：

```
@Test
public void testLogAddAppender() {
  Logger logger = Logger.getLogger("MyLogger");
  logger.removeAllAppenders();
  logger.addAppender(new ConsoleAppender(
    new PatternLayout("%p %t %m%n"),
```

[①] [BeckTDD], pp. 136-137。

```
        ConsoleAppender.SYSTEM_OUT));
    logger.info("hello");
}
```

这回行了，"hello"字样的日志信息出现在控制台上！必须告知`ConsoleAppender`，让它往控制台写字，看起来有点儿奇怪。

很有趣，当我们移除`ConsoleAppender.SystemOut`参数时，那个"hello"字样仍然输出到屏幕上。但如果取走`PatternLayout`，就会出现关于没有输出流的错误信息。这实在太古怪了。

再仔细看看文档，我们看到默认的`ConsoleAppender`构造器是"未配置"的，这看起来并不明显或没什么用，反而像是`log4j`的一个缺陷，或者至少是前后不太一致。

再搜索、阅读、测试，最终我们得到代码清单8-1。我们极大地发掘了`log4j`的工作方式，也将得到的知识融入了一系列简单的单元测试中。

代码清单 8-1　LogTest.java

```java
public class LogTest {
    private Logger logger;

    @Before
    public void initialize() {
        logger = Logger.getLogger("logger");
        logger.removeAllAppenders();
        Logger.getRootLogger().removeAllAppenders();
    }
    @Test
    public void basicLogger() {
        BasicConfigurator.configure();
        logger.info("basicLogger");
    }

    @Test
    public void addAppenderWithStream() {
        logger.addAppender(new ConsoleAppender(
            new PatternLayout("%p %t %m%n"),
            ConsoleAppender.SYSTEM_OUT));
        logger.info("addAppenderWithStream");
    }

    @Test
    public void addAppenderWithoutStream() {
        logger.addAppender(new ConsoleAppender(
            new PatternLayout("%p %t %m%n")));
        logger.info("addAppenderWithoutStream");
    }
}
```

现在我们知道如何初始化一个简单的控制台日志器，也能把这些知识封装到自己的日志类中，好将应用程序的其他部分与`log4j`的边界接口隔离开来。

8.4 学习性测试的好处不只是免费

学习性测试毫无成本。无论如何我们都得学习要使用的 API，而编写测试则是获得这些知识的容易而不会影响其他工作的途径。学习性测试是一种精确试验，帮助我们增进对 API 的理解。

学习性测试不光免费，还在投资上有正向的回报。当第三方程序包发布了新版本，我们可以运行学习性测试，看看程序包的行为有没有改变。

学习性测试确保第三方程序包按照我们想要的方式工作。一旦整合进来，就不能保证第三方代码总与我们的需求兼容。原作者不得不修改代码来满足他们自己的新需求。他们会修正缺陷、添加新功能。风险伴随新版本而来。如果第三方程序包的修改与测试不兼容，我们也能马上发现。

无论你是否需要通过学习性测试来学习，都要有一系列与生产代码中调用方式一致的输出测试来支持整洁的边界。如果不使用这些边界测试来减轻迁移的劳力，我们可能就会超出应有时限，长久地绑在旧版本上面。

8.5 使用尚不存在的代码

还有另一种边界，那种将已知和未知分隔开的边界。在代码中总有许多地方是我们的知识未及之处。有时，边界那边就是未知的（至少目前未知）。有时，我们并不往边界那边看过去。

好多年以前，我曾在一个开发无线通信系统软件的团队中工作。该系统有一个子系统 `Transmitter`（发送机）。我们对 `Transmitter` 知之甚少，而该子系统的开发者还没有对接口进行定义。我们不想受这种事阻碍，就从距未知那部分代码很远处开始工作。

对于我们的世界如何结束、新世界如何开始，我们有许多好主意。工作时，我们偶尔会跨越那道边界。尽管云雾遮挡了我们看向边界那边的视线，但是我们还是从工作中了解到我们想要的边界接口是什么样的。我们想要告知发送机一些事：

> 将发送机置于指定频率，并发出自这个流得到的数据的模拟表示。

我们不知这会如何做到，因为 API 还没设计出来。所以，我们决定过后再编写细节代码。

为了不受阻碍，我们定义了自己使用的接口。我们给它起了个好记的名字，比如

`Transmitter`。我们给它写了个名为 `transmit` 的方法，获取频率参数和数据流。这就是我们希望得到的接口。

编写我们想得到的接口的好处之一是它在我们控制之下。这有助于保持客户代码更可读，且集中于它该完成的工作。

从图 8-2 中可以看到，我们将 `CommunicationsController` 类从发送器 API（Transmitter API，该 API 不受我们控制，而且还没定义）中隔离出来。通过使用符合应用程序的接口，`CommunicationsController` 代码整洁且足以表达其意图。一旦发送器 API 被定义出来，我们就编写 `TransmitterAdapter` 来跨接。ADAPTER[①]封装了与 API 的互动，也提供了一个当 API 发生变动时唯一需要改动的地方。

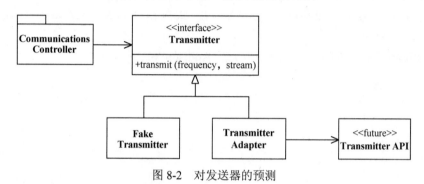

图 8-2　对发送器的预测

这套设计方案为测试提供了一种极为方便的接缝[②]。使用适当的 `FakeTransmitter`，我们就能测试 `CommunicationsController` 类。在拿到 TransmitterAPI 时，我们也能创建确保正确使用 API 的边界测试。

8.6　整洁的边界

边界上会发生有趣的事。改动是其中之一。如果有良好的软件设计，则无须巨大投入和重写即可进行修改。在使用我们控制不了的代码时，必须加倍小心保护投资，确保未来的修改不至于代价太大。

边界上的代码需要清晰的分割和定义了期望的测试。应该避免我们的代码过多地了解第三方代码中的特定信息。依靠你能控制的东西，好过依靠你控制不了的东西，免得日后受它控制。

---

① 见[GOF]中的 Adapter 模式。
② 在[WELC]中可查阅更多关于接缝（seam）的信息。

我们通过代码中少数几处引用第三方边界接口的位置来管理第三方边界。可以像我们对待 Map 那样包装它们，也可以使用 ADAPTER 模式将我们的接口转换为第三方提供的接口。采用这两种方式，代码都能更好地与我们沟通，如果在边界两边推动内部一致的用法，当第三方代码有改动时修改点就会更少。

8.7 文献

[BeckTDD]：*Test Driven Development,* Kent Beck, Addison-Wesley, 2003.

[GOF]：*Design Patterns: Elements of Reusable Object Oriented Software,* Gamma et al., Addison-Wesley, 1996.

[WELC]：*Working Effectively with Legacy Code,* Addison-Wesley, 2004.

第 9 章

单元测试

过去十年以来，编程专业领域进步很大。1997 年时，没人听说过测试驱动开发。对于我们之中的大多数人来说，单元测试是那种用来确保程序"可运行"的用过即扔的短代码。我们辛勤地编写类和方法，再弄出一些特殊代码来测试它们。通常这些代码会是一种简单的驱动式程序，让我们能够手工与自己编写的程序交互。

我记得在 20 世纪 90 年代曾为一套嵌入式实时系统编写过 C++ 程序。该程序是一个简单的计时器，有如下签名：

```
void Timer::ScheduleCommand(Command* theCommand, int milliseconds)
```

想法很简单：到达指定毫秒数时，在一个新线程中执行 `Command` 的 `execute` 方法。问题在于如何测试它。

我随便写了个简单的驱动式程序，聆听来自键盘的动作。键盘输入一个字符时，它就安排 5 秒之后输出同样的字符。我输入了一句带节奏的歌词，然后等着 5 秒之后它在屏幕上重现出来。

"I . . . want-a-girl . . . just . . . like-the-girl-who-marr . . . ied . . . dear . . . old . . . dad."[①]

在按下那些"."键时，我真的在哼着那段旋律，当那些句点出现在屏幕上时，我又哼了一次。

那就是我的测试！我看到这法子可行，演示给同事们看，然后就把代码扔掉了。

如前文所述，我们的专业领域进步甚多。如今，我会编写测试，确保代码中每个犄角旮旯都如我所愿地工作。我会将代码和操作系统隔离开，而不是直接调用标准计时功能。我会伪造一套计时函数，这样就能全面控制时间。我会安排一些设置布尔值标识的命令，往前步进时间，查看这些标识，确保它们在我将时间调到正确值时由 `false` 变为 `true`。

有了一套运行通过的测试，我会确保任何需要用到代码的人都能方便地使用这些测试。我会确保测试和代码一起签入同一个代码包。

对，我们进步甚多，但还有很长的路要走。敏捷和 TDD 运动鼓舞了许多程序员编写自动化单元测试，每天还有更多人加入这个行列。但是，在争先恐后将测试加入规程中时，许多程序员遗漏了一些关于编写好测试的更细微但却重要的要点。

9.1 TDD 三定律

谁都知道 TDD 要求我们在编写生产代码前先编写单元测试。但这条规则只是冰山之巅。看看下列 3 条定律[②]。

第一定律 在编写不能通过的单元测试前，不可编写生产代码。
第二定律 只可编写刚好无法通过的单元测试，不能编译也算不通过。
第三定律 只可编写刚好足以通过当前失败测试的生产代码。

这 3 条定律将你限制在大概 30 秒一个的循环中。测试与生产代码一起编写，测试只比生产代码早写几秒。

这样写程序，我们每天就会编写数十个测试，每个月编写数百个测试，每年编写数千个测试。这样写程序，测试将覆盖所有生产代码。测试代码量足以匹敌生产代码量，导致令人生畏的管理问题。

[①] "I want a girl just like the girl who married dear old dad" 是 20 世纪初 American Quartet 四重唱乐队的歌曲名，也是歌词中的一句。——译者注

[②] Professionalism and Test-Driven Development, Robert C. Martin, Object Mentor, IEEE Software, May/June 2007 (Vol. 24, No. 3) pp. 32-36。

9.2 保持测试整洁

几年前,有人请我去指导一个开发团队。那个团队认定,测试代码的维护不应遵循生产代码的质量标准。他们彼此默许在单元测试中破坏规矩。"速而不周"成了团队格言,即变量命名不用很好,测试函数不必短小和具有描述性,测试代码不必做良好设计和仔细划分,只要测试代码还能工作,只要还覆盖着生产代码,就足够好了。

有些读者可能会同意这种做法。或许,在很久以前,你也用过我为那个 `Timer` 类写测试的方法。从编写那种用后即扔的测试到编写全套自动化单元测试是一大进步。所以,就像那个我指导过的团队一样,你或许也会认为脏测试好于没测试。

这个团队没有意识到的是,脏测试等同于——如果不是坏的话——没测试。问题在于,测试必须随生产代码的演进而修改。测试越脏,就越难修改。测试代码越缠结,你就越有可能花更多时间塞进新测试,而不是编写新的生产代码。修改生产代码后,旧测试就会开始失败,而测试代码中乱七八糟的东西将阻碍代码再次通过。于是,测试变得就像是不断翻番的债务。

随着版本递进,团队维护测试代码组的代价也在上升。最终,这样的代价变成了开发者最大的抱怨对象。当经理们问及为何超支如此巨大,开发者们就归咎于测试。最后,他们只能扔掉整个测试代码组。

但是,没有了测试代码组,他们就失去了确保对代码的改动能如愿工作的能力。没有了测试代码组,他们就无法确保对系统某个部分的修改不会影响系统的其他部分。故障率开始上升。随着并非出自有意的故障越来越多,他们开始害怕做改动。他们不再清理生产代码,因为他们害怕修改带来的损害多于收益。生产代码开始腐坏。最后,他们只剩下没有测试、纷乱而缺陷缠身的生产代码,沮丧的客户,还有对测试的失望。

在某种意义上,他们说对了。测试的确让他们失望。不过是他们自己决定让测试变得乱七八糟的,而那正是失败的根源。如果他们保持测试整洁,测试就不会令他们失望,我可以拍着胸脯这么说,因为我曾经参与并指导了多个凭借整洁单元测试获得成功的团队。

故事的寓意很简单:测试代码和生产代码一样重要。测试代码可不是二等公民,它需要被思考、被设计和被照料,它该像生产代码一般保持整洁。

测试带来一切好处

如果测试不能保持整洁,你就会失去它们。没有了测试,你就会失去保证生产代码可扩展的一切要素。你没看错,正是单元测试让你的代码可扩展、可维护、可复用。原因很简单。有了测试,你就不用担心对代码的修改!没有测试,每次修改都可能带来缺陷。无论架构多有扩展性，无论设计划分得有多好，如果没有了测试，你就很难做改动，因为你担忧改动会引入不可预知的缺陷。

有了测试，愁云一扫而空。测试覆盖率越高，你就越不用担心。哪怕是对于那种架构并不优秀、设计晦涩纠缠的代码，你也能近乎没有后患地做修改。实际上，你甚至能毫无顾虑地改进架构和设计！

所以，覆盖了生产代码的自动化单元测试程序组能尽可能地保持设计和架构的整洁。测试带来了一切好处，因为测试使改动变得可能。

如果测试不干净，你改动自己代码的能力就会有所限制，而你也会开始失去改进代码结构的能力。测试越脏，代码就会变得越脏。最终，你丢失了测试，代码开始腐坏。

9.3 整洁的测试

整洁的测试有哪些要素呢？有 3 个要素：可读性、可读性和可读性。在单元测试中，可读性甚至比在生产代码中还重要。测试如何才能做到可读？和在其他代码中一样：明确，简洁，并有足够的表达力。在测试中，你要以尽可能少的文字表达大量内容。

我们来看看代码清单 9-1 中来自 FitNesse 的代码。这 3 个测试很难读懂，显然有改善空间。首先，其中有数量巨大的重复代码[G5]调用 `addPage` 和 `assertSubString`。更重要的是，代码中充满干扰测试表达力的细节。

代码清单 9-1　SerializedPageResponderTest.java

```java
public void testGetPageHieratchyAsXml() throws Exception
{
  crawler.addPage(root, PathParser.parse("PageOne"));
  crawler.addPage(root, PathParser.parse("PageOne.ChildOne"));
  crawler.addPage(root, PathParser.parse("PageTwo"));

  request.setResource("root");
  request.addInput("type", "pages");
  Responder responder = new SerializedPageResponder();
  SimpleResponse response =
    (SimpleResponse) responder.makeResponse(
      new FitNesseContext(root), request);
  String xml = response.getContent();

  assertEquals("text/xml", response.getContentType());
  assertSubString("<name>PageOne</name>", xml);
  assertSubString("<name>PageTwo</name>", xml);
  assertSubString("<name>ChildOne</name>", xml);
}

public void testGetPageHieratchyAsXmlDoesntContainSymbolicLinks()
throws Exception
{
```

```
    WikiPage pageOne = crawler.addPage(root, PathParser.parse("PageOne"));
    crawler.addPage(root, PathParser.parse("PageOne.ChildOne"));
    crawler.addPage(root, PathParser.parse("PageTwo"));

    PageData data = pageOne.getData();
    WikiPageProperties properties = data.getProperties();
    WikiPageProperty symLinks = properties.set(SymbolicPage.PROPERTY_NAME);
    symLinks.set("SymPage", "PageTwo");
    pageOne.commit(data);

    request.setResource("root");
    request.addInput("type", "pages");
    Responder responder = new SerializedPageResponder();
    SimpleResponse response =
      (SimpleResponse) responder.makeResponse(
        new FitNesseContext(root), request);
    String xml = response.getContent();

    assertEquals("text/xml", response.getContentType());
    assertSubString("<name>PageOne</name>", xml);
    assertSubString("<name>PageTwo</name>", xml);
    assertSubString("<name>ChildOne</name>", xml);
    assertNotSubString("SymPage", xml);
  }

  public void testGetDataAsHtml() throws Exception
  {
    crawler.addPage(root, PathParser.parse("TestPageOne"), "test page");

    request.setResource("TestPageOne");
    request.addInput("type", "data");
    Responder responder = new SerializedPageResponder();
    SimpleResponse response =
      (SimpleResponse) responder.makeResponse(
        new FitNesseContext(root), request);
    String xml = response.getContent();

    assertEquals("text/xml", response.getContentType());
    assertSubString("test page", xml);
    assertSubString("<Test", xml);
  }
```

请看对 `PathParser` 的那些调用，它们将字符串转换为供爬虫使用的 `PagePath` 实体。转换与测试毫无关系，徒然混淆了代码的意图。与创建 responder 相关的细节，还有 response 的收集与转换也尽是噪声，此外还有从 resource 和参数构造请求 URL 的笨手段。（这些代码我有幸参与编写，所以可以敞开来批评。）

最终，这段代码不是设计来给人看的。可怜的读者淹没在细节的汪洋大海中，在真正用到测试之前，还得理解这些细节。

现在看看代码清单 9-2 中改进了的测试。这些测试还是做同样的事，不过已经被重构为更整洁和更有表达力的形式。

代码清单 9-2　SerializedPageResponderTest.java（重构后）

```java
public void testGetPageHierarchyAsXml() throws Exception {
  makePages("PageOne", "PageOne.ChildOne", "PageTwo");

  submitRequest("root", "type:pages");

  assertResponseIsXML();
  assertResponseContains(
    "<name>PageOne</name>", "<name>PageTwo</name>", "<name>ChildOne</name>"
  );
}

public void testSymbolicLinksAreNotInXmlPageHierarchy() throws Exception {
  WikiPage page = makePage("PageOne");
  makePages("PageOne.ChildOne", "PageTwo");

  addLinkTo(page, "PageTwo", "SymPage");

  submitRequest("root", "type:pages");

  assertResponseIsXML();
  assertResponseContains(
    "<name>PageOne</name>", "<name>PageTwo</name>", "<name>ChildOne</name>"
  );
  assertResponseDoesNotContain("SymPage");
}

public void testGetDataAsXml() throws Exception {
  makePageWithContent("TestPageOne", "test page");

  submitRequest("TestPageOne", "type:data");

  assertResponseIsXML();
  assertResponseContains("test page", "<Test");
}
```

　　这些测试显然呈现了构造-操作-检验（BUILD-OPERATE-CHECK）模式。每个测试都清晰地拆分为 3 个环节。第一个环节构造测试数据，第二个环节操作测试数据，第三个环节部分检验操作是否得到期望的结果。

　　注意，大部分恼人的细节消失了。测试直达目的，只用到那些真正需要的数据类型和函数。读测试的人应该都能够很快搞清楚状况，而不至于被细节误导或吓倒。

9.3.1　面向特定领域的测试语言

　　代码清单 9-2 中的测试展示了为测试构造一种面向特定领域的语言的技巧。我们没有直接使用程序员用来对系统进行操作的 API，而是打造了一套包装这些 API 的函数和工具代码，这样就能更方便地编写测试，写出来的测试也更便于阅读。那正是一种测试语言，可以帮助程序员编写自己的测试，也可以帮助后来者阅读测试。

这种测试 API 并非起初就设计出来的，而是在对那些充满令人迷惑细节的测试代码进行后续重构时逐渐演进的。如同你看见我将代码清单 9-1 重构为代码清单 9-2 一般，守规矩的开发者也将他们的测试代码重构为更简洁和更具表达力的形式。

9.3.2 双重标准

在某种意义上，本章开始处提到的那个团队的做法是正确的。测试 API 中的代码与生产代码相比，的确有一套不同的工程标准。测试代码应当简单、精悍、足具表达力，但它该和生产代码一般有效。毕竟它是在测试环境而非生产环境中运行的，这两种环境有着截然不同的需求。

请看代码清单 9-3 中的测试。在为某个环境控制系统设计原型时，我写了这个测试。无须深入细节，你就能说出该测试是在"温度太低"时检验温度警报器、加热器和送风机是否全部打开。

代码清单 9-3　EnvironmentControllerTest.java

```java
@Test
public void turnOnLoTempAlarmAtThreashold() throws Exception {
  hw.setTemp(WAY_TOO_COLD);
  controller.tic();
  assertTrue(hw.heaterState());
  assertTrue(hw.blowerState());
  assertFalse(hw.coolerState());
  assertFalse(hw.hiTempAlarm());
  assertTrue(hw.loTempAlarm());
}
```

当然，这里也有许多细节。例如，`tic` 函数是做什么的？实际上，在读测试时你可以不用担心这些问题，你只需考虑是否同意系统最终状态与"温度太低"的情况相符。

当你阅读这个测试时，可以留意到自己的眼光得在被检验的状态的名称与状态的意义之间来回跳转。你看到 `heaterState`，眼光向左滑到 `assertTrue`。你看到 `coolerState`，眼光向左看 `assertFalse`。这个过程既乏味又不可靠，它让测试变得难以阅读。

我大幅改进了测试的可读性，得到代码清单 9-4。

代码清单 9-4　EnvironmentControllerTest.java（重构后）

```java
@Test
public void turnOnLoTempAlarmAtThreashold() throws Exception {
  wayTooCold();
  assertEquals("HBchL", hw.getState());
}
```

当然，我创建了一个 `wayTooCold` 函数，隐藏了 `tic` 函数的细节。不过要注意的是，

assertEquals 中的那个奇怪的字符串，大写表示"打开"，小写表示"关闭"，那些字符遵循以下次序：{heater, blower, cooler, hi-temp-alarm, lo-temp-alarm}。

尽管这破坏了思维映射[①]的规则，但看来它在这种情况下还是适用的。只要你明白其含义，你就能一眼看到那个字符串，并迅速译解出结果，如代码清单 9-5 所示。

代码清单 9-5　EnvironmentControllerTest.java（扩展到更大范围）

```java
@Test
public void turnOnCoolerAndBlowerIfTooHot() throws Exception {
  tooHot();
  assertEquals("hBChl", hw.getState());
}

@Test
public void turnOnHeaterAndBlowerIfTooCold() throws Exception {
  tooCold();
  assertEquals("HBchl", hw.getState());
}

@Test
public void turnOnHiTempAlarmAtThreshold() throws Exception {
  wayTooHot();
  assertEquals("hBCHl", hw.getState());
}

@Test
public void turnOnLoTempAlarmAtThreshold() throws Exception {
  wayTooCold();
  assertEquals("HBchL", hw.getState());
}
```

代码清单 9-6 中给出了 getState 函数，注意，它的代码效率不是非常高。要提升效率，可能应该使用 StringBuffer。

代码清单 9-6　MockControlHardware.java

```java
public String getState() {
  String state = "";
  state += heater ? "H" : "h";
  state += blower ? "B" : "b";
  state += cooler ? "C" : "c";
  state += hiTempAlarm ? "H" : "h";
  state += loTempAlarm ? "L" : "l";
  return state;
}
```

StringBuffer 有点儿丑陋。即便在生产代码中，就算代价较小，我也会避免使用 StringBuffer；而且你可以看到，代码清单 9-6 中代码的代价的确很小。这套应用显然是

[①] 见第 2 章。

嵌入式实时系统，计算机和内存资源都很有限。不过，测试环境大概完全不必做限制。

这就是双重标准。有些事你大概永远不会在生产环境中做，而在测试环境中做却完全没问题。通常这关乎内存或 CPU 效率的问题，不过却永远不会与整洁有关。

9.4 每个测试一个断言

有一个流派认为，JUnit 中每个测试函数都应该有且只有一个断言语句。这条规则看似过于苛刻，但其好处却可以在代码清单 9-5 中看到。这些测试都归结为一个可快速方便地理解的结论。

代码清单 9-2 又如何？我们能将关于输出是 XML 的断言与输出包含某些子字符串的断言轻易地组合到一起，不过这样做看来毫无道理。然而，我们可以将测试分解为两个单独的测试，每个测试都有各自的断言，如代码清单 9-7 所示。

代码清单 9-7　SerializedPageResponderTest.java（单个断言的版本）

```java
public void testGetPageHierarchyAsXml() throws Exception {
  givenPages("PageOne", "PageOne.ChildOne", "PageTwo");

  whenRequestIsIssued("root", "type:pages");

  thenResponseShouldBeXML();
}
public void testGetPageHierarchyHasRightTags() throws Exception {
  givenPages("PageOne", "PageOne.ChildOne", "PageTwo");

  whenRequestIsIssued("root", "type:pages");

  thenResponseShouldContain(
    "<name>PageOne</name>", "<name>PageTwo</name>", "<name>ChildOne</name>"
  );
}
```

注意，我修改了那些函数的名称，以符合 given-when-then[1]约定。这让测试更易阅读。不幸的是，如此分解测试，导致了许多重复代码的出现。

可以利用模板方法（TEMPLATE METHOD）[2]模式，将 given/when 部分放到基类中，将 then 部分放到派生类中，消除代码重复问题。或者，我们也可以创建一个完整的单独测试类，把 given 和 when 部分放到@Before 函数中，把 when 部分放到每个@Test 函数中，但对于这个小问题，这种做法看来有点儿机械。最后，我还是保留了代码清单 9-2 那种多个断言的形式。

---

[1] [RSpec]。
[2] [GOF]。

我认为，单个断言是一个好准则①。我通常都会创建支持这条准则的特定领域测试语言，如代码清单 9-5 所示。不过，我也不害怕在单个测试中放入多于一个的断言。我认为最好的说法是，单个测试中的断言数量应该最小化。

每个测试一个概念

更好一些的规则或许是每个测试函数中只测试一个概念。我们不想要超长的测试函数，测试完这个又测试那个。代码清单 9-8 就是那样一种测试的例子。这个测试应当拆解为 3 个单独测试，因为它测试了 3 件不同的事。如果把 3 件事混到一起，读者就不得不猜想每段代码出现的理由，以及那段代码到底要测试什么。

代码清单 9-8

```java
/**
 * Miscellaneous tests for the addMonths() method.
 */
public void testAddMonths() {
  SerialDate d1 = SerialDate.createInstance(31, 5, 2004);

  SerialDate d2 = SerialDate.addMonths(1, d1);
  assertEquals(30, d2.getDayOfMonth());
  assertEquals(6, d2.getMonth());
  assertEquals(2004, d2.getYYYY());

  SerialDate d3 = SerialDate.addMonths(2, d1);
  assertEquals(31, d3.getDayOfMonth());
  assertEquals(7, d3.getMonth());
  assertEquals(2004, d3.getYYYY());

  SerialDate d4 = SerialDate.addMonths(1, SerialDate.addMonths(1, d1));
  assertEquals(30, d4.getDayOfMonth());
  assertEquals(7, d4.getMonth());
  assertEquals(2004, d4.getYYYY());
}
```

这 3 个测试函数大概应该像下面这个样子。

- 对于某个有 31 天的月份（如 5 月）的最后一天
    - 增加一个该月最末一天为 30 日的月份（如 6 月）时，日期应该是该月的 30 日而非 31 日。
    - 增加最末月有 31 天的两个月时，日期应该是 31 日。
- 对于某个有 30 天的月份（如 6 月）的最后一天
    - 增加一个有 31 天的月份时，日期应该是 30 日而非 31 日。

这样一来，你可以看到，在这些混杂的测试当中，隐藏有一条普遍规则。增加月数时，

---

① "照规矩办"（Keep to the code!）。（这是电影《加勒比海盗》中的一句台词。——译者注）

日期不能大于该月的最末一天。这意味着在 2 月 28 日增加月份数，就会得到 3 月 28 日。而这个测试应该有用，但被遗漏了。

并非是由于代码清单 9-8 中每个段落的多重断言导致问题。问题在于，有多个概念被测试，所以，最佳规则也许是应该尽可能减少每个概念的断言数量，每个测试函数只测试一个概念。

9.5 F.I.R.S.T.[①]

整洁的测试还遵循以下 5 条规则，这 5 条规则的首字母构成了本节标题。

- **快速**（Fast）。测试应该够快。测试应该能快速运行。测试运行缓慢，你就不会想要频繁地运行它。如果你不频繁运行测试，就不能尽早发现问题，也无法轻易修正，从而也不能轻而易举地清理代码。最终，代码就会腐坏。
- **独立**（Independent）。测试应该相互独立。某个测试不应为下一个测试设定条件。你应该可以单独运行每个测试，以及以任何顺序运行测试。当测试互相依赖时，头一个测试没通过就会导致一连串的测试失败，使问题诊断变得困难，隐藏了下级错误。
- **可重复**（Repeatable）。测试应当可以在任何环境中重复通过。你应该能够在生产环境、质检环境中运行测试，也能够在无网络的列车上用笔记本电脑运行测试。如果测试不能在任意环境中重复，你就总会有个解释其失败的借口。当环境条件不具备时，你也会无法运行测试。
- **自足验证**（Self-Validating）。测试应该有布尔值输出。无论是通过或失败，你都不应该通过查看日志文件来确认测试是否通过。你不应该手工对比两个不同的文本文件来确认测试是否通过。如果测试不能自足验证，对失败的判断就会变得依赖主观，而运行测试也需要更长的手工操作时间。
- **及时**（Timely）。测试应及时编写。单元测试应该恰好在使其通过的生产代码之前编写。如果在编写生产代码之后编写测试，你会发现生产代码难以测试。你可能会因为某些生产代码本身难以测试而不去设计可测试的代码。

9.6 小结

我们只是触及了这个话题的表面。实际上，我认为应该为整洁的测试写上一整本书。对于项目的健康度，测试和生产代码同等重要。或许测试更为重要，因为它保证和增强了生产代码的可扩展性、可维护性和可复用性。所以，保持测试整洁吧。让测试具有表达力并短小

---

① 参见 Object Mentor 培训材料。

精悍。发明作为面向特定领域语言的测试 API，帮助自己编写测试。

如果你坐视测试腐坏，那么代码也会跟着腐坏。保持测试整洁吧。

9.7 文献

[RSpec]：*RSpec: Behavior Driven Development for Ruby Programmers*, Aslak Hellesøy, David Chelimsky, Pragmatic Bookshelf, 2008.

[GOF]：*Design Patterns: Elements of Reusable Object Oriented Software*, Gamma et al., Addison-Wesley, 1996.

第10章

类

与 Jeff Langr 合写

到目前为止,本书一直在讨论如何编写良好的代码行和代码块。我们深入研究了函数的恰当构成,以及函数之间如何互相关联。尽管讨论了这么多关于代码语句及由代码语句形成的函数的表达力,但是,除非我们将注意力放到代码组织的更高层面,否则始终不能得到整洁的代码。

10.1 类的组织

遵循标准的 Java 约定，类应该从一组变量列表开始。如果有公共静态常量，应该先出现。然后是私有静态变量，以及私有实体变量。很少会有公共变量。

公共函数应跟在变量列表之后。我们喜欢把由某个公共函数调用的私有工具函数紧随在该公共函数后面。这符合自上向下原则，让程序读起来就像一篇报纸文章。

封装

我们喜欢保持变量和工具函数的私有性，但并不执着于此。有时，我们也需要用到受保护的（protected）变量或工具函数，好让测试可以访问到。对我们来说，测试说了算。若同一程序包内的某个测试需要调用一个函数或变量，我们就会将该函数或变量置为受保护或在整个程序包内可访问。然而，我们首先会想办法使之保有隐私。放松封装总是下策。

10.2 类应该短小

关于类的第一条规则是类应该短小。第二条规则是还要更短小。不，我们并不是要重谈"函数"一章的论调。就像函数一样，在设计类时，首要规则就是要更短小。和函数一样，马上有个问题出现，那就是"多小合适呢？"

对于函数，我们通过计算代码行数衡量其大小。对于类，我们采用不同的衡量方法，即计算其权责（responsibility）①。

代码清单 10-1 给出了某个类的轮廓。`SuperDashboard` 类曝露大概 70 个公共方法，其中大多数开发者都会同意，这实在是太长了。有些开发者或许会将 `SuperDashboard` 类称为"神类"。

代码清单 10-1 权责太多

```
public class SuperDashboard extends JFrame implements MetaDataUser
    public String getCustomizerLanguagePath()
    public void setSystemConfigPath(String systemConfigPath)
    public String getSystemConfigDocument()
    public void setSystemConfigDocument(String systemConfigDocument)
    public boolean getGuruState()
    public boolean getNoviceState()
    public boolean getOpenSourceState()
```

---

① [RDD]。

```java
public void showObject(MetaObject object)
public void showProgress(String s)
public boolean isMetadataDirty()
public void setIsMetadataDirty(boolean isMetadataDirty)
public Component getLastFocusedComponent()
public void setLastFocused(Component lastFocused)
public void setMouseSelectState(boolean isMouseSelected)
public boolean isMouseSelected()
public LanguageManager getLanguageManager()
public Project getProject()
public Project getFirstProject()
public Project getLastProject()
public String getNewProjectName()
public void setComponentSizes(Dimension dim)
public String getCurrentDir()
public void setCurrentDir(String newDir)
public void updateStatus(int dotPos, int markPos)
public Class[] getDataBaseClasses()
public MetadataFeeder getMetadataFeeder()
public void addProject(Project project)
public boolean setCurrentProject(Project project)
public boolean removeProject(Project project)
public MetaProjectHeader getProgramMetadata()
public void resetDashboard()
public Project loadProject(String fileName, String projectName)
public void setCanSaveMetadata(boolean canSave)
public MetaObject getSelectedObject()
public void deselectObjects()
public void setProject(Project project)
public void editorAction(String actionName, ActionEvent event)
public void setMode(int mode)
public FileManager getFileManager()
public void setFileManager(FileManager fileManager)
public ConfigManager getConfigManager()
public void setConfigManager(ConfigManager configManager)
public ClassLoader getClassLoader()
public void setClassLoader(ClassLoader classLoader)
public Properties getProps()
public String getUserHome()
public String getBaseDir()
public int getMajorVersionNumber()
public int getMinorVersionNumber()
public int getBuildNumber()
public MetaObject pasting(
  MetaObject target, MetaObject pasted, MetaProject project)
public void processMenuItems(MetaObject metaObject)
public void processMenuSeparators(MetaObject metaObject)
public void processTabPages(MetaObject metaObject)
public void processPlacement(MetaObject object)
public void processCreateLayout(MetaObject object)
public void updateDisplayLayer(MetaObject object, int layerIndex)
public void propertyEditedRepaint(MetaObject object)
public void processDeleteObject(MetaObject object)
public boolean getAttachedToDesigner()
```

```
    public void processProjectChangedState(boolean hasProjectChang)
    public void processObjectNameChanged(MetaObject object)
    public void runProject()
    public void setAllowDragging(boolean allowDragging)
    public boolean allowDragging()
    public boolean isCustomizing()
    public void setTitle(String title)
    public IdeMenuBar getIdeMenuBar()
    public void showHelper(MetaObject metaObject, String propertyName)
    // ... many non-public methods follow ...
}
```

如果 SuperDashboard 类只包括代码清单 10-2 中的方法呢?

代码清单 10-2　足够短小了吗?

```
public class SuperDashboard extends JFrame implements MetaDataUser{
    public Component getLastFocusedComponent()
    public void setLastFocused(Component lastFocused)
    public int getMajorVersionNumber()
    public int getMinorVersionNumber()
    public int getBuildNumber()
}
```

5 个方法不算多,在这里,虽然方法数量较少,但是 SuperDashboard 还是拥有太多权责。

类的名称应当描述其权责。实际上,命名正是帮助判断类的长度的第一个手段。如果无法为某个类命以精确的名称,那么这个类大概就太长了。类名越含糊,该类越有可能拥有过多权责。例如,如果类名中包含含义模糊的词,如 Processor、Manager 或 Super,那么这种现象往往说明有不恰当的权责聚集情况存在。

我们也应该能够用大概 25 个单词简要描述一个类,且不用"若"(if)"与"(and)"或"(or)或者"但"(but)等词汇。我们该如何描述 SuperDashboard 类呢?"SuperDashboard 类提供了对最后拥有焦点的组件的访问能力,我们还能通过它跟踪版本号和构建序列号。""还能"二字恰好提示了 SuperDashboard 类有太多权责。

10.2.1　单一权责原则

单一权责原则(SRP)[①]认为,类或模块应该有且只有一条加以修改的理由。该原则既给出了权责的定义,又是关于类的长度的指导方针。类只应有一个权责——只有一条修改的理由。

代码清单 10-2 中貌似很小的 SuperDashboard 类有两条加以修改的理由。第一,它跟踪大概会随软件每次发布而更新的版本信息。第二,它管理 Java Swing 组件(派生自 JFrame,顶层 GUI 窗口的 Swing 表现形态)。每次修改 Swing 代码时,无疑都要更新版本号,但反之未必可行:也可能依据系统中其他代码的修改而更新版本信息。

---

① 你可在[PPP]中读到更多信息。

鉴别权责（修改的理由）常常帮助我们在代码中认识到并创建出更好的抽象。可以轻易地将全部 3 个处理版本信息的 `SuperDashboard` 方法拆解到名为 `Version` 的类中（如代码清单 10-3 所示）。`Version` 类是个极有可能在其他应用程序中得到复用的构造！

代码清单 10-3　单一权责类

```
public class Version {
  public int getMajorVersionNumber()
  public int getMinorVersionNumber()
  public int getBuildNumber()
}
```

SRP 是面向对象设计中最为重要的概念之一，也是较为容易理解和遵循的概念之一。奇怪的是 SRP 往往也是最容易被破坏的类设计原则。经常会遇到做太多事的类，为什么呢？

让软件能工作和让软件保持整洁，是两种截然不同的工作。我们中的大多数人脑力有限，只能更多地把精力放在让代码能工作上，而不是放在保持代码有组织和整洁上，这全然正确。分而治之在编程行为中的重要程度，等同于其在程序中的重要程度。

问题是太多人在程序能工作时就以为万事大吉了。我们没能把思维转向有关代码组织和整洁的部分。我们直接转向了下一个问题，而没能回过头将臃肿的类切分为只有单一权责的去耦式单元。

与此同时，许多开发者害怕数量巨大的短小单一目的类会导致难以一目了然抓住全局。他们认为，要搞清楚一件较大工作如何完成，就得在类与类之间找来找去。

然而，有大量短小类的系统并不比有少量庞大类的系统拥有更多移动部件，其数量大致相等。问题是：你是想把工具归置到有许多抽屉、每个抽屉中装有定义和标记良好的组件的工具箱中呢？还是想要少数几个能随便把所有东西扔进去的抽屉？

每个达到一定规模的系统都会包括大量逻辑和复杂性。管理这种复杂性的首要目标就是加以组织，以便开发者知道到哪儿能找到东西，并且在某个特定时间只需要理解直接相关的复杂性。反之，拥有巨大、多目的类的系统，总是让我们在目前并不需要了解的一大堆东西中艰难跋涉。

再强调一下：系统应该由许多短小的类而不是少量巨大的类组成，每个小类封装一个权责，只有一个修改的原因，并与少数其他类一起协同达成期望的系统行为。

10.2.2　内聚

类应该只有少数实体变量。类中的每个方法都应该操作一个或多个这种变量。通常而言，方法操作的变量越多，就越黏聚到类上。如果一个类中的每个变量都被每个方法所使用，则该类具有最大的内聚性。

一方面，一般来说，创建这种极大化内聚类是既不可取也不可能的；另一方面，我们希望内聚性保持在较高位置。内聚性高，意味着类中的方法和变量互相依赖、互相结合成一个逻辑整体。

看看代码清单 10-4 中一个 Stack 类的实现方式。这个类内聚性非常高。在 3 个方法中，只有 size() 方法没有使用所有两个变量。

代码清单 10-4　Stack.java（一个内聚类）

```java
public class Stack {
  private int topOfStack = 0;
  List<Integer> elements = new LinkedList<Integer>();

  public int size() {
    return topOfStack;
  }

  public void push(int element) {
    topOfStack++;
    elements.add(element);
  }

  public int pop() throws PoppedWhenEmpty {
    if (topOfStack == 0)
      throw new PoppedWhenEmpty();
    int element = elements.get(--topOfStack);
    elements.remove(topOfStack);
    return element;
  }
}
```

保持函数和参数列表短小的策略，有时会导致为一组子集方法所用的实体变量数量增加。出现这种情况时，往往意味着至少有一个类要从大类中挣扎出来。你应当尝试将这些变量和方法分拆到两个或多个类中，让新的类更为内聚。

10.2.3　保持内聚性就会得到许多短小的类

仅仅是将较大的函数切割为小函数，就将导致更多的类出现。想想看一个有许多变量的大函数。你想把该函数中某一小部分拆解成单独的函数。不过，你想要拆出来的代码使用了该函数中声明的 4 个变量。是否必须将这 4 个变量都作为参数传递到新函数中去呢？

完全没必要！只要将 4 个变量提升为类的实体变量，完全无须传递任何变量就能拆解代码了。应该很容易将函数拆分为小块。

可惜这也意味着类丧失了内聚性，因为堆积了越来越多只为允许少量函数共享而存在的实体变量。等一下！如果有些函数想要共享某些变量，为什么不让它们拥有自己的类呢？当类丧失了内聚性，就拆分它！

所以，将大函数拆分为许多小函数时，往往也是将类拆分为多个小类的时机。程序会更加有组织，也会拥有更为透明的结构。

为了说明我的意思，不如从 Knuth 的名著 *Literate Programming*[1]中摘取一个经过时间考验的例子。代码清单 10-5 展示了 Knuth 的 `PrintPrimes` 程序的 Java 版本。为示公平，以下程序并非 Knuth 原版，而是用他的 Web 工具输出的版本。采用它作为例子的目的，是因为它是展示如何将较大的函数分解为多个较小的函数和类的极好入手点。

代码清单 10-5　PrintPrimes.java

```java
package literatePrimes;

public class PrintPrimes {
  public static void main(String[] args) {
    final int M = 1000;
    final int RR = 50;
    final int CC = 4;
    final int WW = 10;
    final int ORDMAX = 30;
    int P[] = new int[M + 1];
    int PAGENUMBER;
    int PAGEOFFSET;
    int ROWOFFSET;
    int C;
    int J;
    int K;
    boolean JPRIME;
    int ORD;
    int SQUARE;
    int N;
    int MULT[] = new int[ORDMAX + 1];

    J = 1;
    K = 1;
    P[1] = 2;
    ORD = 2;
    SQUARE = 9;

    while (K < M) {
      do {
        J = J + 2;
        if (J == SQUARE) {
          ORD = ORD + 1;
          SQUARE = P[ORD] * P[ORD];
          MULT[ORD - 1] = J;
        }
        N = 2;
        JPRIME = true;
        while (N < ORD && JPRIME) {
          while (MULT[N] < J)
            MULT[N] = MULT[N] + P[N] + P[N];
          if (MULT[N] == J)
            JPRIME = false;
```

---

[1] [Knuth92]。

```
          N = N + 1;
        }
      } while (!JPRIME);
      K = K + 1;
      P[K] = J;
    }
    {
      PAGENUMBER = 1;
      PAGEOFFSET = 1;
      while (PAGEOFFSET <= M) {
        System.out.println("The First " + M +
                           " Prime Numbers --- Page " + PAGENUMBER);
        System.out.println("");
        for (ROWOFFSET = PAGEOFFSET; ROWOFFSET < PAGEOFFSET + RR; ROWOFFSET++){
          for (C = 0; C < CC;C++)
            if (ROWOFFSET + C * RR <= M)
              System.out.format("%10d", P[ROWOFFSET + C * RR]);
          System.out.println("");
        }
        System.out.println("\f");
        PAGENUMBER = PAGENUMBER + 1;
        PAGEOFFSET = PAGEOFFSET + RR * CC;
      }
    }
  }
}
```

该程序只有一个大函数，简直一团糟，它拥有很深的缩进结构，冗余的变量和紧密耦合的结构。至少应该将其拆分为数个较小的函数。

从代码清单 10-6 到代码清单 10-8，展示了将代码清单 10-5 中的代码拆分为较小的类和函数，并为这些类、函数和变量起个好名字后的结果。

代码清单 10-6　PrimePrinter.java（重构后）

```
package literatePrimes;

public class PrimePrinter {
  public static void main(String[] args) {
    final int NUMBER_OF_PRIMES = 1000;
    int[] primes = PrimeGenerator.generate(NUMBER_OF_PRIMES);

    final int ROWS_PER_PAGE = 50;
    final int COLUMNS_PER_PAGE = 4;
    RowColumnPagePrinter tablePrinter =
      new RowColumnPagePrinter(ROWS_PER_PAGE,
                               COLUMNS_PER_PAGE,
                               "The First " + NUMBER_OF_PRIMES +
                               " Prime Numbers");
    tablePrinter.print(primes);
  }
}
```

代码清单 10-7　RowColumnPagePrinter.java

```java
package literatePrimes;

import java.io.PrintStream;

public class RowColumnPagePrinter {
  private int rowsPerPage;
  private int columnsPerPage;
  private int numbersPerPage;
  private String pageHeader;
  private PrintStream printStream;

  public RowColumnPagePrinter(int rowsPerPage,
                              int columnsPerPage,
                              String pageHeader) {
    this.rowsPerPage = rowsPerPage;
    this.columnsPerPage = columnsPerPage;
    this.pageHeader = pageHeader;
    numbersPerPage = rowsPerPage * columnsPerPage;
    printStream = System.out;
  }

  public void print(int data[]) {
    int pageNumber = 1;
    for (int firstIndexOnPage = 0;
         firstIndexOnPage < data.length;
         firstIndexOnPage += numbersPerPage) {
      int lastIndexOnPage =
        Math.min(firstIndexOnPage + numbersPerPage - 1,
                 data.length - 1);
      printPageHeader(pageHeader, pageNumber);
      printPage(firstIndexOnPage, lastIndexOnPage, data);
      printStream.println("\f");
      pageNumber++;
    }
  }

  private void printPage(int firstIndexOnPage,
                         int lastIndexOnPage,
                         int[] data) {
    int firstIndexOfLastRowOnPage =
      firstIndexOnPage + rowsPerPage - 1;
    for (int firstIndexInRow = firstIndexOnPage;
         firstIndexInRow <= firstIndexOfLastRowOnPage;
         firstIndexInRow++) {
      printRow(firstIndexInRow, lastIndexOnPage, data);
      printStream.println("");
    }
  }

  private void printRow(int firstIndexInRow,
                        int lastIndexOnPage,
                        int[] data) {
    for (int column = 0; column < columnsPerPage; column++) {
```

```java
        int index = firstIndexInRow + column * rowsPerPage;
        if (index <= lastIndexOnPage)
          printStream.format("%10d", data[index]);
      }
    }

    private void printPageHeader(String pageHeader,
                                 int pageNumber) {
      printStream.println(pageHeader + " --- Page " + pageNumber);
      printStream.println("");
    }

    public void setOutput(PrintStream printStream) {
      this.printStream = printStream;
    }
  }
```

代码清单 10-8　PrimeGenerator.java

```java
  package literatePrimes;

  import java.util.ArrayList;

  public class PrimeGenerator {
    private static int[] primes;
    private static ArrayList<Integer> multiplesOfPrimeFactors;

    protected static int[] generate(int n) {
      primes = new int[n];
      multiplesOfPrimeFactors = new ArrayList<Integer>();
      set2AsFirstPrime();
      checkOddNumbersForSubsequentPrimes();
      return primes;
    }

    private static void set2AsFirstPrime() {
      primes[0] = 2;
      multiplesOfPrimeFactors.add(2);
    }

    private static void checkOddNumbersForSubsequentPrimes() {
      int primeIndex = 1;
      for (int candidate = 3;
           primeIndex < primes.length;
           candidate += 2) {
        if (isPrime(candidate))
          primes[primeIndex++] = candidate;
      }
    }

    private static boolean isPrime(int candidate) {
      if (isLeastRelevantMultipleOfNextLargerPrimeFactor(candidate)) {
        multiplesOfPrimeFactors.add(candidate);
        return false;
```

```
    return isNotMultipleOfAnyPreviousPrimeFactor(candidate);
  }
  private static boolean
  isLeastRelevantMultipleOfNextLargerPrimeFactor(int candidate) {
    int nextLargerPrimeFactor = primes[multiplesOfPrimeFactors.size()];
    int leastRelevantMultiple = nextLargerPrimeFactor * nextLargerPrimeFactor;
    return candidate == leastRelevantMultiple;
  }
  private static boolean
  isNotMultipleOfAnyPreviousPrimeFactor(int candidate) {
    for (int n = 1; n < multiplesOfPrimeFactors.size(); n++) {
      if (isMultipleOfNthPrimeFactor(candidate, n))
        return false;
    }
    return true;
  }
  private static boolean
  isMultipleOfNthPrimeFactor(int candidate, int n) {
    return
      candidate == smallestOddNthMultipleNotLessThanCandidate(candidate, n);
  }
  private static int
  smallestOddNthMultipleNotLessThanCandidate(int candidate, int n) {
    int multiple = multiplesOfPrimeFactors.get(n);
    while (multiple < candidate)
      multiple += 2 * primes[n];
    multiplesOfPrimeFactors.set(n, multiple);
    return multiple;
  }
}
```

你可能注意到的第一件事就是程序比原来长了许多，从 1 页多增加到了将近 3 页。这有几个原因：其一，重构后的程序采用了更长、更有描述性的变量名；其二，重构后的程序将函数和类声明当作是给代码添加注释的一种手段；其三，我们采用了空格和格式技巧让程序更可读。

注意程序是如何被拆分为 3 个主要权责的。`PrimePrinter` 类中只有主程序。主程序的权责是处理执行环境。如果调用方式改变，它也会随之改变。例如，如果程序被转换为 SOAP 服务，则该类也会受影响。

`RowColumnPagePrinter` 类懂得如何将数字列表格式化到有着固定行数、列数的页面上。若输出格式需要改动，则该类也会受影响。

`PrimeGenerator` 类懂得如何生成素数列表。注意，这并不意味着要实体化为对象。该类就是一个有用的作用域，在其中声明并隐藏变量。如果计算素数的算法发生改动，则该类也会改动。

这并不算是重写！因为我们没有从头开始写一遍程序。实际上，如果你仔细看上述两个不同的程序，就会发现它们采用了同样的算法和机制来完成工作。

我们通过编写验证第一个程序的精确行为的用例来实现修改。然后，我们做了许多小改动，每次改动一处。每改动一次，就执行一次，确保程序的行为没有变化。一小步接着一小步，第一个程序就这样被逐渐清理和转换为第二个程序。

10.3 为了修改而组织

对于多数系统，修改将一直持续。每处修改都让我们冒着系统其他部分不能如期望般工作的风险。在整洁的系统中，我们对类加以组织，以降低修改的风险。

代码清单 10-9 中的 `Sql` 类用来生成提供恰当元数据的 SQL 格式化字符串。这个类还没写完，所以暂时不支持 update 语句等 SQL 功能。当需要 `Sql` 类支持 update 语句时，我们就得"打开"这个类进行修改。打开类带来的问题是风险随之而来。对类的任何修改都有可能破坏类中的其他代码。因此必须全面重新测试。

代码清单 10-9 一个必须打开修改的类

```java
public class Sql {
    public Sql(String table, Column[] columns)
    public String create()
    public String insert(Object[] fields)
    public String selectAll()
    public String findByKey(String keyColumn, String keyValue)
    public String select(Column column, String pattern)
    public String select(Criteria criteria)
    public String preparedInsert()
    private String columnList(Column[] columns)
    private String valuesList(Object[] fields, final Column[] columns)
    private String selectWithCriteria(String criteria)
    private String placeholderList(Column[] columns)
}
```

当增加一种新类型语句时，就要修改 `Sql` 类。改动单个语句类型时，也要进行修改，比如，打算让 select 功能支持子查询。如果存在两个修改的理由，就说明 `Sql` 违反了 SRP 原则。

可以从简单的组织性观点发现对 SRP 的违反。`Sql` 的方法概述显示，存在类似于 `selectWithCriteria` 等只与 select 语句有关的私有方法。

出现了只与类的一小部分有关的私有方法行为，就意味着存在改进空间。然而，展开行动的基本动因却应该是系统的变动。若我们认为 `Sql` 类在逻辑上已具足，则无须担心对权责的拆分。如果在可预见的未来无须增加 update 功能，就不该去修改 `Sql` 类。不过，一旦打开了类，就应当修正设计方案。

代码清单 10-10 中的解决方式如何呢？代码清单 10-9 中 `Sql` 类的每个接口方法都重构到从 `Sql` 类派生出来的类中了。注意那些私有方法，如 `valuesList`，直接移到了需要用它们的地方。公共私有行为被划分到独立的两个工具类 Where 和 ColumnList 中。

代码清单 10-10　一组封闭类

```java
abstract public class Sql {
  public Sql(String table, Column[] columns)
  abstract public String generate();
}

public class CreateSql extends Sql {
  public CreateSql(String table, Column[] columns)
  @Override public String generate()
}

public class SelectSql extends Sql {
  public SelectSql(String table, Column[] columns)
  @Override public String generate()
}

public class InsertSql extends Sql {
  public InsertSql(String table, Column[] columns, Object[] fields)
  @Override public String generate()
  private String valuesList(Object[] fields, final Column[] columns)
}

public class SelectWithCriteriaSql extends Sql {
  public SelectWithCriteriaSql(
    String table, Column[] columns, Criteria criteria)
  @Override public String generate()
}

public class SelectWithMatchSql extends Sql {
  public SelectWithMatchSql(
    String table, Column[] columns, Column column, String pattern)
  @Override public String generate()
}

public class FindByKeySql extends Sql{
  public FindByKeySql(
    String table, Column[] columns, String keyColumn, String keyValue)
  @Override public String generate()
}

public class PreparedInsertSql extends Sql {
  public PreparedInsertSql(String table, Column[] columns)
  @Override public String generate(){
  private String placeholderList(Column[] columns)
}

public class Where {
  public Where(String criteria)
  public String generate()
}
public class ColumnList {
  public ColumnList(Column[] columns)
```

```
    public String generate()
}
```

每个类中的代码都变得极为简单。理解每个类花费的时间缩减到近乎为零。函数对其他函数造成毁坏的风险也变得几近于无。从测试的角度看,验证方案中每一处逻辑都成了极为简单的任务,因为类与类之间相互隔离了。

当需要增加 update 语句时,现存类无须做任何修改,这也同样重要!我们在 Sql 类的新子类 UpdateSql 中构建 update 语句的逻辑,系统中的其他代码都不会因为这个修改而被破坏。

重新架构的 Sql 逻辑百利而无一弊。它符合 SRP,也符合其他面向对象设计的关键原则,如开放闭合原则(OCP)[①]:类应当对扩展开放,对修改封闭。通过子类化手段,重新架构的 Sql 类对添加新功能是开放的,而且可以同时不触及其他类,只要将 UpdateSql 类放置到位就行了。

我们希望精心组织系统,从而在添加或修改特性时尽可能少惹麻烦。在理想系统中,我们通过扩展系统而非修改现有代码来添加新特性。

隔离修改

需求会改变,所以代码也会改变。在 OO 101[②]中,我们学习到,具体类包含实现细节(代码),而抽象类则只呈现概念。依赖具体细节的客户类,当细节改变时,就会有风险。我们可以借助接口和抽象类来隔离这些细节带来的影响。

对具体细节的依赖给对系统的测试带来了挑战。如果我们构建一个依赖外部 TokyoStockExchange API 的 Portfolio 类以代表投资组合的价值,则测试用例就会受到价值查询的连带影响。如果每隔 5 分钟就有新说法,就很难写出测试来。

与其设计直接依赖 TokyoStockExchange 的 Portfolio 类,不如创建 StockExchange 接口,其中只声明一个方法:

```
public interface StockExchange {
  Money currentPrice(String symbol);
}
```

我们设计 TokyoStockExchange 类来实现这个接口。我们还要确保 Portfolio 的构造器接受作为参数的 StockExchange 引用:

```
public Portfolio {
  private StockExchange  exchange;
  public  Portfolio(StockExchange  exchange) {
    this.exchange = exchange;
  }
  // ...
}
```

---

① [PPP]。
② 即面向对象入门知识。——译者注

现在就可以为 `StockExchange` 接口创建可测试的尝试性实现了，该尝试性实现将返回固定的现值。如果测试中购买了 5 股微软股票，则尝试性实现总是返回每股 100 美元的现值。对于 `StockExchange` 接口的尝试性实现简化为简单的表格查找。然后再编写一个总投资价值为 500 美元的测试。

```java
public class PortfolioTest {
  private FixedStockExchangeStub exchange;
  private Portfolio portfolio;

  @Before
  protected void setUp() throws Exception {
    exchange = new FixedStockExchangeStub();
    exchange.fix("MSFT", 100);
    portfolio = new Portfolio(exchange);
  }

  @Test
  public void GivenFiveMSFTTotalShouldBe500() throws Exception {
    portfolio.add(5, "MSFT");
    Assert.assertEquals(500, portfolio.value());
  }
}
```

如果系统解耦到这样测试的程度，也就更加灵活，更加可复用。部件之间的解耦代表着系统中的元素互相隔离得很好。隔离也让对系统每个元素的理解变得更加容易。

通过降低连接度，我们的类就遵循了另一条类设计原则，即依赖倒置原则（Dependency Inversion Principle，DIP）①。本质而言，DIP 认为类应当依赖抽象而不是依赖具体细节。

我们的 `Portfolio` 类不再依赖 `TokyoStockExchange` 类的实现细节，而是依赖 `StockExchange` 接口。`StockExchange` 接口呈现的是有关询问某只股票价格的抽象概念，这种抽象隔离了所有询价的特定细节，其中包括价格数据来自何处的类。

10.4 文献

[RDD]：*Object Design: Roles, Responsibilities, and Collaborations,* Rebecca Wirfs-Brock et al., Addison-Wesley, 2002.

[PPP]：*Agile Software Development*: *Principles, Patterns, and Practices,* Robert C. Martin, Prentice Hall, 2002.

[Knuth92]：*Literate Programming,* Donald E. Knuth, Center for the Study of language and Information, Leland Stanford Junior University, 1992.

---

① [PPP]。

第 11 章

系 统

Kevin Dean Wampler 博士

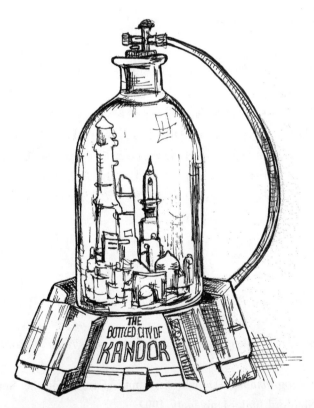

"复杂要人命。它消磨开发者的生命,让产品难以规划、构建和测试。"

——Ray Ozzie,微软公司首席技术官

11.1 如何建造一个城市

你能自己掌管一切细节吗？大概不行。即便是管理一个既存的城市，也是靠单人能力无法做到的。不过，城市还是在运转（多数时候）。这是因为每个城市都有各种组织管理不同的部分，如供水系统、供电系统、交通、执法、立法，诸如此类。有些人负责全局，有些人负责细节。

城市能运转，还因为它演化出恰当的抽象等级和模块，好让个人和其所管理的"组件"即便在不了解全局时也能有效地运转。

尽管软件团队往往也是这样组织起来的，但他们所致力的工作却常常没有同样的关注面切分及抽象层级，而整洁的代码可以帮助我们在较低的抽象层级上达成这一目标。本章将讨论如何在较高的抽象层级（系统层级）上保持整洁。

11.2 将系统的构造与使用分开

首先，构造与使用是非常不一样的过程。当我走笔至此，投目窗外的芝加哥，看到有一家酒店正在建设。今天，那只是个框架结构，起重机和升降机布设在外面。忙碌的人们身穿工作服，头戴安全帽。大概一年之后，酒店就将建成。起重机和升降机都会消失无踪。建筑物变得整洁，覆盖着玻璃幕墙和漂亮的漆色。在其中工作和住宿的人，会看到完全不同的景象。

软件系统应将起始过程和起始过程之后的运行时逻辑分离开，在起始过程中构建应用对象，也会存在互相缠结的依赖关系。

每个应用程序都该留意起始过程。那也是本章中我们首先要考虑的问题。将关注的方面分离开，是软件技能中最古老也最重要的设计技能。

不幸的是，多数应用程序都没有做分离处理。起始过程代码很特殊，被混杂到运行时逻辑中，下例就是典型的这种情形：

```
public Service getService() {
  if (service == null)
    service = new MyServiceImpl(...);  // Good enough default for most cases?
  return service;
}
```

这就是所谓延迟初始化/赋值，它有一些好处，即在真正用到对象之前，无须操心这种架空构造，起始时间也会更短，而且还能保证永远不会返回 `null` 值。

然而，我们也得到了 `MyServiceImpl` 及其构造器所需一切（我省略了那些代码）的硬编码依赖。不搞好这些被依赖项，程序就无法编译，即便在运行时永不使用这种类型的对象！

测试也会是问题。如果 `MyServiceImpl` 是一个重型对象，则我们必须确保在单元测试调用该方法之前，就给服务指派恰当的测试替身（TEST DOUBLE）[①]或仿制对象（MOCK OBJECT）。由于构造逻辑与运行过程相混杂，我们必须测试所有的执行路径（例如，null 值测试及其代码块）。有了这些权责，说明方法做了不止一件事，这样就略微违反了单一权责原则。

最糟糕的大概是，我们不知道 `MyServiceImpl` 在所有情形中是否都是正确的对象。我在代码注释中做了暗示。为什么该方法所属类必须知道全局情景？我们是否真能知道在这里要用到的正确对象？是否真有可能存在一种放之四海而皆准的类型？

当然，仅出现一次的延迟初始化不算是严重问题。不过，在应用程序中往往有许多种类似的情况出现。于是，全局设置策略（如果有的话）在应用程序中四散分布，缺乏模块组织性，通常也会有许多重复代码。

如果我们勤于打造有着良好格式并且强固的系统，就不该让这类就手小技巧破坏模块组织性，对象构造的起始和设置过程也不例外。应当将这个过程从正常的运行时逻辑中分离出来，确保拥有解决主要依赖问题的全局性一贯策略。

11.2.1 分解 main

将构造与使用分开的方法之一是将全部构造过程搬迁到 main 或被称之为 main 的模块中，设计系统的其余部分时，假设所有对象都已正确构造和设置（如图 11-1 所示）。

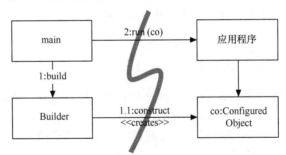

图 11-1　将构造分解到 main() 中

控制流程很容易理解。main 函数创建系统所需的对象，再传递给应用程序，应用程序只管使用。注意，看横贯 main 与应用程序之间隔离的依赖箭头的方向，它们都从 main 函数向外走，这表示应用程序对 main 或者构造过程一无所知，它只是简单地指望一切已齐备。

---

① [Mezzaros07]。

11.2.2 工厂

当然,有时应用程序也要负责确定何时创建对象。例如,在某个订单处理系统中,应用程序必须创建 `LineItem` 实体,添加到 `Order` 对象。在这种情况下,我们可以使用抽象工厂模式[1]让应用自行控制何时创建 `LineItem`,但构造的细节却隔离于应用程序代码之外,如图 11-2 所示。

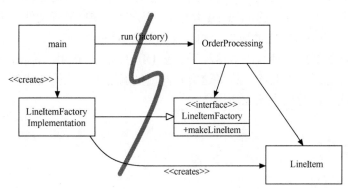

图 11-2 使用工厂分离构造过程

再注意一下,所有依赖都是从 `main` 指向 `OrderProcessing` 应用程序的,这表示应用程序与如何构建 `LineItem` 的细节是分离开来的。其中构建能力由 `LineItemFactoryImplementation` 持有,而 `LineItemFactoryImplementation` 又是在 `main` 这一边的,但应用程序能完全控制 `LineItem` 实体何时构建,甚至能传递应用程序特定的构造器参数。

11.2.3 依赖注入

有一种强大的机制可以实现分离构造与使用,那就是依赖注入(Dependency Injection,DI)。控制反转(Inversion of Control,IoC)在依赖管理中的一种应用手段[2]。控制反转将第二权责从对象中拿出来,转移到另一个专注于此的对象中,从而遵循了单一权责原则。在依赖管理情景中,对象不应负责实体化对自身的依赖,反之,它应当将这份权责移交给其他"有权力"的机制,从而实现控制的反转。因为初始设置是一种全局问题,所以通常这种授权机制要么是 `main` 例程,要么是有特定目的的容器。

JNDI 查找是 DI 的一种"部分"实现。在 JNDI 中,对象请求目录服务器提供一种符合某个特定名称的"服务"。

---

[1] [GOF]。
[2] 可参见[Fowler]。

```
MyService myService = (MyService)(jndiContext.lookup("NameOfMyService"));
```

调用对象并不控制真正返回对象的类别（当然前提是它实现了恰当的接口），但调用对象仍然主动解决了依赖问题。

真正的依赖注入还要更进一步。类并不直接解决依赖问题，而是保持完全被动。它提供可用于注入依赖的赋值器方法或构造器参数（或二者皆有）。在构造过程中，DI 容器实体化需要的对象（通常按需创建），并使用构造器参数或赋值器方法将依赖连接到一起。至于哪个依赖对象真正得到使用，是通过配置文件或在一个有特殊目的的构造模块中编程决定的。

Spring 框架提供了最有名的 Java DI 容器[①]。用户在 XML 配置文件中定义互相关联的对象，然后用 Java 代码请求特定的对象。稍后我们就会看到相关例子。

但延后初始化的好处是什么呢？这种手段在 DI 中也有其作用。首先，多数 DI 容器在需要对象之前并不构造对象。其次，许多这类容器提供调用工厂或构造代理的机制，而这种机制可为延迟赋值或类似的优化处理所用[②]。

11.3 扩容

城市由城镇而来，城镇由乡村而来。一开始，道路狭窄，几乎无人涉足，随后逐渐拓宽。小型建筑和空地渐渐被更大的建筑所取代，一些地方最终矗立起摩天大楼。

一开始，供电、供水、下水、互联网（哇！）等服务全部欠奉。随着人口和建筑密度的增加，这些服务也开始出现。

这种成长并非全无阵痛。想想看你有多少次开着车，艰难穿行一个"道路改善"工程时，是否问过自己："他们为什么不一开始就修条够宽的路呢？！"

不过那无论如何不可能实现。谁敢打包票说在小镇里修建一条六车道的公路并不浪费呢？谁会想要这么一条穿过他们小镇的路呢？

"一开始就做对系统"纯属神话。反之，我们应该只去实现今天的用户故事，然后重构，明天再扩展系统、实现新的用户故事。这就是迭代和增量敏捷的精髓所在。测试驱动开发、重构以及它们打造出的整洁代码，在代码层面保证了这个过程的实现。

但在系统层面又如何呢？难道系统架构不需要预先做好计划吗？系统理所当然不可能从简单递增到复杂，它能行吗？

> 与物理系统相比，软件系统比较独特。软件系统的架构可以递增式地增长，只要我们持续将关注面恰当地切分。

---

① 见[Spring]。另外也有一个 Spring.NET 框架。
② 别忘记延迟初始化/赋值只是一种优化手段，而且可能是一种不成熟的手段。

如我们将见到的那样，软件系统短生命周期的本质使这一切变得可行。我们先来看一个没有充分隔离关注问题的架构反例。

初始的 EJB1 和 EJB2 架构没有恰当地切分关注面，从而给有机增长压上了不必要的负担。比如一个持久 Bank 类的 Entity Bean。Entity Bean 是关系数据在内存中的体现，换言之，是表格的一行。

首先，你要定义一个本地（进程内）或远程（分离的 JVM）接口，供客户代码使用。代码清单 11-1 就是一种可能的本地接口：

代码清单 11-1　Bank EJB 的 EJB2 本地接口

```
package com.example.banking;
import java.util.Collections;
import javax.ejb.*;

public interface BankLocal extends java.ejb.EJBLocalObject {
  String getStreetAddr1() throws EJBException;
  String getStreetAddr2() throws EJBException;
  String getCity() throws EJBException;
  String getState() throws EJBException;
  String getZipCode() throws EJBException;
  void setStreetAddr1(String street1) throws EJBException;
  void setStreetAddr2(String street2) throws EJBException;
  void setCity(String city) throws EJBException;
  void setState(String state) throws EJBException;
  void setZipCode(String zip) throws EJBException;
  Collection getAccounts() throws EJBException;
  void setAccounts(Collection accounts) throws EJBException;
  void addAccount(AccountDTO accountDTO) throws EJBException;
}
```

上面列出了银行地址的几个属性，和一组该银行拥有的账户，其中每个账户的数据都由单独的 Account EJB 所持有。代码清单 11-2 展示了 Bank Bean 的相应实现类。

代码清单 11-2　相应的 EJB2 Entity Bean 实现

```
package com.example.banking;
import java.util.Collections;
import javax.ejb.*;

public abstract class Bank implements javax.ejb.EntityBean {
  // Business logic...
  public abstract String getStreetAddr1();
  public abstract String getStreetAddr2();
  public abstract String getCity();
  public abstract String getState();
  public abstract String getZipCode();
  public abstract void setStreetAddr1(String street1);
  public abstract void setStreetAddr2(String street2);
  public abstract void setCity(String city);
  public abstract void setState(String state);
```

```
    public abstract void setZipCode(String zip);
    public abstract Collection getAccounts();
    public abstract void setAccounts(Collection accounts);
    public void addAccount(AccountDTO accountDTO) {
      InitialContext context = new InitialContext();
      AccountHomeLocal accountHome = context.lookup("AccountHomeLocal");
      AccountLocal account = accountHome.create(accountDTO);
      Collection accounts = getAccounts();
      accounts.add(account);
    }
    // EJB container logic
    public abstract void setId(Integer id);
    public abstract Integer getId();
    public Integer ejbCreate(Integer id) { ... }
    public void ejbPostCreate(Integer id) { ... }
    // The rest had to be implemented but were usually empty:
    public void setEntityContext(EntityContext ctx) {}
    public void unsetEntityContext() {}
    public void ejbActivate() {}
    public void ejbPassivate() {}
    public void ejbLoad() {}
    public void ejbStore() {}
    public void ejbRemove() {}
}
```

我没有列出对应的 LocalHome 接口，该接口基本上是用来创建对象的，也没有列出你可能添加的 Bank 查找器（查询）。

最后，你要编写一个或多个 XML 部署说明，将对象相关映射细节指定给某个持久化存储空间，说明期望的事物行为、安全约束等。

业务逻辑与 EJB2 应用"容器"紧密耦合。你必须子类化容器类型，必须提供许多个该容器所需要的生命周期方法。

由于存在这种与重量级容器的紧耦合，隔离单元测试就很困难。有必要模拟出容器（这很难），或者花费大量时间在真实服务器上部署 EJB 和测试。也由于耦合的存在，在 EJB2 架构之外的复用实际上变得不可能。

最终，连面向对象编程本身也被侵蚀。bean 不能继承自另一个 bean。留意添加新账号的逻辑。在 EJB2 bean 中，定义一种本质上是无行为 struct 的"数据传输对象"（DTO）很常见。这往往会导致拥有同样数据的冗余类型出现，而且也需要在对象之间复制数据的八股式代码。

横贯式关注面

在某些领域，EBJ2 架构已经很接近真正的关注面切分。例如，在与源代码分离的部署描述中声明了期待的事务、安全及部分持久化行为。

注意，持久化之类关注面倾向于横贯某个领域的天然对象边界。你会想用同样的策略来持久化所有对象，例如，使用 DBMS[①]而非平面文件，表名和列名遵循某种命名约定，采用一致的事务语义，等等。

原则上，你可以从模块、封装的角度推理持久化策略。但在实践上，你却不得不将实现了持久化策略的代码铺展到许多对象中。我们用术语横贯式关注面（Cross-Cutting Concern）来形容这类情况。同样，对于持久化框架和领域逻辑，如果我们孤立地看也可以是模块化的。问题在于横贯这些领域的情形。

实际上，EJB 架构处理持久化、安全和事务的方法要早于面向方面编程（aspect-oriented programming，AOP）[②]，而 AOP 是一种恢复横贯式关注面模块化的普适手段。

在 AOP 中，被称为方面（aspect）的模块构造说明了系统中哪些点的行为会以某种一致的方式被修改，从而支持某种特定的场景。这种说明是用某种简洁的声明或编程机制来实现的。

以持久化为例，可以声明哪些对象和属性（或其模式）应当被持久化，然后将持久化任务委托给持久化框架。行为的修改由 AOP 框架以无损方式[③]在目标代码中进行。下面来看看 Java 中的 3 种方面或类似方面的机制。

11.4 Java 代理

Java 代理适用于简单的情况，例如在单独的对象或类中包装方法调用。然而，JDK 提供的动态代理仅能与接口协同工作。对于代理类，你得使用字节码操作库，比如 CGLIB、ASM 或 Javassist[④]。

代码清单 11-3 展示了为我们的 Bank 应用程序提供持久化支持的 JDK 代理，代码仅覆盖设置和取得账号列表的方法。

代码清单 11-3　JDK 代理范例

```
// Bank.java (suppressing package names...)
import java.util.*;

// The abstraction of a bank.
public interface Bank {
  Collection<Account> getAccounts();
  void setAccounts(Collection<Account> accounts);
}

// BankImpl.java
import java.utils.*;
```

[①] 数据库管理系统。
[②] 查阅[AOSD]获取有关方面的一般信息，查阅[AspectJ]和[Colyer]获取有关 AspectJ 的信息。
[③] 即无须手工修改源代码。
[④] 见[CGLIB]、[ASM]和[Javassist]。

```java
// The "Plain Old Java Object" (POJO) implementing the abstraction.
public class BankImpl implements Bank {
  private List<Account> accounts;

  public Collection<Account> getAccounts() {
    return accounts;
  }

  public void setAccounts(Collection<Account> accounts) {
    this.accounts = new ArrayList<Account>();
    for (Account account: accounts) {
      this.accounts.add(account);
    }
  }
}

// BankProxyHandler.java
import java.lang.reflect.*;
import java.util.*;

// "InvocationHandler" required by the proxy API.
public class BankProxyHandler implements InvocationHandler {
  private Bank bank;

  public BankHandler (Bank bank) {
    this.bank = bank;
  }

  // Method defined in InvocationHandler
  public Object invoke(Object proxy, Method method, Object[] args)
      throws Throwable {
    String methodName = method.getName();
    if (methodName.equals("getAccounts")) {
      bank.setAccounts(getAccountsFromDatabase());
      return bank.getAccounts();
    } else if (methodName.equals("setAccounts")) {
      bank.setAccounts((Collection<Account>) args[0]);
      setAccountsToDatabase(bank.getAccounts());
      return null;
    } else {
      ...
    }
  }

  // Lots of details here:
  protected Collection<Account> getAccountsFromDatabase() { ... }
  protected void setAccountsToDatabase(Collection<Account> accounts) { ... }
}

// Somewhere else...

Bank bank = (Bank) Proxy.newProxyInstance(
  Bank.class.getClassLoader(),
```

```
new Class[] { Bank.class },
new BankProxyHandler(new BankImpl()));
```

我们定义了将被代理包装起来的接口 `Bank`，还有旧式的 Java 对象（Plain-Old Java Object，POJO）`BankImpl`，该对象实现业务逻辑（稍后再来看 POJO）。

Proxy API 需要一个 `InvocationHandler` 对象，用来实现对代理的全部 `Bank` 方法调用。`BankProxyHandler` 使用 Java 反射 API 将一般方法调用映射到 `BankImpl` 中的对应方法，以此类推。

即便对于这样简单的例子，也有许多相对复杂的代码[①]。使用那些字节操作类库也同样具有挑战性。代码量和复杂度是代理的两大弱点，创建整洁代码变得很难！另外，代理也没有提供在系统范围内指定执行点的机制，而那正是真正的 AOP 解决方案所必需的[②]。

11.5 纯 Java AOP 框架

幸运的是，编程工具能自动处理大多数代理模板代码。在数个 Java 框架中，代理都是内嵌的，如 Spring AOP 和 JBoss AOP 等，从而能够以纯 Java 代码实现面向方面编程[③]。在 Spring 中，你将业务逻辑编码为旧式 Java 对象。POJO 自扫门前雪，并不依赖于企业框架（或其他域），因此，它在概念上更简单、更易于测试驱动，相对简单，也较易于保证正确地实现相应的用户故事，并为未来的用户故事维护和改进代码。

通过使用描述性配置文件或 API，你可以把需要的应用程序构架组合起来，包括持久化、事务、安全、缓存、恢复等横贯性问题。在许多情况下，你实际上只是指定 Spring 或 Jboss 类库，框架以对用户透明的方式处理使用 Java 代理或字节代码库的机制。这些声明驱动了依赖注入（DI）容器，DI 容器再实体化主要对象，并按需将对象连接起来。

代码清单 11-4 展示了 Spring V2.5 配置文件 app.xml 的典型片段。

代码清单 11-4 Spring 2.x 的配置文件

```
<beans>
  ...
  <bean id="appDataSource"
    class="org.apache.commons.dbcp.BasicDataSource"
    destroy-method="close"
    p:driverClassName="com.mysql.jdbc.Driver"
    p:url="jdbc:mysql://localhost:3306/mydb"
    p:username="me"/>
```

---

① 要想了解更多关于 Proxy API 及其用法，请参阅[Goetz]。
② AOP 有时会与实现它的技术相混淆，例如方法拦截和通过代理做的"封包"。AOP 系统的真正价值在于用简洁和模块化的方式指定系统行为。
③ 见[Spring]和[JBoss]。"纯 Java"表示不使用 AspectJ。

```xml
<bean id="bankDataAccessObject"
  class="com.example.banking.persistence.BankDataAccessObject"
  p:dataSource-ref="appDataSource"/>

<bean id="bank"
  class="com.example.banking.model.Bank"
  p:dataAccessObject-ref="bankDataAccessObject"/>
...
</beans>
```

每个 bean 就像是嵌套"俄罗斯套娃"中的一个，每个由数据存取器对象（DAO）代理（包装）的 Bank 都有一个域对象，而 bean 本身又是由 JDBC 驱动程序数据源代理（如图 11-3 所示）。

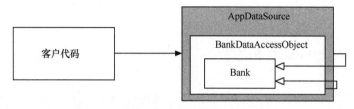

图 11-3 "俄罗斯套娃"式的油漆工模式

客户代码以为调用的是 Bank 对象的 getAccount() 方法，其实它是在与一组扩展 Bank POJO 基础行为的 DECORATOR[①] 对象中最外面的那个沟通。

在应用程序中，只添加了少数几行代码，用来向 DI 容器请求系统中的顶层对象，如 XML 文件中所定义的那样。

```
XmlBeanFactory bf =
  new XmlBeanFactory(new ClassPathResource("app.xml", getClass()));
Bank bank = (Bank) bf.getBean("bank");
```

只需区区几行与 Spring 相关的 Java 代码，应用程序就几乎与 Spring 完全分离了，消除了 EJB2 之类系统中那种紧耦合问题。

尽管 XML 可能会冗长且难以阅读[②]，配置文件中定义的"策略"还是要比那种隐藏在幕后自动创建的复杂的代理和方面逻辑来得简单。这种类型的架构是如此引人注目，Spring 之类的框架最终导致了 EJB 标准在第 3 版的彻底变化。使用 XML 配置文件和/或 Java 5 注解，EJB3 很大程度上遵循了 Spring 通过描述性手段支持横贯式关注面的模型。

代码清单 11-5 展示了用 EJB3 重写的 Bank 对象。

代码清单 11-5 EJB3 版本的 Bank

```
package com.example.banking.model;
import javax.persistence.*;
```

---

① [GOF]。
② 可以使用遵循"约定胜于配置"的机制和 Java 5 annotation 来减少外露的连接逻辑，从而简化这个例子。

```java
import java.util.ArrayList;
import java.util.Collection;

@Entity
@Table(name = "BANKS")
public class Bank implements java.io.Serializable {
  @Id @GeneratedValue(strategy=GenerationType.AUTO)
  private int id;

  @Embeddable // An object "inlined" in Bank's DB row
  public class Address {
    protected String streetAddr1;
    protected String streetAddr2;
    protected String city;
    protected String state;
    protected String zipCode;
  }

  @Embedded
  private Address address;

  @OneToMany(cascade = CascadeType.ALL, fetch = FetchType.EAGER,
             mappedBy = "bank")
  private Collection<Account> accounts = new ArrayList<Account>();

  public int getId() {
    return id;
  }

  public void setId(int id) {
    this.id = id;
  }

  public void addAccount(Account account) {
    account.setBank(this);
    accounts.add(account);
  }

  public Collection<Account> getAccounts() {
    return accounts;
  }

  public void setAccounts(Collection<Account> accounts) {
    this.accounts = accounts;
  }
}
```

上述代码要比原本的 EJB2 代码整洁多了。有些实体细节仍然在注解中存在。不过，因为没有任何信息超出注解之外，代码依然整洁、清晰，也因此而易于测试驱动、易于维护。

如果愿意的话，注解中的一些或全部持久化信息可以转移到 XML 部署描述中，只留下真正的纯 POJO。如果持久化映射细节不会频繁改动，许多团队可能会选择保留注解，但与 EJB2 那种侵害性相比还是少了很多问题。

11.6　AspectJ 的方面

通过方面来实现关注面切分的功能最全的工具是 AspectJ 语言[①]，它提供"一流的"将方面作为模块构造处理支持的 Java 扩展。在 80%～90%用到方面特性的情况下，Spring AOP 和 JBoss AOP 提供的纯 Java 实现手段足够使用。然而，AspectJ 却提供了一套用以切分关注面的丰富而强有力的工具。AspectJ 的弱势在于，需要采用几种新工具，学习新语言构造和使用方式。

借由 AspectJ 近期引入的"annotation form"（使用 Java 5 注解定义纯 Java 代码的方面），采用新工具的问题大大减少。另外，Spring 框架也有一些让拥有较少 AspectJ 经验的团队更容易组合基于注解的方面的特性。

关于 AspectJ 的全面探讨已经超出本书范围。更多信息可参见[AspectJ]、[Colyer]和[Spring]。

11.7　测试驱动系统架构

通过方面式的手段切分关注面的威力不可低估。假使你能用 POJO 编写应用程序的领域逻辑，在代码层面与架构关注面分离开，就有可能真正地用测试来驱动架构。采用一些新技术，就能将架构按需从简单演化到精细。没必要先做大设计（Big Design Up Front，BDUF）[②]。实际上，BDUF 甚至是有害的，它阻碍改进，因为心理上会抵制丢弃既成之事，也因为架构上的方案选择影响后续的设计思路。

建筑设计师不得不做 BDUF，因为一旦建造过程开始，就不可能对大型物理建筑的结构做根本性改动[③]。尽管软件也有物理[④]的一面，但只要软件的构架有效切分了各个关注面，还是有可能做根本性改动的。

这意味着我们可以从"简单自然"但切分良好的架构开始做软件项目，快速交付可开展工作的用户故事，随着规模的增加添加更多基础架构。有些世界上最大的网站采用了精密的数据缓存、安全、虚拟化等技术，获得了极高的可用性和性能，在每个抽象层和范围之内，那些最小化耦合的设计都简单到位，效率和灵活性也随之而来。

当然，这不是说要毫无准备地进入一个项目。对于总的覆盖范围、目标、项目进度和最

---

① 参见[AspectJ]和[Colyer]。
② BDUF 是一种预先设计好一切实现的方式，不能与先做设计（up-front design）的良好实践手段相混淆。
③ 即便在构建开始之后，也会有大量迭代式的考察和细节讨论。
④ "软件物理"一词最早由[Kolence]提出。

终系统的总体构架，我们会有所预期。不过，我们必须有能力随机应变。

EJB 早期架构就是一种著名的过度工程化而没能有效切分关注面的 API。在没能真正得到使用时，设计得再好的 API 也等于是杀鸡用牛刀。优秀的 API 在大多数时间都该在视线之外，这样团队才能将创造力集中在要实现的用户故事上。否则，架构上的约束就会妨碍向客户交付优化价值的软件。

概言之，

> 最佳的系统架构由模块化的关注面领域组成，每个关注面均用纯 Java（或其他语言）对象实现。不同的领域之间用最不具有侵害性的方面或类方面工具整合起来。这种架构能测试驱动，就像代码一样。

11.8 优化决策

模块化和关注面切分成就了分散化管理和决策。在巨大的系统中，不管是一座城市或一个软件项目，无人能做所有决策。

众所周知，对于决策最好是授权给最有资格的人。但我们常常忘记了，延迟决策至最后一刻也是好手段，这不是懒惰或不负责，而是让我们能够基于最有可能的信息做出选择。提前决策是一种预备知识不足的决策。如果决策太早，就会缺少太多客户反馈、关于项目的思考和实施经验。

> 拥有模块化关注面的 POJO 系统提供的敏捷能力，允许我们基于最新的知识做出优化的、时机刚好的决策。决策的复杂性也降低了。

11.9 明智使用添加了可论证价值的标准

建筑构造大有可观，既因为新建筑的构建过程（即便是在隆冬季节），也因为那些现今科技所能实现的超凡设计。建筑业是一个成熟行业，有着高度优化的部件、方法和久经岁月历练的标准。

即便是轻量级和更直截了当的设计已足敷使用，许多团队也还是采用了 EJB2 架构，这是因为 EJB2 是一个标准。我见过一些团队，纠缠于这个或那个名声大噪的标准，却丧失了对为客户实现价值的关注。

有了标准，就更易复用想法和组件、雇用拥有相关经验的人才、封装好点子，以及将组件连接起来。不过，创立标准的过程有时却漫长到行业等不及的程度，有些标准没能与它要服务的采用者的真实需求相结合。

11.10　系统需要领域特定语言

建筑，与大多数其他领域一样，发展出一套丰富的语言，有词汇、熟语和清晰而简洁地表达基础信息的句式[①]。在软件领域，领域特定语言（Domain-Specific Language，DSL）[②]最近重受关注。DSL 是一种单独的小型脚本语言或以标准语言写就的 API，领域专家可以用它编写读起来像是组织严谨的散文一般的代码。

优秀的 DSL 填平了领域概念和实现领域概念的代码之间的"壕沟"，就像敏捷实践优化了开发团队和甲方之间的沟通一样。如果你用与领域专家使用的同一种语言来实现领域逻辑，就会降低不正确地将领域翻译为实现的风险。

DSL 在有效使用时能提升代码惯用法和设计模式之上的抽象层次，它允许开发者在恰当的抽象层级上直指代码的初衷。

> 领域特定语言允许所有抽象层级和应用程序中的所有领域，从高级策略到底层细节，使用 POJO 来表达。

11.11　小结

系统也应该是整洁的。侵害性架构会湮灭领域逻辑，冲击敏捷能力。如果领域逻辑受到困扰，质量就会堪忧，因为缺陷更易隐藏，用户故事更难实现。当敏捷能力受到损害时，生产力也会降低，TDD 的好处遗失殆尽。

在所有的抽象层级上，意图都应该清晰可辨。只有在编写 POJO 并使用类方面的机制来无损地组合其他关注面时，这种事情才会发生。

无论是设计系统还是单独的模块，别忘了使用大概可开展工作的最简单方案。

11.12　文献

[Alexander]: Christopher Alexander, *A Timeless Way of Building,* Oxford University Press, New York, 1979.

[AOSD]: Aspect-Oriented Software Development port.

---

① [Alexander]的著作对软件社区影响至深。
② 参见[DSL]。[JMock]是创建 DSL 的 Java API 的优秀范例。

[ASM]：ASM Home Page.
[AspectJ]：Eclipse Foundation AspectJ.
[CGLIB]：Code Generation Library.
[Colyer]：Adrian Colyer, Andy Clement, George Hurley, Mathew Webster, *Eclipse AspectJ,* Person Education, Inc., Upper Saddle River, NJ, 2005.
[DSL]：Domain-specific programming language.
[Fowler]：Inversion of Control Containers and the Dependency Injection pattern.
[Goetz]：Brian Goetz, *Java Theory and Practice: Decorating with Dynamic Proxies*.
[Javassist]：Javassist Home Page.
[JBoss]：JBoss Home Page.
[JMock]：JMock—A Lightweight Mock Object Library for Java.
[Kolence]：Kenneth W. Kolence, Software physics and computer performance measurements, *Proceedings of the ACM annual conference—Volume 2*, Boston, Massachusetts, pp. 1024-1040, 1972.
[Spring]：*The Spring Framework*.
[Mezzaros07]：*XUnit Patterns*, Gerard Mezzaros, Addison-Wesley, 2007.
[GOF]：*Design Patterns: Elements of Reusable Object Oriented Software*, Gamma et al., Addison-Wesley, 1996.

第12章

迭 进

Jeff Langr

12.1 通过迭进设计达到整洁目的

假使有4条简单的规则,跟着做就能帮助你创建优良的设计,会如何?假使遵循这些规则,你就能洞见代码的结构和设计,更能轻易地应用 SRP 和 DIP 之类的原则,便会如何?假使这4条规则有利于良好的设计"浮现"出来,又会如何?

我们中的许多人认为，Kent Beck 关于简单设计[①]的 4 条规则，对于创建具有良好设计的软件有着莫大的帮助。

据 Kent 所述，只要遵循以下规则，设计就能变得"简单"：

- 运行所有测试；
- 不可重复；
- 表达了程序员的意图；
- 尽可能减少类和方法的数量。

以上规则按其重要程度排列。

12.2 简单设计规则 1：运行所有测试

设计必须制造出如预期一般工作的系统，这是首要因素。系统也许有一套绝佳设计，但如果缺乏验证系统是否真按预期那样工作的简单方法，那就无异于纸上谈兵。

全面测试并持续通过所有测试的系统，就是可测试的系统。这看似浅显，却很重要。不可测试的系统同样不可验证。不可验证的系统，绝不应部署。

幸运的是，只要系统可测试，就会导向保持类短小且目的单一的设计方案。遵循 SRP 的类，测试起来较为简单。测试编写得越多，就越能持续走向编写较易测试的代码。所以，确保系统完全可测试能帮助我们创建更好的设计。

紧耦合的代码难以编写测试。同样，编写测试越多，就越会遵循 DIP 之类的规则，从而越会使用依赖注入、接口和抽象等工具尽可能减少耦合，如此一来，设计就会有长足进步。

遵循有关编写测试并持续运行测试的简单、明确的规则，系统就会更贴近面向对象低耦合度、高内聚度的目标。编写测试将会引致更好的设计。

12.3 简单设计规则 2~4：重构

有了测试，就能保持代码和类的整洁，方法就是递增式地重构代码。添加了几行代码后，就要暂停，琢磨一下变化了的设计。设计退步了吗？如果是，就要清理它，并且运行测试，保证没有破坏任何东西。**测试消除了对清理代码就会破坏代码的恐惧。**

在重构过程中，可以应用有关优秀软件设计的一切知识，提升内聚性，降低耦合度，切分关注面，模块化系统性关注面，缩小函数和类的尺寸，选用更好的名称，如此等等。这也是应用简单设计后 3 条规则的地方：消除重复，保证表达力，尽可能减少类和方法的数量。

---

① [XPE]。

12.4 不可重复

重复是拥有良好设计的系统的大敌。它代表着额外的工作、额外的风险和额外且不必要的复杂度。重复有多种表现。极其雷同的代码行当然是重复。类似的代码往往可以调整得更相似，这样就能更容易地进行重构。重复也有实现上的重复等其他一些形态，例如，在某个群集类中可能会有以下两个方法：

```
int size()       {}
boolean isEmpty()   {}
```

这两个方法可以分别实现。isEmpty 方法跟踪一个布尔值，而 size 方法则跟踪一个计数器。或者，也可以通过在 isEmpty 方法中使用 size 方法来消除重复：

```
boolean isEmpty() {
  return 0 == size();
}
```

要想创建整洁的系统，需要有消除重复的意愿，即便对于短短几行也是如此。例如，以下代码：

```
public void scaleToOneDimension(
    float desiredDimension, float imageDimension) {
  if (Math.abs(desiredDimension - imageDimension) < errorThreshold)
    return;
  float scalingFactor = desiredDimension / imageDimension;
  scalingFactor = (float)(Math.floor(scalingFactor * 100) * 0.01f);

  RenderedOp newImage = ImageUtilities.getScaledImage(
      image, scalingFactor, scalingFactor);
  image.dispose();
  System.gc();
  image = newImage;
}
public synchronized void rotate(int degrees) {
  RenderedOp newImage = ImageUtilities.getRotatedImage(
    image, degrees);
  image.dispose();
  System.gc();
  image = newImage;
}
```

要保持系统整洁，应该消除 scaleToOneDimension 方法和 rotate 方法里面的少量重复：

```
public void scaleToOneDimension(
    float desiredDimension, float imageDimension) {
  if (Math.abs(desiredDimension - imageDimension) < errorThreshold)
```

```
        return;
    float scalingFactor = desiredDimension / imageDimension;
    scalingFactor = (float)(Math.floor(scalingFactor * 100) * 0.01f);
    replaceImage(ImageUtilities.getScaledImage(
      image, scalingFactor, scalingFactor));
}

public synchronized void rotate(int degrees) {
    replaceImage(ImageUtilities.getRotatedImage(image, degrees));
}

private void replaceImage(RenderedOp newImage) {
    image.dispose();
    System.gc();
    image = newImage;
}
```

做了一点点共性抽取后,我们意识到已经违反了 SRP 原则。所以,可以把一个新方法分解到另外的类中,从而提升其可见性。团队中的其他成员也许会发现进一步抽象新方法的机会,并且在其他场景中复用之。"小规模复用"可大量降低系统复杂性。要想实现大规模复用,必须理解如何实现小规模复用。

模板方法模式(Template Method)[1]是一种移除高层级重复的通用技巧。例如:

```
public class VacationPolicy {
  public void accrueUSDivisionVacation() {
    // code to calculate vacation based on hours worked to date
    // ...
    // code to ensure vacation meets US minimums
    // ...
    // code to apply vaction to payroll record
    // ...
  }

   public void accrueEUDivisionVacation() {
     // code to calculate vacation based on hours worked to date
     // ...
     // code to ensure vacation meets EU minimums
     // ...
     // code to apply vaction to payroll record
     // ...
   }
}
```

除了计算法定最少数量假期的部分,accrueUSDivisionVacation 和 accrueEuropeanDivisionVacation 中有大量代码雷同。那部分的算法,依据员工类型而变。

可以通过应用模板方法模式来消除明显的重复。

```
abstract public class VacationPolicy {
   public void accrueVacation() {
```

---

① [GOF]。

```
    calculateBaseVacationHours();
    alterForLegalMinimums();
    applyToPayroll();
  }

  private void calculateBaseVacationHours() { /* ... */ };
  abstract protected void alterForLegalMinimums();
  private void applyToPayroll() { /* ... */ };
}

public class USVacationPolicy extends VacationPolicy {
  @Override protected void alterForLegalMinimums() {
    // US specific logic
  }
}

public class EUVacationPolicy extends VacationPolicy {
  @Override protected void alterForLegalMinimums() {
    // EU specific logic
  }
}
```

子类填充了 accrueVacation 算法中的"空洞",提供不重复的信息。

12.5 表达力

我们中的大多数人都经历过费解代码的纠缠。我们中的许多人自己就编写过费解的代码。写出自己能理解的代码很容易,因为在写这些代码时,我们正深入于要解决的问题中。代码的其他维护者不会那么深入,也就不易理解代码。

软件项目的主要成本在于长期的维护。为了在修改时尽量降低出现缺陷的可能性,很有必要理解系统是做什么的。当系统变得越来越复杂,开发者就需要越来越多的时间来理解它,而且也极有可能误解。所以,代码应当清晰地表达其作者的意图。作者把代码写得越清晰,其他人花在理解代码上的时间也就越少,从而减少缺陷,缩减维护成本。

可以通过选用好名称来表达。我们想要听到好类名和好函数名,而且在查看其权责时不会大吃一惊。

也可以通过保持函数和类尺寸短小来表达。短小的类和函数通常易于命名,易于编写,易于理解。

还可以通过采用标准命名法来表达。例如,设计模式很大程度上关乎沟通和表达。通过在实现这些模式的类的名称中采用标准模式名,如 COMMAND 或 VISITOR,就能充分地向其他开发者描述你的设计。

编写良好的单元测试也具有表达性。测试的主要目的之一就是通过实例起到文档的作用。读到测试的人应该能很快理解某个类是做什么的。

不过，做到有表达力的最重要方式却是尝试。有太多时候，我们一旦写出能工作的代码，就转移到下一个问题上，而没有下足功夫调整代码，让后来者易于阅读。记住，下一位读代码的人最有可能是你自己。

所以，多少尊重一下你的手艺吧。花一点点时间在每个函数和类上。选用较好的名称，将大函数切分为小函数，时时照拂自己创建的东西。用心是最珍贵的资源。

12.6 尽可能少的类和方法

即便是消除重复、代码表达力和 SRP 等最基础的概念，也会被过度使用。为了保持类和函数短小，我们可能会造出太多的细小类和方法。所以这条规则也主张函数和类的数量要少。

类和方法的数量太多，有时是由毫无意义的教条主义导致的。例如，某个编码标准就坚称应当为每个类创建接口。也有开发者认为，字段和行为必须切分到数据类和行为类中。应该抵制这类教条，采用更实用的手段。

我们的目标是在保持函数和类短小的同时，保持整个系统短小精悍。不过要记住，这在关于简单设计的 4 条规则里面是优先级最低的一条。所以，尽管使类和函数的数量尽量少是很重要的，但更重要的却是测试、消除重复和表达力。

12.7 小结

有没有能替代经验的一套简单实践手段呢？当然不会有。另外，本章中写到的实践来自本书作者数十年经验的精练总结。遵循简单设计的实践手段，开发者不必经年学习就能掌握好的原则和模式。

12.8 文献

[XPE]：*Extreme Programming Explained: Embrace Change*, Kent Beck, Addison-Wesley, 1999.

[GOF]：*Design Patterns: Elements of Reusable Object Oriented Software*, Gamma et al., Addison-Wesley, 1996.

第13章

并发编程

Brett L.Schuchert

"对象是过程的抽象。线程是调度的抽象。"

——James O Coplien[①]

---

① 来自私人邮件。

编写整洁的并发程序很难——非常难,而编写在单线程中执行的代码却简单得多。编写表面上看似不错、深入进去却支离破碎的多线程代码也简单,但是系统一旦遭受压力,这种代码就扛不住了。

本章将讨论并发编程的需求及其困难之处,并给出一些应对这些难点、编写整洁的并发代码的建议。最后,我们将讨论与测试并发代码有关的问题。

整洁的并发编程是个复杂话题,值得用一整本书来讨论。本书只做概览,并在附录 A 中提供更详细的指引。如果你只是对并发好奇,阅读本章就足够了。如果你需要更深入地理解并发,就应读完整个附录 A。

13.1 为什么要并发

并发是一种解耦策略。它帮助我们把做什么(目的)和何时做(时机)分解开。在单线程应用中,目的与时机紧密耦合,很多时候只要查看栈追踪即可断定应用程序的状态。调试这种系统的程序员可以设定断点或者断点序列,通过查看到达哪个断点来了解系统状态。

解耦目的与时机能明显地改进应用程序的吞吐量和结构。从结构的角度来看,应用程序看起来更像是许多台协同工作的计算机,而不是一个大循环。系统因此会更被易于理解,给出了许多切分关注面的有力手段。

例如,Web 应用的 Servlet 标准模式。这类系统运行于 Web 或 EJB 容器的保护伞之下,Web 或 EJB 为你部分地处理并发问题。当有 Web 请求时,servlet 就会异步执行。servlet 程序员无须管理所有的请求。原则上,每次 servlet 是在自己的小世界中执行,与其他 servlet 的执行是分离的。

当然,如果只是那么简单,也就没必要写这一章了。实际上,Web 容器提供的解耦手段离完美还差得远。servlet 程序员得非常警惕、非常小心地保证并发程序不出错。同样,servlet 模式的结构性好处还是很明显的。

但结构并非采用并发的唯一动机。有些系统对响应时间和吞吐量有要求,需要手工编写并发解决方案。例如,考虑一个单线程信息聚合程序,它从许多 Web 站点获取信息,再合并写入日志中。因为该系统是单线程的,它会逐个访问 Web 站点,在开始下一个访问之前等待当前站点访问完毕。每天的执行时间必须少于 24 小时。然而,随着要访问的站点越来越多,采集所有数据花费的时间也越来越多,最终超过了 24 小时的限制。单线程程序许多时间花在等待 Web 套接字 I/O 结束上面。通过采用同时访问多个站点的多线程算法,就能改进性能。

或者,考虑某个每次花费 1 秒处理一个用户请求的系统。该系统在用户量较少的时候响应及时,但随着用户数增加,系统的响应时间也增加了。没人想排在 150 个人后面!通过并发处理多个用户请求,就能改善系统响应时间。

再或者,考虑某个解释大量数据集,但只在处理完全部数据后给出一个完整解决方案的系统。或许可以在独立的计算机上处理每个数据集,那样的话许多数据集就能并行地得到处理。

迷思与误解

看来有足够的理由采用并发方案。然而，如前文所述，并发编程**很难**。如果你不那么细心，就会搞出不堪入目的东西来。看看以下常见的迷思和误解：

（1）并发总能改进性能。并发有时能改进性能，但只在多个线程或处理器之间能分享大量等待时间的时候管用。事情没那么简单。

（2）编写并发程序无须修改设计。事实上，并发算法的设计有可能与单线程系统的设计极不相同。目的与时机的解耦往往会对系统结构产生巨大影响。

（3）在采用 Web 或 EJB 容器的时候，理解并发问题并不重要。实际上，你最好了解容器在做什么，了解如何应对本章后面将提到的并发更新、死锁等问题。

下面是一些有关编写并发软件的中肯说法：

- 并发会在性能和编写额外代码上增加一些开销；
- 正确的并发是复杂的，即便对于简单的问题也是如此；
- 并发缺陷并非总能重现，所以常被看作偶发事件[①]而忽略，未被当作真的缺陷看待；
- 并发常常需要对设计策略做根本性修改。

13.2 挑战

并发编程为何如此之难？来看看下面这个小型类：

```
public class X {
  private int lastIdUsed;

  public int getNextId() {
    return ++lastIdUsed;
  }
}
```

例如，创建 X 的一个实体，将 `lastIdUsed` 设置为 42，在两个线程中共享这个实体。假设这两个线程都调用 `getNextId()` 方法，结果可能有以下 3 种输出：

- 线程一得到值 43，线程二得到值 44，`lastIdUsed` 为 44；
- 线程一得到值 44，线程二得到值 43，`lastIdUsed` 为 44；
- 线程一得到值 43，线程二得到值 43，`lastIdUsed` 为 43。

第三种结果令人惊异[②]，当两个线程相互影响时就会出现这种情况。这是因为线程在执

[①] 宇宙射线、狼来了等。（作者在这里开了个小玩笑。程序员常把不能复现的程序错误的原因归结为宇宙射线等偶发性和无法修正的问题。）——译者注

[②] 见附录 A.2.2。

行那行 Java 代码时有许多可能的路径，有些路径会产生错误的结果。有多少种不同路径呢？要真正回答这个问题，需要理解 Just-In-Time 编译器如何对待生成的字节码，还要理解 Java 内存模型认为什么东西具有原子性。

简单回答一下，就生成的字节码而言，对于在 getNextId 方法中执行的那两个线程，有 12 870 种不同的可能执行路径[①]。如果 lastIdUsed 的类型从 int 变为 long，则可能路径的数量将增至 2 704 156 种。当然，多数路径都能得到正确结果，问题是其中一些不能得到正确结果。

13.3 并发防御原则

下面给出一系列防御并发代码问题的原则和技巧。

13.3.1 单一权责原则

单一权责原则（SRP）[②]认为，方法/类/组件应当只有一个修改的理由。并发设计足够复杂，以至于这种复杂足以成为其需要修改的理由，所以确实应该从其他代码中分离出来。不幸的是，并发实现细节常常直接嵌入其他生产代码中。下面是要考虑的一些问题：
- 并发相关代码有自己的开发、修改和调优生命周期；
- 并发相关代码有自己要应对的挑战，它和非并发相关代码不同，而且往往更为困难；
- 即便没有增加周边应用程序的负担，写得不好的并发代码可能的出错方式数量也已经足具挑战性。

建议：分离并发相关代码与其他代码[③]。

13.3.2 推论：限制数据作用域

如我们所见，两个线程修改共享对象的同一字段时，可能会互相干扰，导致未预期的行为。解决方案之一是采用关键字 synchronized 在代码中保护一块使用共享对象的临界区（critical section）。限制临界区的数量很重要。更新共享数据的地方越多，就越可能：
- 你会忘记保护一个或多个临界区——破坏了修改共享数据的代码；
- 得多花力气保证一切都受到有效防护（违反了 DRY 原则[④]）；

---

① 见附录 A.2.1。
② [PPP]。
③ 见附录 A.1。
④ [PRAG]。

- 很难找到错误源，也很难判断错误源。

建议：谨记数据封装；严格限制对可能被共享的数据的访问。

13.3.3 推论：使用数据副本

避免共享数据的好方法之一是，从开始就避免共享数据。在某些情形下，有可能复制对象并以只读方式对待。在另外的情况下，有可能复制对象，从多个线程收集所有副本的结果，并在单个线程中合并这些结果。

如果有避免共享数据的简易手段，结果代码就会大大减少导致错误的可能。你可能会关心创建额外对象的成本，值得试验一下看看那是否真是个问题。然而，假使使用对象副本能避免代码同步执行，则因避免了锁定而省下的价值有可能补偿得上额外的创建成本和垃圾收集开销。

13.3.4 推论：线程应尽可能地独立

让每个线程在自己的世界中存在，而不与其他线程共享数据。每个线程处理一个客户端请求，从不共享的源头接纳所有请求数据，存储为本地变量。这样一来，每个线程都像是世界中的唯一线程，没有同步需要。

例如，`HttpServlet` 的子类接收所有以参数形式传递给 `doGet` 方法和 `doPost` 方法的信息。每个 `Servlet` 都像拥有独立虚拟机一般运行。只要 `Servlet` 中的代码只使用本地变量，`Servlet` 就不会导致同步问题。当然，多数使用 `Servlet` 的应用程序最终都还是会用到类似数据库连接之类的共享资源。

建议：尝试将数据分解为可被独立线程（可能在不同处理器上）操作的独立子集。

13.4 了解 Java 库

与之前的版本相比，Java 5 提供了许多并发开发方面的改进。在用 Java 5 编写线程代码时，要注意以下几点：
- 使用类库提供的线程安全群集；
- 使用 executor 框架执行无关任务；
- 尽可能使用非锁定解决方案；
- 有几个类并不是线程安全的。

线程安全群集

当 Java 还年轻时，Doug Lea 编写了《Java 并发编程》（*Concurrent Programming in Java*）

教程[1]，同时开发了几个线程安全群集，这些代码后来成为 JDK 中 `java.util.concurrent` 包的一部分。该代码包中的群集对于多线程解决方案是安全的，执行良好。实际上，在几乎所有情况下，ConcurrentHashMap 实现都比 HashMap 表现得好，它还支持同步并发读写，也拥有支持非线程安全的合成操作的方法。如果部署环境是 Java 5，可以采用 ConcurrentHashMap。

还有几个支持高级并发设计的类。以下是其中的一小部分，如表 13-1 所示。

表 13-1　　　　　　　　　　　支持高级并发设计的类（部分）

类	作　用
`ReentrantLock`	可在一个方法中获取、在另一个方法中释放的锁
`Semaphore`	经典的"信号"的一种实现，有计数器的锁
`CountDownLatch`	在释放所有等待的线程之前，等待指定数量事件发生的锁。这样，所有线程都平等地几乎同时启动

建议：检读可用的类。对于 Java，需要掌握 `java.util.concurrent`、`java.util.concurrent.atomic` 和 `java.util.concurrent.locks`。

13.5　了解执行模型

有几种在并发应用中切分行为的途径。要讨论这些途径，我们需要理解一些基础定义，如表 13-2 所示。

表 13-2　　　　　　　　　　　　　基础定义

定　义	描　述
限定资源	并发环境中有着固定尺寸或数量的资源。例如，数据库连接和固定尺寸读/写缓存等
互斥	每一时刻仅有一个线程能访问共享数据或共享资源
线程饥饿	一个或一组线程在很长时间内或永久被禁止。例如，总是让执行得快的线程先运行，假如执行得快的线程没完没了，则执行时间长的线程就会"挨饿"
死锁	两个或多个线程互相等待执行结束。每个线程都拥有其他线程需要的资源，如果得不到其他线程拥有的资源，就无法终止
活锁	执行次序一致的线程，每个都想要起步，但发现其他线程已经"在路上"。由于竞步的原因，线程会持续尝试起步，但在很长时间内却无法如愿，甚至永远无法启动

有了这些定义，我们就能讨论在并发编程中用到的几种执行模型了。

---

[1] [Lea99]。

13.5.1 生产者-消费者模型

一个或多个生产者线程创建某些工作，并置于缓存或队列中。一个或多个消费者线程从队列中获取并完成这些工作。生产者和消费者之间的队列是一种限定资源。

13.5.2 读者-作者模型

当存在一个主要为读者线程提供信息源，但只偶尔被作者线程更新的共享资源，吞吐量就会是个问题。增加吞吐量，会导致线程饥饿和过时信息的累积。更新会影响吞吐量。协调读者线程，不去读作者线程正在更新的信息（反之亦然），这是一种辛苦的平衡工作。作者线程倾向于长期锁定许多读者线程，从而导致吞吐量问题。

挑战之处在于平衡读者线程和作者线程的需求，实现正确操作，提供合理的吞吐量，避免线程饥饿。

13.5.3 宴席哲学家

想象一下，一群哲学家环坐在圆桌旁。每个哲学家的左手边放了一把叉子。桌面中央摆着一大碗意大利面。哲学家们思索良久，直至肚子饿了。每个人都要拿起叉子吃饭。但除非手上有两把叉子，否则就没法进食。如果左边或右边的哲学家已经取用一把叉子，中间这位就得等到别人吃完、放回叉子。每位哲学家吃完后，就将两把叉子放回桌面，直到肚子再饿。

用线程代替哲学家，用资源代替叉子，就变成了许多企业级应用中进程竞争资源的情形。如果没有用心设计，这种竞争式系统就会遭遇死锁、活锁、吞吐量和效率降低等问题。

你可能遇到的并发问题，大多数都是以上 3 种模型的变种。请研究并使用这些算法，这样，遇到并发问题时你就能有解决问题的准备了。

建议：学习这些基础算法，理解其解决方案。

13.6 警惕同步方法之间的依赖

同步方法之间的依赖会导致并发代码中的狡猾缺陷。Java 语言有 `synchronized` 概念，可以用来保护单个方法。然而，如果在同一共享类中有多个同步方法，系统就可能写得不太正确了[①]。

[①] 见附录 A.4。

建议：避免使用一个共享对象的多个方法。

有时必须使用一个共享对象的多个方法。在这种情况发生时，有以下 3 种写对代码的手段：

- **基于客户端的锁定**——客户端代码在调用第一个方法前锁定服务端，确保锁的范围覆盖了调用最后一个方法的代码；
- **基于服务端的锁定**——在服务端内创建锁定服务端的方法，调用所有方法，然后解锁。让客户端代码调用新方法；
- **适配服务端**——创建执行锁定的中间层。这是一种基于服务端的锁定的例子，但不修改原始服务端代码。

13.7 保持同步区域微小

关键字 `synchronized` 制造了锁。同一个锁维护的所有代码区域在任一时刻保证只有一个线程执行。锁是昂贵的，因为它们带来了延迟和额外开销。所以我们不愿将代码扔给 `synchronized` 语句了事。另外，临界区[1]应该被保护起来。所以，应该尽可能少地设计临界区。

有些天真的程序员想通过扩大临界区面积达到这个目的。然而，如果将同步延展到最小临界区范围之外，会加剧资源争用，降低执行效率[2]。

建议：尽可能减小同步区域。

13.8 很难编写正确的关闭代码

编写永远运行的系统，与编写运行一段时间后平静地关闭的系统是两码事。

平静关闭很难做到。常见问题与死锁[3]有关，线程一直等待永远不会到来的信号。

例如，想象一个系统中有一个父线程分裂出数个子线程，父线程等待所有子线程结束后才会释放资源并关闭。如果其中一个子线程发生死锁会怎样？父线程将一直等待下去，而系统就永远不能关闭。

或者，考虑一个被指示关闭的类似系统。父线程告知全体子线程放弃任务并结束。如果其中两个子线程正以生产者-消费者模型操作会怎样呢？假设生产者线程从父线程处接收到信号，并迅速关闭，而消费者线程可能还在等待生产者线程发来消息，于是就被锁定在无法

---

[1] 临界区是为了确保程序正确而要阻止同时使用的代码区域。
[2] 见附录 A.5。
[3] 见附录 A.6。

接收到关闭信号的状态中，它会死等生产者线程，永不结束，从而导致父线程也无法结束。

这类情形并非不常见。如果你要编写涉及平静关闭的并发代码，就多预留一些时间搞对关闭过程。

建议：尽早考虑关闭问题，尽早令其工作正常。这会花费比你预期的更多的时间。检视既有算法，因为这可能会比想象中难得多。

13.9 测试线程代码

证明代码的正确性不切实际。测试并不能确保正确性，然而，好的测试却能尽量降低风险。这对于所有单线程解决方案都是对的。当有两个或多个线程使用同一代码段和共享数据，事情就变得非常复杂了。

建议：编写有潜力曝露问题的测试，在不同的编程配置、系统配置和负载条件下频繁运行。如果测试失败，就跟踪错误，别因为后来测试通过了后来的运行就忽略失败。

有一大堆问题要考虑。下面是一些精练的建议：

- 将伪失败看作可能的线程问题；
- 先使非线程代码可工作；
- 编写可插拔的线程代码；
- 编写可调整的线程代码；
- 运行多于处理器数量的线程；
- 在不同平台上运行；
- 调整代码并强迫错误发生。

13.9.1 将伪失败看作可能的线程问题

线程代码导致"不可能失败的"失败。多数开发者缺乏有关线程如何与其他代码（可能由其他作者编写）互动的直觉。线程代码中的缺陷可能在一千或一百万次执行中才会显现一次。重复执行想要复现问题令人沮丧。所以开发者常常会将失败归咎于宇宙射线、硬件错误或其他"偶发事件"。最好假设这种偶发事件根本不存在。"偶发事件"被忽略得越久，代码就越有可能搭建于不完善的基础之上。

建议：不要将系统错误归咎于偶发事件。

13.9.2 先使非线程代码可工作

这看起来太浅显，但强调一下不无益处。确保线程之外的代码可工作，通常，这意味着创建由线程调用的 POJO。POJO 与线程无涉，所以可在线程环境之外测试。能放进 POJO 中的代码越多越好。

建议：不要同时追踪非线程缺陷和线程缺陷。确保代码在线程之外可工作。

13.9.3 编写可插拔的线程代码

编写可在数个配置环境下运行的线程代码：
- 单线程与多个线程在执行时不同的情况；
- 线程代码与实物或测试替身互动；
- 用运行快速、缓慢和有变动的测试替身执行；
- 将测试配置为能运行一定数量的迭代。

建议：编写可插拔的线程代码，这样就能在不同的配置环境下运行。

13.9.4 编写可调整的线程代码

要获得良好的线程平衡，常常需要反复试验。一开始，在不同的配置环境下监测系统性能。要允许线程数量可调整。在系统运行时允许线程发生变动。允许线程依据吞吐量和系统使用率自我调整。

13.9.5 运行多于处理器数量的线程

系统在切换任务时会发生一些事。为了促使任务交换的发生，运行多于处理器或处理器核心数量的线程。任务交换越频繁，越有可能找到错过临界区或导致死锁的代码。

13.9.6 在不同平台上运行

2007 年，我们开发了一套关于并发编程的课程。该课程主要在 OS X 下开发，在运行于虚拟机的 Windows XP 上展示。对于用于演示的测试失败条件，在 OS X 上要比在 XP 上失败得更频繁。

被测试的代码已知是不正确的。这正强调了不同操作系统有着不同线程策略的事实，不同的线程策略影响了代码的执行。在不同环境中，多线程代码的行为也不一样[1]。应该在所有可能部署的环境中运行测试。

---

[1] 你是否知道，Java 的线程模型并不保证线程抢先？现代操作系统支持抢先线程，所以你可以"免费"获得这一特性。即便如此，JVM 也没有做出保证。

建议：尽早并经常地在所有目标平台上运行线程代码。

13.9.7 装置试错代码

并发代码中藏有缺陷，这并不罕见。简单的测试往往无法曝露这些缺陷。实际上，缺陷经常隐藏于一般处理过程中，可能好几个小时、好几天甚至好几个星期才会跳出来一次！

线程中的缺陷之所以如此不频繁、偶发、难以重现，是因为在几千个穿过脆弱区域的可能路径当中，只有少数路径会真的导致失败。经过会导致失败的路径的可能性惊人地低。所以，监测与调试也非常难。

怎么才能增加捕捉住如此罕见之物的机会？可以装置代码，增加对 `Object.wait()`、`Object.sleep()`、`Object.yield()` 和 `Object.priority()` 等方法的调用，改变代码执行顺序。

这些方法都会影响执行顺序，从而增加监测到缺陷的可能性。有问题的代码最好尽早测试，尽可能多地使其通不过测试。

有两种装置代码的方法：

- 硬编码；
- 自动化。

13.9.8 硬编码

你可以手工向代码中插入 `wait()`、`sleep()`、`yield()` 和 `priority()` 的调用。在测试某段棘手的代码时，正当如此操作。

下面是一个例子：

```
public synchronized String nextUrlOrNull() {
  if(hasNext()) {
    String url = urlGenerator.next();
    Thread.yield(); // inserted for testing.
    updateHasNext();
    return url;
  }
  return null;
}
```

插入对 `yield()` 的调用，将改变代码的执行路径，由此可能导致代码在以前未失败过的地方失败。如果代码的确出错，那并非是因为你插入了 `yield()` 方法调用[①]。代码出错了，才是失败的原因。

[①] 严格说来并非如此。JVM 不保证抢占线程，故在不抢占线程的系统上，某个特殊的算法可能一直能工作。反之亦然，但会有其他的原因影响。

这种方法有许多毛病：
- 你得手工找到合适的地方来插入方法调用；
- 你怎么知道在哪里插入调用、插入什么调用？
- 不必要地在产品环境中留下这类代码，将拖慢代码执行速度；
- 这是一种无的放矢的手段。你可能找不到缺陷。实际上，这不在你的把握之中。

我们所需要的，是一种在测试中但不在生产中实现的手段。我们还需要为多次运行方便地调整配置，从而增加总的发现错误的机会。

无疑，如果将系统分解为对线程及控制线程的类一无所知的 POJO，就能更容易地找到装置代码的位置，而且，还能创建许多个以不同方式调用 sleep、yield 等方法的 POJO 测试。

13.9.9　自动化

可以使用 Aspect-Oriented Framework、CGLIB 或 ASM 之类的工具，通过编程来装置代码。例如，可以使用有单个方法的类：

```
public class ThreadJigglePoint {
  public static void jiggle() {
  }
}
```

可以在代码的不同位置调用这个方法：

```
public synchronized String nextUrlOrNull() {
  if(hasNext()) {
    ThreadJigglePoint.jiggle();
    String url = urlGenerator.next();
    ThreadJigglePoint.jiggle();
    updateHasNext();
    ThreadJigglePoint.jiggle();
    return url;
  }
  return null;
}
```

如此，你就得到了一个随机选择无所作为、睡眠或让步的方面。

或者，想象 ThreadJigglePoint 类有两种实现。第一种实现 jiggle 什么都不做，在生产环境中使用。第二种实现生成一个随机数，在睡眠、让步或径直执行间做选择。如果上千次地做这种随机测试，大概就能找到一些缺陷的根源。假如测试都通过了，至少你可以说自己已谨慎对待。这种方法看似过于简单，却是替代复杂工具的一种可选方案。

有一种叫作 ConTest 的工具，由 IBM 公司开发，能做类似的事情，但做法稍微复杂些。

要点是让代码发生"异动"，从而使线程以不同次序执行。编写良好的测试与"异动"组合，能有效地增加发现错误的机会。

建议：使用异动策略搜出错误。

13.10 小结

并发代码很难写正确。加入多线程和共享数据后，简单的代码也会变成噩梦。要编写并发代码，就得严格地编写整洁的代码，否则将面临细微和不频繁发生的失败。

第一要诀是遵循单一权责原则。将系统切分为分离了线程相关代码和线程无关代码的 POJO。确保在测试线程相关代码时只是在测试，而没有做其他事情。线程相关代码应该保持短小和目的集中。

了解并发问题的可能原因：对共享数据的多线程操作，或使用了公共资源池。类似平静关闭或停止循环之类的边界情况尤其棘手。

学习类库，了解基本算法。理解类库提供的与基础算法类似的解决问题的特性。

学习如何找到必须锁定的代码区域并锁定之。不要锁定不必锁定的代码。避免从锁定区域中调用其他锁定区域。这需要深刻理解某物是否已共享。尽可能减少共享对象和共享范围。修改对象的设计，向客户代码提供共享数据，而不是迫使客户代码管理共享状态。

问题会跳出来。那种在早期没跳出来的问题往往是偶发的。这种所谓的偶发问题，通常仅在高负载下出现或者偶然出现。所以，你需要在不同平台上，以不同配置持续重复运行线程代码。跟随 TDD 三要则而来的可测试性意味着某种程度的可插拔性，从而提供了在大量不同配置下运行代码的必要支持。

如果花点时间装置代码，就能极大地提升发现错误代码的机会。可以手工做，也可以使用某种自动化技术。尽早这么做。在将线程代码投入生产环境前，就要尽可能多地运行它。

只要采用了整洁的做法，做对的可能性就会有翻天覆地的提高。

13.11 文献

[Lea99]：*Concurrent Programming in Java: Design Principles and Patterns*, 2d. ed., Doug Lea, Prentice Hall, 1999.

[PPP]：*Agile Software Development: Principles, Patterns, and Practices*, Robert C. Martin, Prentice Hall, 2002.

[PRAG]：*The Pragmatic Programmer*, Andrew Hunt, Dave Thomas, Addison-Wesley, 2000.

第 14 章

逐步改进

对一个命令行参数解析程序的案例研究

本章研究一个逐步改进的案例。你将看到一个开始还不错，但规模扩大后即出问题的模块。你还将看到这个模块是如何被重构得整洁起来的。

我们中的大多数人都会遇到解析命令行参数的情况。如果没有就手的工具，就得遍历传入 main 函数的字符串数组。有一些不同来源的好工具，但没有一个是最符合要求的。所以，我当然要自己写一个，我把它叫作 Args。

Args 非常易于使用。你只要简单地用输入参数和格式化字符串构造 Args 类，再向 Args 实体询问参数值即可。看看代码清单 14-1 中给出的简单例子。

代码清单 14-1　Args 的简单用法

```java
public static void main(String[]args) {
  try {
    Args arg = new Args("l,p#,d*", args);
    boolean logging = arg.getBoolean('l');
    int port = arg.getInt('p');
    String directory = arg.getString('d');
    executeApplication(logging, port, directory);
  } catch (ArgsException e) {
    System.out.printf("Argument error:%s\n", e.errorMessage());
  }
}
```

可以看到这有多简单。我们只是用两个参数创建了 Args 类的一个实体。第一个参数是格式字符串，或范式字符串 "l,p#,d*"。它定义了 3 个命令行参数。第一个，-l，是一个布尔值参数。第二个，-p，是一个整数参数。第三个，-d，是一个字符串参数。向 Args 构造器传入的第二个参数就是向 main 传入的命令行参数数组。

如果构造器正常返回，没有抛出 ArgsException 异常，则命令行参数已传入，Args 实体随时待命。使用 getBoolean、getInteger 和 getString 等方法，可以用参数名称获得参数值。

不管是格式化字符串还是命令行参数出现问题，都会抛出一个 ArgsException 异常。可以从该异常的 errorMessage 中获得关于错误的描述。

14.1　Args 的实现

代码清单 14-2 是 Args 类的实现。请仔细阅读。我在代码风格和结构上花了大力气，使之值得仿效。

代码清单 14-2　Args.java

```java
package com.objectmentor.utilities.args;

import static com.objectmentor.utilities.args.ArgsException.ErrorCode.*;
import java.util.*;

public class Args {
  private Map<Character, ArgumentMarshaler> marshalers;
```

14.1 Args 的实现

```java
  private Set<Character> argsFound;
  private ListIterator<String> currentArgument;

public Args(String schema, String[] args) throws ArgsException {
  marshalers = new HashMap<Character, ArgumentMarshaler>();
  argsFound = new HashSet<Character>();

  parseSchema(schema);
  parseArgumentStrings(Arrays.asList(args));
}

private void parseSchema(String schema) throws ArgsException {
  for (String element : schema.split(","))
    if (element.length() > 0)
      parseSchemaElement(element.trim());
}

private void parseSchemaElement(String element) throws ArgsException {
  char elementId = element.charAt(0);
  String elementTail = element.substring(1);
  validateSchemaElementId(elementId);
  if (elementTail.length() == 0)
    marshalers.put(elementId, new BooleanArgumentMarshaler());
  else if (elementTail.equals("*"))
    marshalers.put(elementId, new StringArgumentMarshaler());
  else if (elementTail.equals("#"))
    marshalers.put(elementId, new IntegerArgumentMarshaler());
  else if (elementTail.equals("##"))
    marshalers.put(elementId, new DoubleArgumentMarshaler());
  else if (elementTail.equals("[*]"))
    marshalers.put(elementId, new StringArrayArgumentMarshaler());
  else
    throw new ArgsException(INVALID_ARGUMENT_FORMAT, elementId, elementTail);
}

private void validateSchemaElementId(char elementId) throws ArgsException {
  if (!Character.isLetter(elementId))
    throw new ArgsException(INVALID_ARGUMENT_NAME, elementId, null);
}

private void parseArgumentStrings(List<String> argsList) throws ArgsException
{
  for (currentArgument = argsList.listIterator(); currentArgument.hasNext();)
  {
    String argString = currentArgument.next();
    if (argString.startsWith("-")) {
      parseArgumentCharacters(argString.substring(1));
    } else {
      currentArgument.previous();
      break;
    }
  }
}
```

```
  private void parseArgumentCharacters(String argChars) throws ArgsException {
    for (int i = 0; i < argChars.length(); i++)
      parseArgumentCharacter(argChars.charAt(i));
  }

  private void parseArgumentCharacter(char argChar) throws ArgsException {
    ArgumentMarshaler m = marshalers.get(argChar);
    if (m == null) {
      throw new ArgsException(UNEXPECTED_ARGUMENT, argChar, null);
    } else {
      argsFound.add(argChar);
      try {
        m.set(currentArgument);
      } catch (ArgsException e) {
        e.setErrorArgumentId(argChar);
        throw e;
      }
    }
  }

  public boolean has(char arg) {
    return argsFound.contains(arg);
  }

  public int nextArgument() {
    return currentArgument.nextIndex();
  }

  public boolean getBoolean(char arg) {
    return BooleanArgumentMarshaler.getValue(marshalers.get(arg));
  }

  public String getString(char arg) {
    return StringArgumentMarshaler.getValue(marshalers.get(arg));
  }

  public int getInt(char arg) {
    return IntegerArgumentMarshaler.getValue(marshalers.get(arg));
  }

  public double getDouble(char arg) {
    return DoubleArgumentMarshaler.getValue(marshalers.get(arg));
  }

  public String[] getStringArray(char arg) {
    return StringArrayArgumentMarshaler.getValue(marshalers.get(arg));
  }
}
```

注意，你可以从上到下阅读这些代码，不用跳来跳去，也不用先看后面的部分。唯一需要先看的是 ArgumentMarshaler 的定义，这部分我有意省略了。仔细看这段代码，你应该能理解 ArgumentMarshaler 接口是什么，其派生类做什么。下面我将向你展示一部分（如代码清单 14-3～代码清单 14-6 所示）。

代码清单 14-3　ArgumentMarshaler.java

```java
public interface ArgumentMarshaler {
  void set(Iterator<String> currentArgument) throws ArgsException;
}
```

代码清单 14-4　BooleanArgumentMarshaler.java

```java
public class BooleanArgumentMarshaler implements ArgumentMarshaler {
  private boolean booleanValue = false;

  public void set(Iterator<String> currentArgument) throws ArgsException {
    booleanValue = true;
  }

  public static boolean getValue(ArgumentMarshaler am) {
    if (am != null && am instanceof BooleanArgumentMarshaler)
      return ((BooleanArgumentMarshaler) am).booleanValue;
    else
      return false;
  }
}
```

代码清单 14-5　StringArgumentMarshaler.java

```java
import static com.objectmentor.utilities.args.ArgsException.ErrorCode.*;

public class StringArgumentMarshaler implements ArgumentMarshaler {
  private String stringValue = "";

  public void set(Iterator<String> currentArgument) throws ArgsException {
    try {
      stringValue = currentArgument.next();
    } catch (NoSuchElementException e) {
      throw new ArgsException(MISSING_STRING);
    }
  }

  public static String getValue(ArgumentMarshaler am) {
    if (am != null && am instanceof StringArgumentMarshaler)
      return ((StringArgumentMarshaler) am).stringValue;
    else
      return "";
  }
}
```

代码清单 14-6　IntegerArgumentMarshaler.java

```java
import static com.objectmentor.utilities.args.ArgsException.ErrorCode.*;
```

```java
public class IntegerArgumentMarshaler implements ArgumentMarshaler {
  private int intValue = 0;

  public void set(Iterator<String> currentArgument) throws ArgsException {
    String parameter = null;
    try {
      parameter = currentArgument.next();
      intValue = Integer.parseInt(parameter);
    } catch (NoSuchElementException e) {
      throw new ArgsException(MISSING_INTEGER);
    } catch (NumberFormatException e) {
      throw new ArgsException(INVALID_INTEGER, parameter);
    }
  }

  public static int getValue(ArgumentMarshaler am) {
    if (am != null && am instanceof IntegerArgumentMarshaler)
      return ((IntegerArgumentMarshaler) am).intValue;
    else
      return 0;
  }
}
```

ArgumentMarshaler 的其他派生类以同样的模式处理 double 和 String 数组，一一列出反而阻碍行文。你可以练习自己实现它们。

还有些信息可能会困扰你：错误码常量的定义。这些是在 ArgsException 类（代码清单 14-7）中定义的。

代码清单 14-7　ArgsException.java

```java
import static com.objectmentor.utilities.args.ArgsException.ErrorCode.*;

public class ArgsException extends Exception {
  private char errorArgumentId = '\0';
  private String errorParameter = null;
  private ErrorCode errorCode = OK;

  public ArgsException() {}

  public ArgsException(String message) { super(message);}

  public ArgsException(ErrorCode errorCode) {
    this.errorCode = errorCode;
  }

  public ArgsException(ErrorCode errorCode, String errorParameter) {
    this.errorCode = errorCode;
    this.errorParameter = errorParameter;
  }

  public ArgsException(ErrorCode errorCode,
                       char errorArgumentId,String errorParameter) {
```

```java
    this.errorCode = errorCode;
    this.errorParameter = errorParameter;
    this.errorArgumentId = errorArgumentId;
  }

  public char getErrorArgumentId() {
    return errorArgumentId;
  }

  public void setErrorArgumentId(char errorArgumentId) {
    this.errorArgumentId = errorArgumentId;
  }

  public String getErrorParameter() {
    return errorParameter;
  }

  public void setErrorParameter(String errorParameter) {
    this.errorParameter = errorParameter;
  }

  public ErrorCode getErrorCode() {
    return errorCode;
  }

  public void setErrorCode(ErrorCode errorCode) {
    this.errorCode = errorCode;
  }

  public String errorMessage() {
    switch (errorCode) {
      case OK:
        return "TILT: Should not get here.";
      case UNEXPECTED_ARGUMENT:
        return String.format("Argument -%c unexpected.", errorArgumentId);
      case MISSING_STRING:
        return String.format("Could not find string parameter for -%c.",
                             errorArgumentId);
      case INVALID_INTEGER:
        return String.format("Argument -%c expects an integer but was '%s'.",
                             errorArgumentId, errorParameter);
      case MISSING_INTEGER:
        return String.format("Could not find integer parameter for -%c.",
                             errorArgumentId);
      case INVALID_DOUBLE:
        return String.format("Argument -%c expects a double but was '%s'.",
                             errorArgumentId, errorParameter);
      case MISSING_DOUBLE:
        return String.format("Could not find double parameter for -%c.",
                             errorArgumentId);
      case INVALID_ARGUMENT_NAME:
        return String.format("'%c' is not a valid argument name.",
                             errorArgumentId);
```

```
        case INVALID_ARGUMENT_FORMAT:
          return String.format("'%s' is not a valid argument format.",
                               errorParameter);
      }
      return "";
    }

    public enum ErrorCode {
      OK, INVALID_ARGUMENT_FORMAT, UNEXPECTED_ARGUMENT, INVALID_ARGUMENT_NAME,
      MISSING_STRING,
      MISSING_INTEGER, INVALID_INTEGER,
      MISSING_DOUBLE, INVALID_DOUBLE
    }
```

为了充实这么一个简单概念的细节，需要如此多代码，这很值得注意。原因之一是我们使用了 Java 这种唠叨型语言。作为一种静态类型语言，需要大量语句才能满足类型系统的要求。在 Ruby、Python 或 Smalltalk 等语言中，程序会短很多[1]。

请再次阅读这段代码。特别注意命名方式、函数大小和代码格式。如果你是经验丰富的程序员，可能会对风格或结构有着这样或那样的不同观点。不过，希望你认为这段程序总体上编写良好，有着整洁的结构。

例如，如何增加新参数类型，如日期或复杂数字参数。其实现手段很清楚，而且只需要花一点点力气即可。简言之，只需要从 ArgumentMarshaler 派生一个新类，写一个新的 `getXXX` 函数，在 `parseSchemaElement` 函数中添加一个新的 `case` 语句。可能还需要添加新的 `ArgsException.ErrorCode` 和新错误信息。

我怎么做的

先放松一下神经。这段程序并非从一开始就写成了现在的样子。更重要的是，我也没指望你能够一次就写出整洁、漂亮的程序。如果说我们从过去几十年里学到什么东西的话，那就是编程是一种技艺甚于科学的东西。要编写整洁代码，必须先写肮脏代码，然后再清理它。

你应该不会对此感到惊讶。我们在小学就学过这条真理了。那时，老师（通常是徒劳地）努力让我们写作文草稿。他们告诉我们，我们应该先写草稿，再写二稿，一次又一次地草撰，直至写出终稿。他们尽力告诉我们，写出好作文是一个逐步改进的过程。

多数新手程序员（就像多数小学生一样）没有特别认真地遵循这个建议。他们相信，首要任务是写出能工作的程序。只要程序"能工作"，就转移到下一个任务上，而那个"能工作"的程序就留在了最后那个所谓"能工作"的状态。多数有经验的程序员都知道，这是一种自毁行为。

---

[1] 最近我用 Ruby 语言重写了这个模块。大概只有 Java 版本的 1/7 大小，而且结构也稍好一些。

14.2 Args：草稿

代码清单 14-8 展示了 Args 类的一个早期版本。它"能工作"，但却很烂。

代码清单 14-8　Args.java（初稿）

```java
import java.text.ParseException;
import java.util.*;

public class Args {
  private String schema;
  private String[] args;
  private boolean valid = true;
  private Set<Character> unexpectedArguments = new TreeSet<Character>();
  private Map<Character, Boolean> booleanArgs =
    new HashMap<Character, Boolean>();
  private Map<Character, String> stringArgs = new HashMap<Character, String>();
  private Map<Character, Integer> intArgs = new HashMap<Character, Integer>();
  private Set<Character> argsFound = new HashSet<Character>();
  private int currentArgument;
  private char errorArgumentId = '\0';
  private String errorParameter = "TILT";
  private ErrorCode errorCode = ErrorCode.OK;

  private enum ErrorCode {
    OK, MISSING_STRING, MISSING_INTEGER, INVALID_INTEGER, UNEXPECTED_ARGUMENT}

  public Args(String schema, String[] args) throws ParseException {
    this.schema = schema;
    this.args = args;
    valid = parse();
  }

  private boolean parse() throws ParseException {
    if (schema.length() == 0 && args.length == 0)
      return true;
    parseSchema();
    try {
      parseArguments();
    } catch (ArgsException e) {
    }
    return valid;
  }

  private boolean parseSchema() throws ParseException {
    for (String element : schema.split(",")) {
      if (element.length() > 0) {
        String trimmedElement = element.trim();
        parseSchemaElement(trimmedElement);
```

```java
        }
      }
      return true;
    }

    private void parseSchemaElement(String element) throws ParseException {
      char elementId = element.charAt(0);
      String elementTail = element.substring(1);
      validateSchemaElementId(elementId);
      if (isBooleanSchemaElement(elementTail))
        parseBooleanSchemaElement(elementId);
      else if (isStringSchemaElement(elementTail))
        parseStringSchemaElement(elementId);
      else if (isIntegerSchemaElement(elementTail)) {
        parseIntegerSchemaElement(elementId);
      } else {
        throw new ParseException(
          (String.format("Argument: %c has invalid format: %s.",
                        elementId,elementTail),0);
      }
    }

    private void validateSchemaElementId(char elementId) throws ParseException {
      if (!Character.isLetter(elementId)) {
        throw new ParseException(
          "Bad character:" + elementId + "in Args format: "+schema,0);
      }
    }

    private void parseBooleanSchemaElement(char elementId) {
      booleanArgs.put(elementId, false);
    }

    private void parseIntegerSchemaElement(char elementId) {
      intArgs.put(elementId, 0);
    }

    private void parseStringSchemaElement(char elementId) {
      stringArgs.put(elementId, "");
    }

    private boolean isStringSchemaElement(String elementTail) {
      return elementTail.equals("*");
    }

    private boolean isBooleanSchemaElement(String elementTail) {
      return elementTail.length() == 0;
    }

    private boolean isIntegerSchemaElement(String elementTail) {
      return elementTail.equals("#");
    }

    private boolean parseArguments() throws ArgsException {
      for (currentArgument = 0; currentArgument < args.length; currentArgument++) {
```

14.2 Args：草稿

```java
      String arg = args[currentArgument];
      parseArgument(arg);
    }
    return true;
  }

  private void parseArgument(String arg) throws ArgsException {
    if (arg.startsWith("-"))
      parseElements(arg);
  }

  private void parseElements(String arg) throws ArgsException {
    for (int i = 1; i < arg.length(); i++)
      parseElement(arg.charAt(i));
  }

  private void parseElement(char argChar) throws ArgsException {
    if (setArgument(argChar))
      argsFound.add(argChar);
    else {
      unexpectedArguments.add(argChar);
      errorCode = ErrorCode.UNEXPECTED_ARGUMENT;
      valid = false;
    }
  }

  private boolean setArgument(char argChar) throws ArgsException {
    if (isBooleanArg(argChar))
      setBooleanArg(argChar, true);
    else if (isStringArg(argChar))
      setStringArg(argChar);
    else if (isIntArg(argChar))
      setIntArg(argChar);
    else
      return false;

    return true;
  }

  private boolean isIntArg(char argChar) {return intArgs.containsKey(argChar);}

  private void setIntArg(char argChar) throws ArgsException {
    currentArgument++;
    String parameter = null;
    try {
      parameter = args[currentArgument];
      intArgs.put(argChar, new Integer(parameter));
    } catch (ArrayIndexOutOfBoundsException e) {
      valid = false;
      errorArgumentId = argChar;
      errorCode = ErrorCode.MISSING_INTEGER;

      throw new ArgsException();
    } catch (NumberFormatException e) {
```

```
      valid = false;
      errorArgumentId = argChar;
      errorParameter = parameter;
      errorCode = ErrorCode.INVALID_INTEGER;
      throw new ArgsException();
    }
  }

  private void setStringArg(char argChar) throws ArgsException {
    currentArgument++;
    try {
      stringArgs.put(argChar, args[currentArgument]);
    } catch (ArrayIndexOutOfBoundsException e) {
      valid = false;
      errorArgumentId = argChar;
      errorCode = ErrorCode.MISSING_STRING;
      throw new ArgsException();
    }
  }

  private boolean isStringArg(char argChar) {
    return stringArgs.containsKey(argChar);
  }

  private void setBooleanArg(char argChar, boolean value) {
    booleanArgs.put(argChar, value);
  }

  private boolean isBooleanArg(char argChar) {
    return booleanArgs.containsKey(argChar);
  }

  public int cardinality() {
    return argsFound.size();
  }

  public String usage() {
    if (schema.length() > 0)
      return "-[" + schema + "]";
    else
      return "";
  }

  public String errorMessage() throws Exception {
    switch (errorCode) {
      case OK:
        throw new Exception("TILT: Should not get here.");
      case UNEXPECTED_ARGUMENT:
        return unexpectedArgumentMessage();
      case MISSING_STRING:
        return String.format("Could not find string parameter for -%c.",
                             errorArgumentId);

      case INVALID_INTEGER:
```

```java
          return String.format("Argument -%c expects an integer but was '%s'.",
                        errorArgumentId, errorParameter);
        case MISSING_INTEGER:
          return String.format("Could not find integer parameter for -%c.",
                        errorArgumentId);
      }
      return "";
    }

    private String unexpectedArgumentMessage() {
      StringBuffer message = new StringBuffer("Argument(s) -");
      for (char c : unexpectedArguments) {
        message.append(c);
      }
      message.append(" unexpected.");

      return message.toString();
    }

    private boolean falseIfNull(Boolean b) {
      return b != null && b;
    }

    private int zeroIfNull(Integer i) {
      return i == null ? 0 : i;
    }

    private String blankIfNull(String s) {
      return s == null ? "" : s;
    }

    public String getString(char arg) {
      return blankIfNull(stringArgs.get(arg));
    }

    public int getInt(char arg) {
      return zeroIfNull(intArgs.get(arg));
    }

    public boolean getBoolean(char arg) {
      return falseIfNull(booleanArgs.get(arg));
    }

    public boolean has(char arg) {
      return argsFound.contains(arg);
    }

    public boolean isValid() {
      return valid;
    }

    private class ArgsException extends Exception {
    }
  }
```

第 14 章　逐步改进

希望你看到这段乱七八糟的代码时，第一反应是"他没就此罢手，真令人高兴！"如果你这么想，不如想想其他人对你留置在草稿形态的代码的想法吧。

实际上，"草稿"大概会是你对这段代码的最高评价。它显然还需打磨。实体变量的数量多到吓人，诸如 `"TILT"` 之类奇怪的字符串、`HashSets` 和 `TreeSets`，还有那些 `try-catch-catch` 代码块，组成了一个烂摊子。

我不想写出一个烂摊子。我也一直想保持一切有序。从函数和变量命名，以及程序的粗略架构中，你可以看出这一点。不过，显然我没能做到。

混乱是逐渐产生的。更早的版本并不如此肮脏。例如，代码清单 14-9 展示了一个早期版本代码，那时只支持 `Boolean` 类型参数。

代码清单 14-9　Args.java（只支持布尔类型）

```java
package com.objectmentor.utilities.getopts;

import java.util.*;

public class Args {
  private String schema;
  private String[] args;
  private boolean valid;
  private Set<Character> unexpectedArguments = new TreeSet<Character>();
  private Map<Character, Boolean> booleanArgs =
    new HashMap<Character, Boolean>();
  private int numberOfArguments = 0;

  public Args(String schema, String[] args) {
    this.schema = schema;
    this.args = args;
    valid = parse();
  }

  public boolean isValid() {
    return valid;
  }

  private boolean parse() {
    if (schema.length() == 0 && args.length == 0)
      return true;
    parseSchema();
    parseArguments();
    return unexpectedArguments.size() == 0;
  }

  private boolean parseSchema() {
    for (String element : schema.split(",")) {
      parseSchemaElement(element);
    }

    return true;
  }
```

14.2 Args：草稿

```java
  private void parseSchemaElement(String element) {
    if (element.length() == 1) {
      parseBooleanSchemaElement(element);
    }
  }

  private void parseBooleanSchemaElement(String element) {
    char c = element.charAt(0);
    if (Character.isLetter(c)) {
      booleanArgs.put(c, false);
    }
  }

  private boolean parseArguments() {
    for (String arg : args)
      parseArgument(arg);
    return true;
  }

  private void parseArgument(String arg) {
    if (arg.startsWith("-"))
      parseElements(arg);
  }

  private void parseElements(String arg) {
    for (int i = 1; i < arg.length(); i++)
      parseElement(arg.charAt(i));
  }

  private void parseElement(char argChar) {
    if (isBoolean(argChar)) {
      numberOfArguments++;
      setBooleanArg(argChar, true);
    } else
      unexpectedArguments.add(argChar);
  }

  private void setBooleanArg(char argChar, boolean value) {
    booleanArgs.put(argChar, value);
  }

  private boolean isBoolean(char argChar) {
    return booleanArgs.containsKey(argChar);
  }

  public int cardinality() {
    return numberOfArguments;
  }

  public String usage() {
    if (schema.length() > 0)
      return "-["+schema+"]";
    else
```

```
      return "";
    }

    public String errorMessage() {
      if (unexpectedArguments.size() > 0) {
        return unexpectedArgumentMessage();
      } else
        return "";
    }

    private String unexpectedArgumentMessage() {
      StringBuffer message = new StringBuffer("Argument(s)  -");
      for (char c : unexpectedArguments) {
        message.append(c);
      }
      message.append(" unexpected.");

      return message.toString();
    }

    public boolean getBoolean(char arg) {
      return booleanArgs.get(arg);
    }
  }
```

尽管你可能对这段代码很不满意，但是其实它并非如此糟糕。它精练、简单，易于理解。然而，在这段代码中很容易找到后面烂摊子的根源，能很清楚看到小问题是如何变成大混乱的。

注意，后面的混乱代码只比这个版本多支持两种参数类型：String 和 integer。只增加两种参数类型支持，就对代码产生了如此巨大的负面影响。它从某种可维护之物变成了满是缺陷的东西。

我逐步添加了对这两种参数类型的支持。首先，我添加对 String 参数的支持，如代码清单 14-10 所示。

代码清单 14-10 Args.java（Boolean 和 String）

```
package com.objectmentor.utilities.getopts;

import java.text.ParseException;
import java.util.*;

public class Args {
  private String schema;
  private String[] args;
  private boolean valid = true;
  private Set<Character> unexpectedArguments = new TreeSet<Character>();
  private Map<Character, Boolean> booleanArgs =
    new HashMap<Character, Boolean>();

  private Map<Character, String> stringArgs =
```

```java
    new HashMap<Character, String>();
  private Set<Character> argsFound = new HashSet<Character>();
  private int currentArgument;
  private char errorArgument = '\0';

  enum ErrorCode {
    OK, MISSING_STRING}

  private ErrorCode errorCode = ErrorCode.OK;

  public Args(String schema, String[] args) throws ParseException {
    this.schema = schema;
    this.args = args;
    valid = parse();
  }

  private boolean parse() throws ParseException {
    if (schema.length() == 0 && args.length == 0)
      return true;
    parseSchema();
    parseArguments();
    return valid;
  }

  private boolean parseSchema() throws ParseException {
    for (String element : schema.split(",")) {
      if (element.length() > 0) {
        String trimmedElement = element.trim();
        parseSchemaElement(trimmedElement);
      }
    }
    return true;
  }

  private void parseSchemaElement(String element) throws ParseException {
    char elementId = element.charAt(0);
    String elementTail = element.substring(1);
    validateSchemaElementId(elementId);
    if (isBooleanSchemaElement(elementTail))
      parseBooleanSchemaElement(elementId);
    else if (isStringSchemaElement(elementTail))
      parseStringSchemaElement(elementId);
  }

  private void validateSchemaElementId(char elementId) throws ParseException {
    if (!Character.isLetter(elementId)) {
      throw new ParseException(
        "Bad  character:" + elementId + "in Args  format: " + schema, 0);
    }

  }

  private void parseStringSchemaElement(char elementId) {
    stringArgs.put(elementId, "");
```

```java
  }

  private boolean isStringSchemaElement(String elementTail) {
    return elementTail.equals("*");
  }

  private boolean isBooleanSchemaElement(String elementTail) {
    return elementTail.length() == 0;
  }

  private void parseBooleanSchemaElement(char elementId) {
    booleanArgs.put(elementId, false);
  }

  private boolean parseArguments() {
    for (currentArgument = 0; currentArgument < args.length; currentArgument++)
    {
      String arg = args[currentArgument];
      parseArgument(arg);
    }
    return true;
  }

  private void parseArgument(String arg) {
    if (arg.startsWith("-"))
      parseElements(arg);
  }

  private void parseElements(String arg) {
    for (int i = 1; i < arg.length(); i++)
      parseElement(arg.charAt(i));
  }

  private void parseElement(char argChar) {
    if (setArgument(argChar))
      argsFound.add(argChar);
    else {
      unexpectedArguments.add(argChar);
      valid = false;
    }
  }

  private boolean setArgument(char argChar) {
    boolean set = true;
    if (isBoolean(argChar))
      setBooleanArg(argChar, true);
    else if (isString(argChar))
      setStringArg(argChar, "");
    else
      set = false;

    return set;
  }

  private void setStringArg(char argChar, String s) {
```

14.2 Args：草稿

```
      currentArgument++;
      try {
        stringArgs.put(argChar, args[currentArgument]);
      } catch (ArrayIndexOutOfBoundsException e) {
        valid = false;
        errorArgument = argChar;
        errorCode = ErrorCode.MISSING_STRING;
      }
    }

    private boolean isString(char argChar) {
      return stringArgs.containsKey(argChar);
    }

    private void setBooleanArg(char argChar, boolean value) {
      booleanArgs.put(argChar, value);
    }

    private boolean isBoolean(char argChar) {
      return booleanArgs.containsKey(argChar);
    }

    public int cardinality() {
      return argsFound.size();
    }

    public String usage() {
      if (schema.length() > 0)
        return "-[" + schema + "]";
      else
        return "";
    }

    public String errorMessage() throws Exception {
      if (unexpectedArguments.size() > 0) {
        return unexpectedArgumentMessage();
      } else
        switch (errorCode) {
          case MISSING_STRING:
            return String.format("Could not find string parameter for -%c.",
                                 errorArgument);
          case OK:
            throw new Exception("TILT: Should not get here.");
        }
      return "";
    }

    private String unexpectedArgumentMessage() {
      StringBuffer message = new StringBuffer("Argument(s) -");
      for (char c : unexpectedArguments) {
        message.append(c);
      }
      message.append(" unexpected.");
```

```
    return message.toString();
  }

  public boolean getBoolean(char arg) {
    return falseIfNull(booleanArgs.get(arg));
  }

  private boolean falseIfNull(Boolean b) {
    return b == null ? false : b;
  }

  public String getString(char arg) {
    return blankIfNull(stringArgs.get(arg));
  }

  private String blankIfNull(String s) {
    return s == null ? "" : s;
  }

  public boolean has(char arg) {
    return argsFound.contains(arg);
  }

  public boolean isValid() {
    return valid;
  }
}
```

你可以看到，代码开始失去控制，这还算不上可怕，可怕的是混乱已经开始生长，已经出现了一堆东西，不过还没烂掉。增加对整数参数类型的支持后，那堆东西才真的变质腐烂了。

14.2.1 所以我暂停了

还有至少两种参数类型要添加，而且情形一定会更加糟糕。如果一味蛮干，大概也能让它工作，不过就会留下一大堆要调整的混乱代码。如果希望代码结构一直可维护，现在正是调整的时机。

所以我暂停添加特性，开始重构。由于刚添加了 `String` 类型和 `integer` 类型参数，我知道每种参数类型都需要在 3 个主要位置增加新代码。首先，每种参数类型都要解析其范式元素，从而为该种类型选择 `HashMap` 的方法。其次，每种参数类型都需要在命令行字符串中解析，然后再转换为真实类型。最后，每种参数类型都需要一个 `getXXX` 方法，并按照其真实类型向调用者返回参数值。

许多种不同类型，类似的方法——听起来像是一个类。`ArgumentMarshaler` 的概念就是这样产生的。

14.2.2 渐进

毁坏程序的最好方法之一就是以改进之名大动其结构。有些程序永远不能从这种所谓"改进"中恢复过来。问题在于，很难让程序以"改进"之前的方式工作。

为了避免这种状况发生，我采用了测试驱动开发的规程，这种方法的核心原则之一是保持系统始终能运行。换言之，采用 TDD，我不会允许做出破坏系统的修改，每次修改都必须保证系统能像以前一样工作。

我需要一套能随需运行，确保系统行为不会改动的自动化测试。在我搞出那个烂摊子的同时，也为 Args 类创建了一套单元测试和验收测试。单元测试用 Java 写成，采用 JUnit 管理。验收测试用 FitNesse 以 wiki 页形式写成。我可以随时运行这些测试，如果测试通过，就能打包票说系统能以我期望的方式工作。

于是我开始做出大量小规模修改。每次修改都将系统结构向 ArgumentMarshaler 概念的方向推动，而且每次修改后，系统都要能正常工作。第一个修改是在烂摊子末尾添加 ArgumentMarshaler 的轮廓，如代码清单 14-11 所示。

代码清单 14-11　向 Args.java 添加 ArgumentMarshaler

```
private class ArgumentMarshaler {
  private boolean booleanValue = false;

  public void setBoolean(boolean value) {
    booleanValue = value;
  }

  public boolean getBoolean() {return booleanValue;}
}

private class BooleanArgumentMarshaler extends ArgumentMarshaler {
}

private class StringArgumentMarshaler extends ArgumentMarshaler {
}

private class IntegerArgumentMarshaler extends ArgumentMarshaler {
}
```

显然，这什么也不会破坏。于是我做了一点最简单的、破坏性尽可能小的修改。我修改了 HashMap，采用 ArgumentMarshaler，使之支持 Boolean 类型参数。

```
private Map<Character, ArgumentMarshaler> booleanArgs =
  new HashMap<Character, ArgumentMarshaler>();
```

这个修改影响少数语句，我很快就修正了。

```
    ...
    private void parseBooleanSchemaElement(char elementId) {
      booleanArgs.put(elementId, new BooleanArgumentMarshaler());
    }
    ...
    private void setBooleanArg(char argChar, boolean value) {
      booleanArgs.get(argChar).setBoolean(value);
    }
    ...
    public boolean getBoolean(char arg) {
      return falseIfNull(booleanArgs.get(arg).getBoolean());
    }
```

注意，这些修改正是在我之前提到的那些区域之内所做的：参数类型的 `parse`、`set` 和 `get` 操作。不幸的是，即便修改如此细微，有些测试也还是会失败。仔细看 `getBoolean`，可以看到如果用 y 去调用，但没有 y 这个参数，则 `booleanArgs.get('y')` 就会返回 `null` 值，函数将抛出 `NullPointerException` 异常。函数 `falseIfNull` 用以防止这种状况发生，但我做出的修改却导致该函数无所作为。

渐进主义要求我在做其他修改之前迅速修正这个问题。修正并不费劲。我只是把对 null 值的检查移了个位置。再也不用检测 `bollean` 是否为 `null`，而是检查 `ArgumentMarshaler` 是否为 `null`。

首先，我移除了 `getBoolean` 函数中的 `falseIfNull` 调用。现在它没什么用了，所以我也删除了这个函数。测试还是以同样的方式失败，所以我确定没有引入新的错误。

```
    public boolean getBoolean(char arg) {
      return booleanArgs.get(arg).getBoolean();
    }
```

下一步，我把函数拆解为两行，并把 `ArgumentMarshaler` 放到它自己的名为 `argumentMarshaler` 的变量中①。我不在意变量名太长，但它确实有点啰唆，把函数搞得支离破碎。所以我把变量名缩短为 am[N5]。

```
    public boolean getBoolean(char arg) {
      Args.ArgumentMarshaler am = booleanArgs.get(arg);
      return am.getBoolean();
    }
```

然后再放入检测 null 值的逻辑。

```
    public boolean getBoolean(char arg) {
      Args.ArgumentMarshaler am = booleanArgs.get(arg);
      return am != null && am.getBoolean();
    }
```

---

① 即创建一个类型为 ArgumentMarshaler 的对象实体。——译者注

14.3 字符串类型参数

添加 `String` 类型参数和添加 `boolean` 类型参数非常像。我要修改 `HashMap`，让 `parse`、`set` 和 `get` 函数能工作。跟着就是按部就班，但我似乎该把所有的 marshalling（编组）实现放到 `ArgumentMarshaler` 基类而不是派生类中。

```java
  private Map<Character, ArgumentMarshaler> stringArgs =
    new HashMap<Character, ArgumentMarshaler>();
...
  private void parseStringSchemaElement(char elementId) {
    stringArgs.put(elementId, new StringArgumentMarshaler());
  }
...
  private void setStringArg(char argChar) throws ArgsException {
    currentArgument++;
    try {
      stringArgs.get(argChar).setString(args[currentArgument]);
    } catch (ArrayIndexOutOfBoundsException e) {
      valid = false;
      errorArgumentId = argChar;
      errorCode = ErrorCode.MISSING_STRING;
      throw new ArgsException();
    }
  }
...
  public String getString(char arg) {
    Args.ArgumentMarshaler am = stringArgs.get(arg);
    return am == null ? "" : am.getString();
  }
...
  private class ArgumentMarshaler {
    private boolean booleanValue = false;
    private String stringValue;

    public void setBoolean(boolean value) {
      booleanValue = value;
    }

    public boolean getBoolean() {
      return booleanValue;
    }

    public void setString(String s) {
      stringValue = s;
    }

    public String getString() {
```

```
      return stringValue == null ? "" : stringValue;
    }
  }
```

同样，也是每次修改一个地方，持续运行测试。如果测试出错，则在做下一个修改前确保通过。

现在你应该明白我的意图了。一旦我将当前的编组行为放到 ArgumentMarshaler 基类中，就会开始往派生类推入该行为。这样，在我逐渐修改程序的形状时，还能保持一切正常。

下一步显而易见，把 int 参数的相关功能放到 ArgumentMarshaler 里面。同样，也是照方抓药。

```
  private Map<Character, ArgumentMarshaler> intArgs =
    new HashMap<Character, ArgumentMarshaler>();
...
  private void parseIntegerSchemaElement(char elementId) {
    intArgs.put(elementId, new IntegerArgumentMarshaler());
  }
...
  private void setIntArg(char argChar) throws ArgsException {
    currentArgument++;
    String parameter = null;
    try {
      parameter = args[currentArgument];
      intArgs.get(argChar).setInteger(Integer.parseInt(parameter));
    } catch (ArrayIndexOutOfBoundsException e) {
      valid = false;
      errorArgumentId = argChar;
      errorCode = ErrorCode.MISSING_INTEGER;
      throw new ArgsException();
    } catch (NumberFormatException e) {
      valid = false;
      errorArgumentId = argChar;
      errorParameter = parameter;
      errorCode = ErrorCode.INVALID_INTEGER;
      throw new ArgsException();
    }
  }
...
  public int getInt(char arg) {
    Args.ArgumentMarshaler am = intArgs.get(arg);
    return am == null ? 0 : am.getInteger();
  }
...
  private class ArgumentMarshaler {
    private boolean booleanValue = false;
    private String stringValue;
    private int integerValue;

    public void setBoolean(boolean value) {
      booleanValue = value;
```

```
  }

  public boolean getBoolean() {
    return booleanValue;
  }

  public void setString(String s) {
    stringValue = s;
  }

  public String getString() {
    return stringValue == null ? "" : stringValue;
  }

  public void setInteger(int i) {
    integerValue = i;
  }

  public int getInteger() {
    return integerValue;
  }
}
```

当所有的编组操作都放到了 ArgumentMarshaler 中，我开始向派生类移植功能。第一步是把 setBoolean 函数放到 BooleanArgumentMarshaler 中，确保它能被正确调用。所以我创建了一个抽象的 set 方法。

```
private abstract class ArgumentMarshaler {
  protected boolean booleanValue = false;
  private String stringValue;
  private int integerValue;

  public void setBoolean(boolean value) {
    booleanValue = value;
  }

  public boolean getBoolean() {
    return booleanValue;
  }

  public void setString(String s) {
    stringValue = s;
  }

  public String getString() {
    return stringValue == null ? "" : stringValue;
  }

  public void setInteger(int i) {
    integerValue = i;
  }

  public int getInteger() {
    return integerValue;
```

```java
  public abstract void set(String s);
}
```

然后在 `BooleanArgumentMarshaler` 中实现 `set` 方法。

```java
private class BooleanArgumentMarshaler extends ArgumentMarshaler {
  public void set(String s) {
    booleanValue = true;
  }
}
```

最后，通过调用 `set`，替换对 `setBoolean` 函数的调用。

```java
private void setBooleanArg(char argChar, boolean value) {
  booleanArgs.get(argChar).set("true");
}
```

测试仍然全部通过。因为这次修改将 `set` 函数放到了 `BooleanArgumentMarshaler` 里面，所以我从 `ArgumentMarshaler` 基类删除了 `setBoolean` 方法。

注意，抽象函数 `set` 有一个 `String` 类型参数，但其在 `BooleanArgumentMarshaler` 中的实现却没有使用这个参数。之所以在这里放这个参数，是因为我知道 `StringArgumentMarshaler` 和 `IntegerArgumentMarshaler` 可能会使用它。

接着，我打算把 `get` 方法放到 `BooleanArgumentMarshaler` 中。这有点难看，因为返回类型必须是 `Object`，且在这里需要转换为 `Boolean` 类型值。

```java
public boolean getBoolean(char arg) {
  Args.ArgumentMarshaler am = booleanArgs.get(arg);
  return am != null && (Boolean)am.get();
}
```

为了编译通过，我把 `get` 函数加到 `ArgumentMarshaler` 中。

```java
private abstract class ArgumentMarshaler {
  ...

  public Object get() {
    return null;
  }
}
```

这样一来，虽然可以编译，但却无法通过测试。只要将 `get` 修改为抽象方法，并在 `BooleanArgumentMarshaler` 中实现，就能重新通过测试。

```java
private abstract class ArgumentMarshaler {
  protected boolean booleanValue = false;
  ...

  public abstract Object get();
}
```

14.3 字符串类型参数

```java
    private class BooleanArgumentMarshaler extends ArgumentMarshaler {
      public void set(String s) {
        booleanValue = true;
      }

      public Object get() {
        return booleanValue;
      }
    }
```

测试又通过了。get 方法和 set 方法都已部署到 BooleanArgumentMarshaler 中！这样我就可以从 ArgumentMarshaler 里面移除旧的 getBoolean 函数，把受保护的 booleanValue 变量向下移动到 BooleanArgumentMarshaler，并将其设置为 private。

对于 String 类型也照此处理，即修改 set 和 get 的部署方式，删除无用的函数，并移动了变量。

```java
  private void setStringArg(char argChar) throws ArgsException {
    currentArgument++;
    try {
      stringArgs.get(argChar).set(args[currentArgument]);
    } catch (ArrayIndexOutOfBoundsException e) {
      valid = false;
      errorArgumentId = argChar;
      errorCode = ErrorCode.MISSING_STRING;
      throw new ArgsException();
    }
  }
...
  public String getString(char arg) {
    Args.ArgumentMarshaler am = stringArgs.get(arg);
    return am == null ? "" : (String) am.get();
  }
...
  private abstract class ArgumentMarshaler {
    private int integerValue;

    public void setInteger(int i) {
      integerValue = i;
    }

    public int getInteger() {
      return integerValue;
    }

    public abstract void set(String s);

    public abstract Object get();
  }

  private class BooleanArgumentMarshaler extends ArgumentMarshaler {
    private boolean booleanValue = false;
```

```java
    public void set(String s) {
      booleanValue = true;
    }

    public Object get() {
      return booleanValue;
    }
  }

  private class StringArgumentMarshaler extends ArgumentMarshaler {
    private String stringValue = "";

    public void set(String s) {
      stringValue = s;
    }

    public Object get() {
      return stringValue;
    }
  }

  private class IntegerArgumentMarshaler extends ArgumentMarshaler {
    public void set(String s) {

    }

    public Object get() {
      return null;
    }
  }
}
```

最后，我为 `integer` 类型参数重复上述过程。这稍稍复杂一点，因为 `integer` 需要解析，而 `parse` 操作会抛出异常。不过结果会更好，因为 `NumberFormatException` 的概念在 `IntegerArgumentMarshaler` 中隐藏了。

```java
  private boolean isIntArg(char argChar) {return intArgs.containsKey(argChar);}

  private void setIntArg(char argChar) throws ArgsException {
    currentArgument++;
    String parameter = null;
    try {
      parameter = args[currentArgument];
      intArgs.get(argChar).set(parameter);
    } catch (ArrayIndexOutOfBoundsException e) {
      valid = false;
      errorArgumentId = argChar;
      errorCode = ErrorCode.MISSING_INTEGER;
      throw new ArgsException();
    } catch (ArgsException e) {
      valid = false;
      errorArgumentId = argChar;
```

```
      errorParameter = parameter;
      errorCode = ErrorCode.INVALID_INTEGER;
      throw e;
    }
  }
  ...
  private void setBooleanArg(char argChar) {
    try {
      booleanArgs.get(argChar).set("true");
    } catch (ArgsException e) {
    }
  }
  ...
  public int getInt(char arg) {
    Args.ArgumentMarshaler am = intArgs.get(arg);
    return am == null ? 0 : (Integer) am.get();
  }
  ...
  private abstract class ArgumentMarshaler {
    public abstract void set(String s) throws ArgsException;
    public abstract Object get();
  }
  ...
  private class IntegerArgumentMarshaler extends ArgumentMarshaler {
    private int intValue = 0;

    public void set(String s) throws ArgsException {
      try {
        intValue = Integer.parseInt(s);
      } catch (NumberFormatException e) {
        throw new ArgsException();
      }
    }

    public Object get() {
      return intValue;
    }
  }
```

测试当然会继续通过。下一步，我要删除算法顶端的 3 种不同映射。这样，整个系统就变得更通用了。不过，如果只是删除它们则无法达到目的，因为那样会破坏系统。取而代之的是，我为 `ArgumentMarshaler` 添加一个新的映射，然后再逐个修改那些方法，让那些方法调用这个新映射。

```
public class Args {
  ...
  private Map<Character, ArgumentMarshaler> booleanArgs =
    new HashMap<Character, ArgumentMarshaler>();
  private Map<Character, ArgumentMarshaler> stringArgs =
    new HashMap<Character, ArgumentMarshaler>();
  private Map<Character, ArgumentMarshaler> intArgs =
    new HashMap<Character,ArgumentMarshaler>();
```

第 14 章 逐步改进

```
    private  Map<Character, ArgumentMarshaler> marshalers =
      new HashMap<Character, ArgumentMarshaler>();
...
    private  void  parseBooleanSchemaElement(char elementId)  {
      ArgumentMarshaler m = new BooleanArgumentMarshaler();
      booleanArgs.put(elementId, m);
      marshalers.put(elementId, m);
    }

    private  void  parseIntegerSchemaElement(char elementId)  {
      ArgumentMarshaler m = new IntegerArgumentMarshaler();
      intArgs.put(elementId, m);
      marshalers.put(elementId, m);
    }

    private  void  parseStringSchemaElement(char elementId)  {
      ArgumentMarshaler m = new StringArgumentMarshaler();
      stringArgs.put(elementId,m);
      marshalers.put(elementId, m);
    }
```

当然，测试还是通过了。接着，我把 isBooleanArg

```
private boolean isBooleanArg(char argChar) {
  return booleanArgs.containsKey(argChar);
}
```

修改成这样：

```
private boolean isBooleanArg(char argChar) {
  ArgumentMarshaler m = marshalers.get(argChar);
  return m instanceof BooleanArgumentMarshaler;
}
```

测试仍然通过。于是我修改了 `isIntArg` 和 `isStringArg`：

```
private boolean isIntArg(char argChar) {
  ArgumentMarshaler m = marshalers.get(argChar);
  return m instanceof IntegerArgumentMarshaler;
}

private boolean isStringArg(char argChar) {
  ArgumentMarshaler m = marshalers.get(argChar);
  return m instanceof StringArgumentMarshaler;
}
```

测试继续通过。我接着删除对 `marshaler.get` 的重复调用：

```
private boolean setArgument(char argChar) throws ArgsException {
  ArgumentMarshaler m = marshalers.get(argChar);
  if (isBooleanArg(m))
    setBooleanArg(argChar);
  else if (isStringArg(m))
    setStringArg(argChar);
```

```
      else if (isIntArg(m))
        setIntArg(argChar);
      else
        return false;

      return true;
  }

  private boolean isIntArg(ArgumentMarshaler m) {
    return m instanceof IntegerArgumentMarshaler;
  }

  private boolean isStringArg(ArgumentMarshaler m) {
    return m instanceof StringArgumentMarshaler;
  }

  private boolean isBooleanArg(ArgumentMarshaler m) {
    return m instanceof BooleanArgumentMarshaler;
  }
```

存在 3 个 `isxxxArg` 方法毫无道理，所以我做了内联修改：

```
  private boolean setArgument(char argChar) throws ArgsException {
    ArgumentMarshaler  m = marshalers.get(argChar);
    if (m instanceof BooleanArgumentMarshaler)
      setBooleanArg(argChar);
    else if (m instanceof StringArgumentMarshaler)
      setStringArg(argChar);
    else if (m instanceof IntegerArgumentMarshaler)
      setIntArg(argChar);
    else
      return false;

    return true;
  }
```

下一步，我开始在 `set` 函数中使用 `marshalers` 映射，停止使用另外 3 个映射。从 boolean 开始：

```
      private boolean setArgument(char argChar) throws ArgsException {
        ArgumentMarshaler m = marshalers.get(argChar);
        if (m instanceof BooleanArgumentMarshaler)
          setBooleanArg(m);
        else if (m instanceof StringArgumentMarshaler)
          setStringArg(argChar);
        else if (m instanceof IntegerArgumentMarshaler)
          setIntArg(argChar);
        else
          return false;

        return true;
      }
...
      private void setBooleanArg(ArgumentMarshaler m) {
```

```
          try {
            m.set("true");  // was: booleanArgs.get(argChar).set("true");
          } catch (ArgsException e) {
          }
        }
```

测试通过，于是我如法炮制 String 类型和 Integer 类型参数。这样我就能把有些丑陋的异常管理代码整合到 setArgument 函数中。

```
    private boolean setArgument(char argChar) throws ArgsException {
      ArgumentMarshaler m = marshalers.get(argChar);
      try {
        if (m instanceof BooleanArgumentMarshaler)
          setBooleanArg(m);
        else if (m instanceof StringArgumentMarshaler)
          setStringArg(m);
        else if (m instanceof IntegerArgumentMarshaler)
          setIntArg(m);
        else
          return false;
      } catch (ArgsException e) {
        valid = false;
        errorArgumentId = argChar;
        throw e;
      }
      return true;
    }

    private void setIntArg(ArgumentMarshaler m) throws ArgsException {
      currentArgument++;
      String parameter = null;
      try {
        parameter = args[currentArgument];
        m.set(parameter);
      } catch (ArrayIndexOutOfBoundsException e) {
        errorCode = ErrorCode.MISSING_INTEGER;
        throw new ArgsException();
      } catch (ArgsException e) {
        errorParameter = parameter;
        errorCode = ErrorCode.INVALID_INTEGER;
        throw e;
      }
    }

    private void setStringArg(ArgumentMarshaler m) throws ArgsException {
      currentArgument++;
      try {
        m.set(args[currentArgument]);
      } catch (ArrayIndexOutOfBoundsException e) {
        errorCode = ErrorCode.MISSING_STRING;
        throw new ArgsException();
      }
    }
```

离彻底删除那 3 个旧映射的时机越来越近了。首先，我需要修改 getBoolean 函数：

```
public boolean getBoolean(char arg) {
  Args.ArgumentMarshaler am = booleanArgs.get(arg);
  return am != null && (Boolean) am.get();
}
```

修改成这样：

```
public boolean getBoolean(char arg) {
  Args.ArgumentMarshaler am = marshalers.get(arg);
  boolean b = false;
  try {
    b = am != null && (Boolean) am.get();
  } catch (ClassCastException e) {
    b = false;
  }
  return b;
}
```

最后这个修改可能令人吃惊。为什么我会突然决定对付 ClassCastException？原因是我有一组单元测试，还有用 FitNesse 编写的一组验收测试。FitNesse 测试确认，如果用非布尔值参数调用 getBoolean，应该返回 false。可单元测试的结果不是这样。而到此时为止，我一直只调用单元测试[①]。

这次修改把另一个对 boolean 映射的使用抽离了：

```
private void parseBooleanSchemaElement(char elementId) {
  ArgumentMarshaler m = new BooleanArgumentMarshaler();
  booleanArgs.put(elementId, m);
  marshalers.put(elementId, m);
}
```

如此我们就能删除 boolean 映射。

```
public class Args {
  ...
  private Map<Character, ArgumentMarshaler> booleanArgs =
    new HashMap<Character, ArgumentMarshaler>();
  private Map<Character, ArgumentMarshaler> stringArgs =
    new HashMap<Character, ArgumentMarshaler>();
  private Map<Character, ArgumentMarshaler> intArgs =
    new HashMap<Character, ArgumentMarshaler>();
  private Map<Character, ArgumentMarshaler> marshalers =
    new HashMap<Character, ArgumentMarshaler>();
  ...
```

接下来，我用同样的方法处理 String 类型和 Integer 类型参数，对 boolean 参数做一点清理工作。

```
  private void parseBooleanSchemaElement(char elementId) {
```

[①] 为了避免这种情况发生，我添加了一个新的单元测试，调用所有 FitNesse 测试。

```java
    marshalers.put(elementId, new BooleanArgumentMarshaler());
  }

  private void parseIntegerSchemaElement(char elementId) {
    marshalers.put(elementId, new IntegerArgumentMarshaler());
  }

  private void parseStringSchemaElement(char elementId) {
    marshalers.put(elementId, new StringArgumentMarshaler());
  }
...
  public String getString(char arg) {
    Args.ArgumentMarshaler am = marshalers.get(arg);
    try {
      return am == null ? "" : (String) am.get();
    } catch (ClassCastException e) {
      return "";
    }
  }

  public int getInt(char arg) {
    Args.ArgumentMarshaler am = marshalers.get(arg);
    try {
      return am == null ? 0 : (Integer) am.get();
    } catch (Exception e) {
      return 0;
    }
  }
...
public class Args {
  ...
  private Map<Character, ArgumentMarshaler> stringArgs =
    new HashMap<Character, ArgumentMarshaler>();
  private Map<Character, ArgumentMarshaler> intArgs =
    new HashMap<Character, ArgumentMarshaler>();
  private Map<Character, ArgumentMarshaler> marshalers =
    new HashMap<Character, ArgumentMarshaler>();
  ...
```

接着，由于那些 parse 方法没有太多事可做，我对它们进行了内联修改：

```java
  private void parseSchemaElement(String element) throws ParseException {
    char elementId = element.charAt(0);
    String elementTail = element.substring(1);
    validateSchemaElementId(elementId);
    if (isBooleanSchemaElement(elementTail))
      marshalers.put(elementId, new BooleanArgumentMarshaler());
    else if (isStringSchemaElement(elementTail))
      marshalers.put(elementId, new StringArgumentMarshaler());
    else if (isIntegerSchemaElement(elementTail)) {
      marshalers.put(elementId, new IntegerArgumentMarshaler());
    } else {
      throw new ParseException(String.format(
        "Argument: %c has invalid format: %s.", elementId, elementTail), 0);
```

    }
}

行了，下面来看看全貌吧。代码清单 14-12 展示了 Args 类的现状。

代码清单 14-12　Args.java（首次重构后）

```java
package com.objectmentor.utilities.getopts;

import java.text.ParseException;
import java.util.*;

public class Args {
  private String schema;
  private String[] args;
  private boolean valid = true;
  private Set<Character> unexpectedArguments = new TreeSet<Character>();
  private Map<Character, ArgumentMarshaler> marshalers =
    new HashMap<Character, ArgumentMarshaler>();
  private Set<Character> argsFound = new HashSet<Character>();
  private int currentArgument;
  private char errorArgumentId = '\0';
  private String errorParameter = "TILT";
  private ErrorCode errorCode = ErrorCode.OK;

  private enum ErrorCode {
    OK, MISSING_STRING, MISSING_INTEGER, INVALID_INTEGER, UNEXPECTED_ARGUMENT}

  public Args(String schema, String[] args) throws ParseException {
    this.schema = schema;
    this.args = args;
    valid = parse();
  }

  private boolean parse() throws ParseException {
    if (schema.length() == 0 && args.length == 0)
      return true;
    parseSchema();
    try {
      parseArguments();
    } catch (ArgsException e) {
    }
    return valid;
  }

  private boolean parseSchema() throws ParseException {
    for (String element : schema.split(",")) {
      if (element.length() > 0) {
        String trimmedElement = element.trim();
        parseSchemaElement(trimmedElement);
      }
    }
    return true;
  }
```

```java
private void parseSchemaElement(String element) throws ParseException {
  char elementId = element.charAt(0);
  String elementTail = element.substring(1);
  validateSchemaElementId(elementId);
  if (isBooleanSchemaElement(elementTail))
    marshalers.put(elementId, new BooleanArgumentMarshaler());
  else if (isStringSchemaElement(elementTail))
    marshalers.put(elementId, new StringArgumentMarshaler());

    else if (isIntegerSchemaElement(elementTail)) {
      marshalers.put(elementId, new IntegerArgumentMarshaler());
    } else {
      throw new ParseException(String.format(
       "Argument: %c has invalid format: %s.", elementId, elementTail), 0);
    }
}

private void validateSchemaElementId(char elementId) throws ParseException {
  if (!Character.isLetter(elementId)) {
    throw new ParseException(
    "Bad character:" + elementId + "in Args format: " + schema, 0);
  }
}

private boolean isStringSchemaElement(String elementTail) {
  return elementTail.equals("*");
}

private boolean isBooleanSchemaElement(String elementTail) {
  return elementTail.length() == 0;
}

private boolean isIntegerSchemaElement(String elementTail) {
  return elementTail.equals("#");
}

private boolean parseArguments() throws ArgsException {
  for (currentArgument=0; currentArgument<args.length; currentArgument++) {
    String arg = args[currentArgument];
    parseArgument(arg);
  }
  return true;
}

private void parseArgument(String arg) throws ArgsException {
  if (arg.startsWith("-"))
    parseElements(arg);
}

private void parseElements(String arg) throws ArgsException {
  for (int i = 1; i < arg.length(); i++)
    parseElement(arg.charAt(i));
}
```

14.3 字符串类型参数

```java
  private void parseElement(char argChar) throws ArgsException {
    if (setArgument(argChar))
      argsFound.add(argChar);
    else {
      unexpectedArguments.add(argChar);
      errorCode = ErrorCode.UNEXPECTED_ARGUMENT;
      valid = false;
    }
  }

  private boolean setArgument(char argChar) throws ArgsException {
    ArgumentMarshaler m = marshalers.get(argChar);
    try {
      if (m instanceof BooleanArgumentMarshaler)
        setBooleanArg(m);
      else if (m instanceof StringArgumentMarshaler)
        setStringArg(m);
      else if (m instanceof IntegerArgumentMarshaler)
        setIntArg(m);
      else
        return false;
    } catch (ArgsException e) {
      valid = false;
      errorArgumentId = argChar;
      throw e;
    }
    return true;
  }

  private void setIntArg(ArgumentMarshaler m) throws ArgsException {
    currentArgument++;
    String parameter = null;
    try {
      parameter = args[currentArgument];
      m.set(parameter);
    } catch (ArrayIndexOutOfBoundsException e) {
      errorCode = ErrorCode.MISSING_INTEGER;
      throw new ArgsException();
    } catch (ArgsException e) {
      errorParameter = parameter;
      errorCode = ErrorCode.INVALID_INTEGER;
      throw e;
    }
  }

  private void setStringArg(ArgumentMarshaler m) throws ArgsException {
    currentArgument++;
    try {
      m.set(args[currentArgument]);
    } catch (ArrayIndexOutOfBoundsException e) {
      errorCode = ErrorCode.MISSING_STRING;
      throw new ArgsException();
    }
```

```
  }

  private void setBooleanArg(ArgumentMarshaler m) {
    try {
      m.set("true");
    } catch (ArgsException e) {
    }
  }

  public int cardinality() {
    return argsFound.size();
  }

  public String usage() {
    if (schema.length() > 0)
      return "-[" + schema + "]";
    else
      return "";
  }

  public String errorMessage() throws Exception {
    switch (errorCode) {
      case OK:
        throw new Exception("TILT: Should not get here.");
      case UNEXPECTED_ARGUMENT:
        return unexpectedArgumentMessage();
      case MISSING_STRING:
        return String.format("Could not find string parameter for -%c.",
                             errorArgumentId);
      case INVALID_INTEGER:
        return String.format("Argument -%c expects an integer but was '%s'.",
                             errorArgumentId, errorParameter);
      case MISSING_INTEGER:
        return String.format("Could not find integer parameter for -%c.",
                             errorArgumentId);
    }
    return "";
  }

  private String unexpectedArgumentMessage() {
    StringBuffer message = new StringBuffer("Argument(s) -");
    for (char c : unexpectedArguments) {
      message.append(c);
    }
    message.append(" unexpected.");

    return message.toString();
  }

  public boolean getBoolean(char arg) {
    Args.ArgumentMarshaler am = marshalers.get(arg);
    boolean b = false;
    try {
```

14.3 字符串类型参数

```java
      b = am != null && (Boolean) am.get();
    } catch (ClassCastException e) {
      b = false;
    }
    return b;
  }

  public String getString(char arg) {
    Args.ArgumentMarshaler am = marshalers.get(arg);
    try {
      return am == null ? "" : (String) am.get();
    } catch (ClassCastException e) {
      return "";
    }
  }


  public int getInt(char arg) {
    Args.ArgumentMarshaler am = marshalers.get(arg);
    try {
      return am == null ? 0 : (Integer) am.get();
    } catch (Exception e) {
      return 0;
    }
  }

  public boolean has(char arg) {
    return argsFound.contains(arg);
  }

  public boolean isValid() {
    return valid;
  }

  private class ArgsException extends Exception {
  }

  private abstract class ArgumentMarshaler {
    public abstract void set(String s) throws ArgsException;
    public abstract Object get();
  }

  private class BooleanArgumentMarshaler extends ArgumentMarshaler {
    private boolean booleanValue = false;

    public void set(String s) {
      booleanValue = true;
    }

    public Object get() {
      return booleanValue;
    }
  }
```

第 14 章　逐步改进

```java
  private class StringArgumentMarshaler extends ArgumentMarshaler {
    private String stringValue = "";

    public void set(String s) {
      stringValue = s;
    }

    public Object get() {
      return stringValue;
    }
  }

  private class IntegerArgumentMarshaler extends ArgumentMarshaler {
    private int intValue = 0;

    public void set(String s) throws ArgsException {
      try {
        intValue = Integer.parseInt(s);
      } catch (NumberFormatException e) {
        throw new ArgsException();
      }
    }

    public Object get() {
      return intValue;
    }
  }
}
```

功夫费尽，还是有点失望。程序结构好了一点，但在代码顶端还是有那一堆变量；在 setArgument 里面还是有那么恐怖的类型转换操作；而且那些 set 函数真的很丑陋。就别提那些错误处理操作了。后面要做的事还很多。

我真是想删掉 setArgument 里面那些类型转换操作[G23]。我希望 setArgument 只简单地调用 ArgumentMarshaler.set。这意味着我需要将 setIntArg、setStringArg 和 setBooleanArg 推到合适的 ArgumentMarshaler 派生类里面。不过这里有一个问题。

仔细看 setIntArg，你会发现，它使用了两个实体变量 args 和 currentArgument。为了把 setIntArg 移到 IntegerArgumentMarshaler 里面，我得把这两个变量都作为函数参数传递过去，这种做法太烂了[F1]，因为我只想传递一个参数。幸运的是，有一个简单的解决方法，可以把 args 数组转换为一个 list，并向 set 函数传递一个 Iterator。这需要经过 10 步，每步都通过了测试。不过我只向你展示结果，你应该能看出每个小修改步骤。

```java
public class Args {
  private String schema;
  private String[] args;
  private boolean valid = true;
  private Set<Character> unexpectedArguments = new TreeSet<Character>();
  private Map<Character, ArgumentMarshaler> marshalers =
    new HashMap<Character, ArgumentMarshaler>();
  private Set<Character> argsFound = new HashSet<Character>();
```

14.3 字符串类型参数

```java
  private Iterator<String> currentArgument;
  private char errorArgumentId = '\0';
  private String errorParameter = "TILT";
  private ErrorCode errorCode = ErrorCode.OK;
  private List<String> argsList;

  private enum ErrorCode {
    OK, MISSING_STRING, MISSING_INTEGER, INVALID_INTEGER, UNEXPECTED_ARGUMENT}
  public Args(String schema, String[] args) throws ParseException {
    this.schema = schema;
    argsList = Arrays.asList(args);
    valid = parse();
  }


  private boolean parse() throws ParseException {
    if (schema.length() == 0 && argsList.size() == 0)
      return true;
    parseSchema();
    try {
      parseArguments();
    } catch (ArgsException e) {
    }
    return valid;
  }
---
  private boolean parseArguments() throws ArgsException {
    for (currentArgument = argsList.iterator();currentArgument.hasNext();) {
      String arg = currentArgument.next();
      parseArgument(arg);
    }

    retun true;
  }
---
  private void setIntArg(ArgumentMarshaler m) throws ArgsException {
    String parameter = null;
    try {
      parameter = currentArgument.next();
      m.set(parameter);
    } catch (NoSuchElementException e) {
      errorCode = ErrorCode.MISSING_INTEGER;
      throw new ArgsException();
    } catch (ArgsException e) {
      errorParameter = parameter;
      errorCode = ErrorCode.INVALID_INTEGER;
      throw e;
    }
  }

  private void setStringArg(ArgumentMarshaler m) throws ArgsException {
    try {
      m.set(currentArgument.next());
```

```
        } catch (NoSuchElementException e) {
          errorCode = ErrorCode.MISSING_STRING;
          throw new ArgsException();
        }
      }
```

是这些简单的修改让测试保持通过。现在我们可以开始把 set 函数移植到合适的派生类中了。第一步，我要在 setArgument 中做以下修改：

```
    private boolean setArgument(char argChar) throws ArgsException {
      ArgumentMarshaler m = marshalers.get(argChar);
      if (m == null)
        return false;
      try {
        if (m instanceof BooleanArgumentMarshaler)
          setBooleanArg(m);
        else if (m instanceof StringArgumentMarshaler)
          setStringArg(m);
        else if (m instanceof IntegerArgumentMarshaler)
          setIntArg(m);
        else
          return false;
      } catch (ArgsException e) {
        valid = false;
        errorArgumentId = argChar;
        throw e;
      }
      return true;
    }
```

这个修改很重要，因为我们想要彻底删除那条 if-else 链。所以，需要把错误条件抽离。

现在可以开始移动 set 函数了。setBooleanArg 函数很小，就从它开始，目标是让 setBooleanArg 函数只与 BooleanArgumentMarshaler 有关。

```
    private boolean setArgument(char argChar) throws ArgsException {
      ArgumentMarshaler m = marshalers.get(argChar);
      if (m == null)
        return false;
      try {
        if (m instanceof BooleanArgumentMarshaler)
          setBooleanArg(m, currentArgument);
        else if (m instanceof StringArgumentMarshaler)
          setStringArg(m);
        else if (m instanceof IntegerArgumentMarshaler)
          setIntArg(m);

      } catch (ArgsException e) {
        valid = false;
        errorArgumentId = argChar;
        throw e;
      }
      return true;
    }
```

```
---
  private void setBooleanArg(ArgumentMarshaler m,
                             Iterator<String> currentArgument)
                             throws ArgsException {
    try {
      m.set("true");
    catch (ArgsException e) {
    }
  }
```

我们不是刚把那个异常处理放进去吗?放进和拿出是重构过程中常见的事。小步幅修改和保持测试通过,意味着你会不断移动各种东西。重构有点儿像是解魔方,需要经过许多小步骤,才能达到较大目标。每一步都是下一步的基础。

为什么要在 setBooleanArg 根本不需要的情况下向其传递 iterator 呢?因为 setIntArg 和 setStringArg 需要!还因为我打算通过 ArgumentMarshaler 中的抽象方法部署这 3 个函数,需要将其传递给 setBooleanArg。

现在 setBooleanArg 没用了。如果 ArgumentMarshaler 中有一个 set 函数,我们就可以直接调用它。是时候打造这个函数了!第一步,在 ArgumentMarshaler 中添加抽象方法。

```
  private abstract class ArgumentMarshaler {
    public abstract void set(Iterator<String> currentArgument)
                        throws ArgsException;
    public abstract void set(String s) throws ArgsException;
    public abstract Object get();
  }
```

当然,这会影响到所有派生类。所以,要逐个实现新方法。

```
private class BooleanArgumentMarshaler extends ArgumentMarshaler {
  private boolean booleanValue = false;

  public void set(Iterator<String> currentArgument) throws ArgsException {
    booleanValue = true;
  }

  public void set(String s) {
    booleanValue = true;
  }

  public Object get() {
    return booleanValue;
  }
}

private class StringArgumentMarshaler extends ArgumentMarshaler {
  private String stringValue = "";

  public void set(Iterator<String> currentArgument) throws ArgsException {
  }
```

```java
  public void set(String s) {
    stringValue = s;
  }

  public Object get() {
    return stringValue;
  }
}

private class IntegerArgumentMarshaler extends ArgumentMarshaler {
  private int intValue = 0;

  public void set(Iterator<String> currentArgument) throws ArgsException {
  }

  public void set(String s) throws ArgsException {
    try {
      intValue = Integer.parseInt(s);
    } catch (NumberFormatException e) {
      throw new ArgsException();
    }
  }

  public Object get() {
    return intValue;
  }
}
```

现在可以删除 setBooleanArg 了!

```java
private boolean setArgument(char argChar) throws ArgsException {
  ArgumentMarshaler m = marshalers.get(argChar);
  if (m == null)
    return false;
  try {
    if (m instanceof BooleanArgumentMarshaler)
      m.set(currentArgument);
    else if (m instanceof StringArgumentMarshaler)
      setStringArg(m);
    else if (m instanceof IntegerArgumentMarshaler)
      setIntArg(m);
  } catch (ArgsException e) {
    valid = false;
    errorArgumentId = argChar;
    throw e;
  }
  return true;
}
```

测试全都通过，而且 set 函数也部署到 BooleanArgumentMarshaler 里面了！现在就能对 String 类型和 Integer 类型参数的处理做同样的修改。

14.3 字符串类型参数

```java
  private boolean setArgument(char argChar) throws ArgsException {
    ArgumentMarshaler m = marshalers.get(argChar);
    if (m == null)
      return false;
    try {
      if (m instanceof BooleanArgumentMarshaler)
        m.set(currentArgument);
      else if (m instanceof StringArgumentMarshaler)
        m.set(currentArgument);
      else if (m instanceof IntegerArgumentMarshaler)
        m.set(currentArgument);

    } catch (ArgsException e) {
      valid = false;
      errorArgumentId = argChar;
      throw e;
    }
    return true;
  }
---
  private class StringArgumentMarshaler extends ArgumentMarshaler {
    private String stringValue = "";

    public void set(Iterator<String> currentArgument) throws ArgsException {
      try {
        stringValue = currentArgument.next();
      } catch (NoSuchElementException e) {
        errorCode = ErrorCode.MISSING_STRING;
        throw new ArgsException();
      }
    }

    public void set(String s) {
    }

    public Object get() {
      return stringValue;
    }
  }

  private class IntegerArgumentMarshaler extends ArgumentMarshaler {
    private int intValue = 0;

  public void set(Iterator<String> currentArgument) throws ArgsException {
    String parameter = null;
    try {
      parameter = currentArgument.next();
      set(parameter);
    } catch (NoSuchElementException e) {
      errorCode = ErrorCode.MISSING_INTEGER;
      throw new ArgsException();
    } catch (ArgsException e) {
      errorParameter = parameter;
```

```
      errorCode = ErrorCode.INVALID_INTEGER;
      throw e;
    }
  }

  public void set(String s) throws ArgsException {
    try {
      intValue = Integer.parseInt(s);
    } catch (NumberFormatException e) {
      throw new  ArgsException();
    }
  }

  public Object get() {
    return intValue;
  }
}
```

最后一击：可以移除类型转换了！看招！

```
private boolean setArgument(char argChar) throws ArgsException {
  ArgumentMarshaler m = marshalers.get(argChar);
  if (m == null)
    return false;
  try {
    m.set(currentArgument);
    return true;
  } catch (ArgsException e) {
    valid = false;
    errorArgumentId = argChar;
    throw e;
  }
}
```

现在可以删除 IntegerArgumentMarshaler 中那些过时的函数，做一下清理了。

```
private class IntegerArgumentMarshaler extends ArgumentMarshaler {
  private int intValue = 0;

  public void set(Iterator<String> currentArgument) throws ArgsException {
    String parameter = null;
    try {
      parameter = currentArgument.next();
      intValue = Integer.parseInt(parameter);
    } catch (NoSuchElementException e) {
      errorCode = ErrorCode.MISSING_INTEGER;
      throw new ArgsException();
    } catch (NumberFormatException e) {
      errorParameter = parameter;
      errorCode = ErrorCode.INVALID_INTEGER;
      throw new ArgsException();
    }
  }

  public Object get() {
    return  intValue;
```

还可以把 ArgumentMarshaler 修改为接口：

```
private interface ArgumentMarshaler {
  void set(Iterator<String> currentArgument) throws ArgsException;
  Object get();
}
```

现在来看看往这个结构中添加新的参数类型有多容易。只需要做少量修改，而且修改是被隔离的。首先，增加一个新的测试用例，检测 double 参数是否正常工作。

```
public void testSimpleDoublePresent() throws Exception {
  Args args = new Args("x##", new String[] {"-x","42.3"});
  assertTrue(args.isValid());
  assertEquals(1, args.cardinality());
  assertTrue(args.has('x'));
  assertEquals(42.3, args.getDouble('x'), .001);
}
```

然后清理范式解析代码，为 double 参数类型添加##监测。

```
private void parseSchemaElement(String element) throws ParseException {
  char elementId = element.charAt(0);
  String elementTail = element.substring(1);
  validateSchemaElementId(elementId);
  if (elementTail.length() == 0)
    marshalers.put(elementId, new BooleanArgumentMarshaler());
  else if (elementTail.equals("*"))
    marshalers.put(elementId, new  StringArgumentMarshaler());
  else if (elementTail.equals("#"))
    marshalers.put(elementId, new  IntegerArgumentMarshaler());
  else if (elementTail.equals("##"))
    marshalers.put(elementId, new DoubleArgumentMarshaler());
  else
    throw new ParseException(String.format(
      "Argument: %c has invalid format: %s.", elementId, elementTail), 0);
}
```

下一步，编写 DoubleArgumentMarshaler 类：

```
private class DoubleArgumentMarshaler implements ArgumentMarshaler {
  private double doubleValue = 0;

  public void set(Iterator<String> currentArgument) throws ArgsException {
    String parameter = null;
    try {
      parameter = currentArgument.next();
      doubleValue = Double.parseDouble(parameter);
    } catch (NoSuchElementException e) {
      errorCode = ErrorCode.MISSING_DOUBLE;
      throw new ArgsException();
    } catch (NumberFormatException e) {
```

```
      errorParameter = parameter;
      errorCode = ErrorCode.INVALID_DOUBLE;
      throw new ArgsException();
    }
  }

  public Object get() {
    return doubleValue;
  }
}
```

然后就得添加一个新的 `ErrorCode`：

```
private enum ErrorCode {
  OK, MISSING_STRING, MISSING_INTEGER, INVALID_INTEGER, UNEXPECTED_ARGUMENT,
  MISSING_DOUBLE, INVALID_DOUBLE}
```

还需要一个 getDouble 函数：

```
public double getDouble(char arg) {
  Args.ArgumentMarshaler am = marshalers.get(arg);
  try {
    return am ==  null ? 0 : (Double) am.get();
  } catch (Exception e) {
    return 0.0;
  }
}
```

全部测试都通过了！完全无痛。再来确保全部错误处理代码正确工作。下一个测试用例用来检测在向##参数传递一个不可解析的字符串时是否会返回错误。

```
public void testInvalidDouble() throws Exception {
  Args args = new  Args("x##", new String[] {"-x","Forty two"});
  assertFalse(args.isValid());
  assertEquals(0, args.cardinality());
  assertFalse(args.has('x'));
  assertEquals(0, args.getInt('x'));
  assertEquals("Argument -x expects a double but was 'Forty two'.",
               args.errorMessage());

}
---
  public String errorMessage() throws Exception {
    switch (errorCode) {
      case OK:
        throw new Exception("TILT: Should not get here.");
      case UNEXPECTED_ARGUMENT:
        return unexpectedArgumentMessage();
      case MISSING_STRING:
        return String.format("Could not find string parameter for -%c.",
                             errorArgumentId);
      case INVALID_INTEGER:
        return String.format("Argument -%c expects an integer but was '%s'.",
                             errorArgumentId, errorParameter);
      case MISSING_INTEGER:
```

```
            return String.format("Could not find integer parameter for -%c.",
                                 errorArgumentId);
        case INVALID_DOUBLE:
            return String.format("Argument -%c expects a double but was '%s'.",
                                 errorArgumentId, errorParameter);
        case MISSING_DOUBLE:
            return String.format("Could not find double parameter for -%c.",
                                 errorArgumentId);
    }
    return "";
}
```

测试通过。下一个测试确保我们正确检测到遗漏的 `double` 参数：

```
public void testMissingDouble() throws Exception {
  Args args = new Args("x##", new String[]{"-x"});
  assertFalse(args.isValid());
  assertEquals(0, args.cardinality());
  assertFalse(args.has('x'));
  assertEquals(0.0, args.getDouble('x'), 0.01);
  assertEquals("Could not find double parameter for -x.",
               args.errorMessage());
}
```

测试如愿通过。我们只是为了保持一切完整而编写这个测试。

异常代码很丑陋，不该在 `Args` 类中存在，我们也抛出 `ParseException`，但那并不真的属于我们自己。因此我们把所有异常都塞到 `ArgsException` 类中，并将其移到它自己的模块里面。

```
public class ArgsException extends Exception {
  private char errorArgumentId = '\0';
  private String errorParameter = "TILT";
  private ErrorCode errorCode = ErrorCode.OK;

  public ArgsException() {}

  public ArgsException(String message) {super(message);}

  public enum ErrorCode {
    OK, MISSING_STRING, MISSING_INTEGER, INVALID_INTEGER, UNEXPECTED_ARGUMENT,
    MISSING_DOUBLE, INVALID_DOUBLE}
}
---
public class Args {
  ...
  private char errorArgumentId = '\0';
  private String errorParameter = "TILT";
  private ArgsException.ErrorCode errorCode = ArgsException.ErrorCode.OK;
  private List<String> argsList;

  public Args(String schema, String[] args) throws ArgsException {
    this.schema = schema;
```

```java
    argsList = Arrays.asList(args);
    valid = parse();
  }

  private boolean parse() throws ArgsException {
    if (schema.length() == 0 && argsList.size() == 0)
      return true;
    parseSchema();
    try {
      parseArguments();
    } catch (ArgsException e) {
    }
    return valid;
  }

  private boolean parseSchema() throws ArgsException {
    ...
  }

  private void parseSchemaElement(String element) throws ArgsException {
    ...
    else
      throw new ArgsException(
        String.format("Argument: %c has invalid format: %s.",
                      elementId,elementTail));
  }

  private void validateSchemaElementId(char elementId) throws ArgsException {
    if (!Character.isLetter(elementId)) {
      throw new ArgsException(
        "Bad character:" + elementId + "in Args format: " + schema);
    }
  }

  ...

  private void parseElement(char argChar) throws ArgsException {
    if (setArgument(argChar))
      argsFound.add(argChar);
    else {
      unexpectedArguments.add(argChar);
      errorCode = ArgsException.ErrorCode.UNEXPECTED_ARGUMENT;
      valid = false;
    }
  }

  ...

  private class StringArgumentMarshaler implements ArgumentMarshaler {
    private String stringValue = "";

    public void set(Iterator<String> currentArgument) throws ArgsException {
      try {
```

14.3 字符串类型参数

```java
      stringValue = currentArgument.next();
    } catch (NoSuchElementException e) {
      errorCode = ArgsException.ErrorCode.MISSING_STRING;
      throw new ArgsException();
    }
  }

  public Object get() {
    return stringValue;
  }
}

private class IntegerArgumentMarshaler implements ArgumentMarshaler {
  private int intValue = 0;

  public void set(Iterator<String> currentArgument) throws ArgsException {
    String parameter = null;
    try {
      parameter = currentArgument.next();
      intValue = Integer.parseInt(parameter);
    } catch (NoSuchElementException e) {
      errorCode = ArgsException.ErrorCode.MISSING_INTEGER;
      throw new ArgsException();
    } catch (NumberFormatException e) {
      errorParameter = parameter;
      errorCode = ArgsException.ErrorCode.INVALID_INTEGER;
      throw new ArgsException();
    }
  }

  public Object get() {
    return intValue;
  }
}

private class DoubleArgumentMarshaler implements ArgumentMarshaler {
  private double doubleValue = 0;

  public void set(Iterator<String> currentArgument) throws ArgsException {
    String parameter = null;
    try {
      parameter = currentArgument.next();
      doubleValue = Double.parseDouble(parameter);
    } catch (NoSuchElementException e) {
      errorCode = ArgsException.ErrorCode.MISSING_DOUBLE;
      throw new ArgsException();
    } catch (NumberFormatException e) {
      errorParameter = parameter;
      errorCode = ArgsException.ErrorCode.INVALID_DOUBLE;
      throw new ArgsException();
    }
  }
```

```java
    public Object get() {
      return doubleValue;
    }
  }
}
```

很好。现在，Args 抛出的唯一一个异常是 `ArgsException`。把 `ArgsException` 移到它自己的模块中，这意味着我们能把大量杂七杂八的错误支持代码从 Args 模块转移到这个模块。

现在我们完全把异常和错误代码从 Args 模块中隔离出来了。（如代码清单 14-13～代码清单 14-16 所示。）为达到这一目标，大概做了 30 次小修改，每次修改都保持测试通过。

代码清单 14-13　ArgsTest.java

```java
package com.objectmentor.utilities.args;

import junit.framework.TestCase;

public class ArgsTest extends TestCase {
  public void testCreateWithNoSchemaOrArguments() throws Exception {
    Args args = new Args("", new String[0]);
    assertEquals(0, args.cardinality());
  }

  public void testWithNoSchemaButWithOneArgument() throws Exception {
    try {
      new Args("", new String[]{"-x"});
      fail();
    } catch (ArgsException e) {
      assertEquals(ArgsException.ErrorCode.UNEXPECTED_ARGUMENT,
                   e.getErrorCode());
      assertEquals('x', e.getErrorArgumentId());
    }
  }

  public void testWithNoSchemaButWithMultipleArguments() throws Exception {
    try {
      new Args("", new String[]{"-x", "-y"});
      fail();
    } catch (ArgsException e) {
      assertEquals(ArgsException.ErrorCode.UNEXPECTED_ARGUMENT,
                   e.getErrorCode());
      assertEquals('x', e.getErrorArgumentId());
    }
  }

  public void testNonLetterSchema() throws Exception {
    try {
      new Args("*", new String[]{});
      fail("Args constructor should have thrown exception");
    } catch (ArgsException e) {
      assertEquals(ArgsException.ErrorCode.INVALID_ARGUMENT_NAME,
```

```java
          e.getErrorCode());
      assertEquals('*', e.getErrorArgumentId());
    }
  }

  public void testInvalidArgumentFormat() throws Exception {
    try {
      new Args("f~", new  String[]{});
      fail("Args constructor should have throws exception");
    } catch (ArgsException e) {
      assertEquals(ArgsException.ErrorCode.INVALID_FORMAT, e.getErrorCode());
      assertEquals('f', e.getErrorArgumentId());
    }
  }

  public void testSimpleBooleanPresent() throws Exception {
    Args args = new  Args("x", new String[]{"-x"});
    assertEquals(1, args.cardinality());
    assertEquals(true, args.getBoolean('x'));
  }

  public void testSimpleStringPresent() throws Exception {
    Args args = new Args("x*", new String[]{"-x", "param"});
    assertEquals(1, args.cardinality());
    assertTrue(args.has('x'));
    assertEquals("param", args.getString('x'));
  }

  public void testMissingStringArgument() throws Exception {
    try {
      new Args("x*", new String[]{"-x"});
      fail();
    } catch (ArgsException e) {
      assertEquals(ArgsException.ErrorCode.MISSING_STRING, e.getErrorCode());
      assertEquals('x', e.getErrorArgumentId());
    }
  }

  public void testSpacesInFormat() throws Exception {
    Args args = new  Args("x, y", new String[]{"-xy"});
    assertEquals(2, args.cardinality());
    assertTrue(args.has('x'));
    assertTrue(args.has('y'));
  }

  public void testSimpleIntPresent() throws Exception {
    Args args = new Args("x#", new String[]{"-x", "42"});
    assertEquals(1, args.cardinality());
    assertTrue(args.has('x'));
    assertEquals(42, args.getInt('x'));
  }

  public void testInvalidInteger() throws Exception {
    try {
      new Args("x#", new String[]{"-x","Forty two"});
```

```java
      fail();
    } catch (ArgsException e) {
      assertEquals(ArgsException.ErrorCode.INVALID_INTEGER, e.getErrorCode());
      assertEquals('x', e.getErrorArgumentId());
      assertEquals("Forty two", e.getErrorParameter());
    }
  }

  public void testMissingInteger() throws Exception {
    try {
      new Args("x#", new String[]{"-x"});
      fail();
    } catch (ArgsException e) {
      assertEquals(ArgsException.ErrorCode.MISSING_INTEGER, e.getErrorCode());
      assertEquals('x', e.getErrorArgumentId());
    }
  }

  public void testSimpleDoublePresent() throws Exception {
    Args args = new Args("x##", new String[]{"-x", "42.3"});
    assertEquals(1, args.cardinality());
    assertTrue(args.has('x'));
    assertEquals(42.3, args.getDouble('x'), .001);
  }

  public void testInvalidDouble() throws Exception {
    try {
      new Args("x##", new  String[]{"-x", "Forty two"});
      fail();
    } catch (ArgsException e) {
      assertEquals(ArgsException.ErrorCode.INVALID_DOUBLE, e.getErrorCode());
      assertEquals('x', e.getErrorArgumentId());
      assertEquals("Forty two", e.getErrorParameter());
    }
  }

  public void testMissingDouble() throws Exception {
    try {
      new Args("x##", new String[]{"-x"});
      fail();
    } catch (ArgsException e) {
      assertEquals(ArgsException.ErrorCode.MISSING_DOUBLE, e.getErrorCode());
      assertEquals('x', e.getErrorArgumentId());
    }
  }
}
```

代码清单 14-14　ArgsExceptionTest.java

```java
public class ArgsExceptionTest extends TestCase {
  public void testUnexpectedMessage() throws Exception {
    ArgsException e =
      new ArgsException(ArgsException.ErrorCode.UNEXPECTED_ARGUMENT,
                        'x', null);
```

```java
    assertEquals("Argument -x unexpected.", e.errorMessage());
}

public void testMissingStringMessage() throws Exception {
    ArgsException e = new ArgsException(ArgsException.ErrorCode.MISSING_STRING,
                                        'x', null);
    assertEquals("Could not find string parameter for -x.", e.errorMessage());
}

public void testInvalidIntegerMessage() throws Exception {
    ArgsException e =
      new ArgsException(ArgsException.ErrorCode.INVALID_INTEGER, 'x', "Forty two");
    assertEquals("Argument -x expects an integer but was 'Forty two'.",
              e.errorMessage());
}

public  void testMissingIntegerMessage() throws Exception {
    ArgsException e =
      new ArgsException(ArgsException.ErrorCode.MISSING_INTEGER, 'x', null);
    assertEquals("Could not find integer parameter for -x.", e.errorMessage());
}

public void testInvalidDoubleMessage() throws Exception {
    ArgsException e = new ArgsException(ArgsException.ErrorCode.INVALID_DOUBLE,
                                        'x', "Forty two");
    assertEquals("Argument -x expects a double but was 'Forty two'.",
              e.errorMessage());
}

public void testMissingDoubleMessage() throws Exception {
    ArgsException e = new ArgsException(ArgsException.ErrorCode.MISSING_DOUBLE,
                                        'x', null);
    assertEquals("Could not find double parameter for -x.", e.errorMessage());
  }
}
```

代码清单 14-15 ArgsException.java

```java
public class ArgsException extends  Exception {
  private char errorArgumentId = '\0';
  private String errorParameter = "TILT";
  private ErrorCode errorCode = ErrorCode.OK;

  public ArgsException() {}

  public ArgsException(String  message) {super(message);}

  public ArgsException(ErrorCode errorCode) {
    this.errorCode = errorCode;
  }

  public ArgsException(ErrorCode errorCode, String errorParameter) {
    this.errorCode = errorCode;
    this.errorParameter = errorParameter;
  }
```

```
public ArgsException(ErrorCode errorCode, char errorArgumentId,
                     String errorParameter) {
  this.errorCode = errorCode;
  this.errorParameter = errorParameter;
  this.errorArgumentId = errorArgumentId;
}

public char getErrorArgumentId() {
  return errorArgumentId;
}

public void setErrorArgumentId(char errorArgumentId) {
  this.errorArgumentId = errorArgumentId;
}

public String getErrorParameter() {
  return errorParameter;
}

public void setErrorParameter(String errorParameter) {
  this.errorParameter = errorParameter;
}

public ErrorCode getErrorCode() {
  return errorCode;
}

public void setErrorCode(ErrorCode errorCode) {
  this.errorCode = errorCode;
}

public String errorMessage() throws Exception {
  switch (errorCode) {
    case OK:
      throw new Exception("TILT: Should not get here.");
    case  UNEXPECTED_ARGUMENT:
      return String.format("Argument -%c unexpected.", errorArgumentId);
    case MISSING_STRING:
      return String.format("Could not find string parameter for -%c.",
                           errorArgumentId);
    case INVALID_INTEGER:
      return  String.format("Argument -%c expects an integer but was '%s'.",
                            errorArgumentId, errorParameter);
    case MISSING_INTEGER:
      return  String.format("Could not find integer parameter for -%c.",
                            errorArgumentId);
    case INVALID_DOUBLE:
      return  String.format("Argument -%c expects a double but was '%s'.",
                            errorArgumentId, errorParameter);


    case MISSING_DOUBLE:
      return  String.format("Could not find double parameter for -%c.",
```

```
                        errorArgumentId);
    }
    return "";
  }

  public enum ErrorCode {
    OK, INVALID_FORMAT, UNEXPECTED_ARGUMENT, INVALID_ARGUMENT_NAME,
    MISSING_STRING,
    MISSING_INTEGER, INVALID_INTEGER,
    MISSING_DOUBLE, INVALID_DOUBLE}
}
```

代码清单 14-16　Args.java

```java
public class Args {
  private String schema;
  private Map<Character, ArgumentMarshaler> marshalers =
    new HashMap<Character, ArgumentMarshaler>();
  private Set<Character> argsFound = new HashSet<Character>();
  private Iterator<String> currentArgument;
  private List<String> argsList;

  public Args(String schema, String[] args) throws ArgsException {
    this.schema = schema;
    argsList = Arrays.asList(args);
    parse();
  }

  private void parse() throws ArgsException {
    parseSchema();
    parseArguments();
  }

  private boolean parseSchema() throws ArgsException {
    for (String element : schema.split(",")) {
      if (element.length() > 0) {
        parseSchemaElement(element.trim());
      }
    }
    return true;
  }

  private void parseSchemaElement(String element) throws ArgsException {
    char elementId = element.charAt(0);
    String elementTail = element.substring(1);
    validateSchemaElementId(elementId);
    if (elementTail.length() == 0)
      marshalers.put(elementId, new BooleanArgumentMarshaler());
    else if (elementTail.equals("*"))
      marshalers.put(elementId, new StringArgumentMarshaler());
    else if (elementTail.equals("#"))
      marshalers.put(elementId, new IntegerArgumentMarshaler());
    else if (elementTail.equals("##"))
      marshalers.put(elementId, new DoubleArgumentMarshaler());
```

```
      else
        throw new ArgsException(ArgsException.ErrorCode.INVALID_FORMAT,
                                elementId, elementTail);
    }

    private void validateSchemaElementId(char elementId) throws ArgsException {
      if (!Character.isLetter(elementId)) {
        throw  new ArgsException(ArgsException.ErrorCode.INVALID_ARGUMENT_NAME,
                                 elementId, null);
      }
    }

    private void parseArguments() throws ArgsException {
      for (currentArgument = argsList.iterator();  currentArgument.hasNext();) {
        String arg = currentArgument.next();
        parseArgument(arg);
      }
    }

    private void parseArgument(String arg) throws ArgsException {
      if (arg.startsWith("-"))
        parseElements(arg);
    }

    private void parseElements(String arg) throws ArgsException {
      for (int i = 1; i < arg.length(); i++)
        parseElement(arg.charAt(i));
    }

    private void parseElement(char argChar) throws ArgsException {
      if (setArgument(argChar))
        argsFound.add(argChar);
      else {
        throw new ArgsException(ArgsException.ErrorCode.UNEXPECTED_ARGUMENT,
                                argChar, null);
      }
    }

    private boolean setArgument(char argChar) throws ArgsException {
      ArgumentMarshaler m = marshalers.get(argChar);
      if (m == null)
        return false;
      try {
        m.set(currentArgument);
        return true;
      } catch (ArgsException e) {
        e.setErrorArgumentId(argChar);
        throw e;
      }
    }


    public int cardinality() {
      return argsFound.size();
    }
```

```java
  public String usage() {
    if (schema.length() > 0)
      return "-[" + schema + "]";
    else
      return "";
  }

  public boolean getBoolean(char arg) {
    ArgumentMarshaler am = marshalers.get(arg);
    boolean b = false;
    try {
      b = am != null && (Boolean) am.get();
    } catch (ClassCastException e) {
      b = false;
    }
    return b;
  }

  public String getString(char arg) {
    ArgumentMarshaler am = marshalers.get(arg);
    try {
      return am == null ? "" : (String) am.get();
    } catch (ClassCastException e) {
      return "";
    }
  }

  public int getInt(char arg) {
    ArgumentMarshaler am = marshalers.get(arg);
    try {
      return am == null ? 0 : (Integer) am.get();
    } catch (Exception e) {
      return 0;
    }
  }


  public double getDouble(char arg) {
    ArgumentMarshaler am = marshalers.get(arg);
    try {
      return am == null ? 0 : (Double) am.get();
    } catch (Exception e) {
      return 0.0;
    }
  }

  public boolean has(char arg) {
    return argsFound.contains(arg);
  }
}
```

对 Args 类所做的最主要的修改是在监测部分。从 Args 里面取出了大量代码，放到 ArgsException 中。很好。我们还把全部 ArgumentMarshaler 转移到了它们自己的文件中。更好！

优秀的软件设计，大都关乎分隔——创建合适的空间放置不同种类的代码。对关注面的分隔让代码更易于理解和维护。

特别有意思的是 `ArgsException` 中的 `errorMessage` 方法。显然，把错误信息格式化操作放在 `Args` 里面，违反了 SRP。`Args` 应该只处理参数，而不该去管错误信息的格式。然而，把错误信息格式化代码放到 `ArgsException` 中是否有道理呢？

实话说，这是一种折中的做法。不打算用 `ArgsException` 提供的错误信息的用户会想自己写错误信息。但如果有备好的错误信息，其方便之处也并非鲜见。

显然，现在我们完成的工作已经非常接近本章开始处所展示的最终解决方案了。最后的工作留给你来练习完成。

14.4 小结

代码能工作还不够。能工作的代码经常会严重崩溃。满足于仅仅让代码能工作的程序员不够专业，他们害怕没时间改进代码的结构和设计，我不敢苟同。没什么能比糟糕的代码给开发项目带来更深远和长期的损害了。进度可以调整，需求可以重新定义，团队可以动态修正，但糟糕的代码只会一直腐败发酵，无情地拖着团队的后腿。我无数次看到开发团队蹒跚前行，就是因为他们匆匆搞出一片代码沼泽，从此之后命运再也不受自己控制。

当然，糟糕的代码可以清理，不过成本高昂。随着代码腐败下去，模块之间互相渗透，出现大量隐藏纠缠的依赖关系，找到和破除陈旧的依赖关系又费时间又费劲。然而，保持代码整洁却相对容易。早晨在模块中制造出一堆混乱，下午就能轻易清理掉。更好的情况是，5 分钟之前制造出混乱，马上就能很容易地清理掉。

所以，解决之道就是保持代码持续整洁和简单，永不让"腐坏"有机会开始。

第 15 章

JUnit 内幕

　　JUnit 是最有名的 Java 框架之一。就像别的框架一样，它概念简单，定义精确，实现优雅。但它的代码是怎样的呢？本章将研判来自 JUnit 框架的一个代码例子。

15.1 JUnit 框架

JUnit 有很多位作者，但它始于 Kent Beck 和 Eric Gamma 一次去亚特兰大的飞行旅程。Kent 想学 Java，而 Eric 则打算学习 Kent 的 Smalltalk 测试框架。"对于两个身处狭窄空间的极客，还有什么会比拿出笔记本电脑开始编码来得更自然呢？[①]"经过 3 小时高海拔工作，他们写出了 JUnit 的基础代码。

我们要查看的模块，是用来帮忙鉴别字符串比较错误的一段聪明代码，该模块被命名为 ComparisonCompactor。对于两个不同的字符串，如 ABCDE 和 ABXDE，该模块将用形如 <...B[X]D...>的字符串来曝露两者的不同之处。

我们可以做进一步解释，但测试用例会更有说服力。看看代码清单 15-1，我们将深入了解到该模块满足的需求，边看代码，边研究该测试的结构，它们能变得更简洁或更明确吗？

代码清单 15-1　ComparisonCompactorTest.java

```java
package junit.tests.framework;

import junit.framework.ComparisonCompactor;
import junit.framework.TestCase;

public class ComparisonCompactorTest extends TestCase {

  public void testMessage() {
    String failure= new ComparisonCompactor(0, "b", "c").compact("a");
    assertTrue("a expected:<[b]> but was:<[c]>".equals(failure));
  }

  public void testStartSame() {
    String failure= new ComparisonCompactor(1, "ba", "bc").compact(null);
    assertEquals("expected:<b[a]> but was:<b[c]>", failure);
  }

  public void testEndSame() {
    String  failure= new ComparisonCompactor(1, "ab", "cb").compact(null);
    assertEquals("expected:<[a]b> but was:<[c]b>", failure);
  }

  public void testSame() {
    String failure= new ComparisonCompactor(1, "ab", "ab").compact(null);
    assertEquals("expected:<ab> but was:<ab>", failure);
  }

  public void testNoContextStartAndEndSame() {
    String failure= new ComparisonCompactor(0, "abc", "adc").compact(null);
```

[①] 摘自 *JUnit Pocket Guide*，Kent Beck，O'Reilly，2004，P.43。

```
    assertEquals("expected:<...[b]...> but was:<...[d]...>", failure);
}

public void testStartAndEndContext() {
    String failure= new ComparisonCompactor(1, "abc", "adc").compact(null);
    assertEquals("expected:<a[b]c> but was:<a[d]c>", failure);
}

public void testStartAndEndContextWithEllipses() {
    String failure=
        new ComparisonCompactor(1, "abcde", "abfde").compact(null);
    assertEquals("expected:<...b[c]d...> but was:<...b[f]d...>", failure);
}

public void testComparisonErrorStartSameComplete() {
    String failure= new ComparisonCompactor(2, "ab", "abc").compact(null);
    assertEquals("expected:<ab[]> but was:<ab[c]>", failure);
}

public void testComparisonErrorEndSameComplete() {
    String failure= new ComparisonCompactor(0, "bc", "abc").compact(null);
    assertEquals("expected:<[]...> but was:<[a]...>", failure);
}

public void testComparisonErrorEndSameCompleteContext() {
    String failure= new ComparisonCompactor(2, "bc", "abc").compact(null);
    assertEquals("expected:<[]bc> but was:<[a]bc>", failure);
}

public void testComparisonErrorOverlapingMatches() {
    String failure= new ComparisonCompactor(0, "abc", "abbc").compact(null);
    assertEquals("expected:<...[]...> but was:<...[b]...>", failure);
}

public void testComparisonErrorOverlapingMatchesContext() {
    String failure= new ComparisonCompactor(2, "abc", "abbc").compact(null);
    assertEquals("expected:<ab[]c> but was:<ab[b]c>", failure);
}

public void testComparisonErrorOverlapingMatches2() {
    String failure= new ComparisonCompactor(0, "abcdde", "abcde").compact(null);
    assertEquals("expected:<...[d]...> but was:<...[]...>", failure);
}

public void testComparisonErrorOverlapingMatches2Context() {
    String failure=
        new ComparisonCompactor(2, "abcdde", "abcde").compact(null);
    assertEquals("expected:<...cd[d]e> but was:<...cd[]e>", failure);
}

public void testComparisonErrorWithActualNull() {
    String failure= new ComparisonCompactor(0, "a", null).compact(null);
    assertEquals("expected:<a> but was:<null>", failure);
```

```java
    }

    public void testComparisonErrorWithActualNullContext() {
        String failure= new ComparisonCompactor(2, "a", null).compact(null);
        assertEquals("expected:<a> but was:<null>", failure);
    }

    public void testComparisonErrorWithExpectedNull() {
        String failure= new ComparisonCompactor(0, null, "a").compact(null);
        assertEquals("expected:<null> but was:<a>", failure);
    }

    public void testComparisonErrorWithExpectedNullContext() {
        String failure= new ComparisonCompactor(2, null, "a").compact(null);
        assertEquals("expected:<null> but was:<a>", failure);
    }

    public void testBug609972() {
        String failure= new ComparisonCompactor(10, "S&P500", "0").compact(null);
        assertEquals("expected:<[S&P50]0> but was:<[]0>", failure);
    }
}
```

我对用到这些测试的 `ComparisonCompactor` 进行了代码覆盖分析，代码被 100% 覆盖了，每行代码、每个 `if` 语句和 `for` 循环都被测试执行了。于是我对代码的工作能力有了极强的信心，也对代码作者们的技艺产生了极高的尊敬之情。

`ComparisonCompactor` 的代码如代码清单 15-2 所示。

代码清单 15-2 ComparisonCompactor.java（原始版本）

```java
package junit.framework;

public class ComparisonCompactor {

    private static final String ELLIPSIS = "...";
    private static final String DELTA_END = "]";
    private static final String DELTA_START = "[";

    private int fContextLength;
    private String fExpected;
    private String fActual;
    private int fPrefix;
    private int fSuffix;

    public ComparisonCompactor(int contextLength,
                               String expected,
                               String actual) {
        fContextLength = contextLength;
        fExpected = expected;
        fActual = actual;
    }

    public String compact(String message) {
        if (fExpected == null || fActual == null || areStringsEqual())
```

```java
      return Assert.format(message, fExpected, fActual);
    findCommonPrefix();
    findCommonSuffix();
    String expected = compactString(fExpected);
    String actual = compactString(fActual);
    return Assert.format(message, expected, actual);
  }

  private String compactString(String source) {
    String result = DELTA_START +
                    source.substring(fPrefix, source.length() -
                         fSuffix + 1) + DELTA_END;
    if (fPrefix > 0)
      result = computeCommonPrefix() + result;
    if (fSuffix > 0)
      result = result + computeCommonSuffix();
    return result;
  }

  private void findCommonPrefix() {
    fPrefix = 0;
    int end = Math.min(fExpected.length(), fActual.length());
    for (; fPrefix < end; fPrefix++) {
      if (fExpected.charAt(fPrefix) != fActual.charAt(fPrefix))
        break;
    }
  }

  private void findCommonSuffix() {
    int expectedSuffix = fExpected.length() - 1;
    int actualSuffix = fActual.length() - 1;
    for (;
        actualSuffix >= fPrefix && expectedSuffix >= fPrefix;
        actualSuffix--, expectedSuffix--) {
      if (fExpected.charAt(expectedSuffix) != fActual.charAt(actualSuffix))
        break;
    }
    fSuffix = fExpected.length() - expectedSuffix;
  }

  private String computeCommonPrefix() {
    return (fPrefix > fContextLength ? ELLIPSIS : "") +
            fExpected.substring(Math.max(0, fPrefix - fContextLength),
                                fPrefix);
  }

  private String computeCommonSuffix() {
    int end = Math.min(fExpected.length() - fSuffix + 1 + fContextLength,
                       fExpected.length());
    return fExpected.substring(fExpected.length() - fSuffix + 1, end) +
          (fExpected.length() - fSuffix + 1 < fExpected.length() -
           fContextLength ? ELLIPSIS : "");
  }
```

```java
    private boolean areStringsEqual() {
      return fExpected.equals(fActual);
    }
  }
```

你可能会对这个模块有所抱怨,例如,里面有些长表达式,有些奇怪的+1 操作,如此等等。不过,总的来说,这个模块很不错,毕竟它原本可能被写成如代码清单 15-3 中的样子。

代码清单 15-3　ComparisonCompactor.java(背离版本)

```java
package junit.framework;

public class ComparisonCompactor {
  private int ctxt;
  private String s1;
  private String s2;
  private int pfx;
  private int sfx;

  public ComparisonCompactor(int ctxt, String s1, String s2) {
    this.ctxt = ctxt;
    this.s1 = s1;
    this.s2 = s2;
  }

  public String compact(String msg) {
    if (s1 == null || s2 == null || s1.equals(s2))
      return Assert.format(msg, s1, s2);

    pfx = 0;
    for (; pfx < Math.min(s1.length(), s2.length()); pfx++) {
      if (s1.charAt(pfx) != s2.charAt(pfx))
        break;
    }
    int sfx1 = s1.length() - 1;
    int sfx2 = s2.length() - 1;
    for (; sfx2 >= pfx && sfx1 >= pfx; sfx2--, sfx1--) {
      if (s1.charAt(sfx1) != s2.charAt(sfx2))
        break;
    }
    sfx = s1.length() - sfx1;
    String cmp1 = compactString(s1);
    String cmp2 = compactString(s2);
    return Assert.format(msg, cmp1, cmp2);
  }

  private String compactString(String s) {
    String result =
      "[" + s.substring(pfx, s.length() - sfx + 1) + "]";
    if (pfx > 0)
      result = (pfx > ctxt ? "..." : "") +
        s1.substring(Math.max(0, pfx - ctxt), pfx) + result;
    if (sfx > 0) {
```

```
      int end = Math.min(s1.length() - sfx + 1 + ctxt, s1.length());
      result = result + (s1.substring(s1.length() - sfx + 1, end) +
        (s1.length() - sfx + 1 < s1.length() - ctxt ? "..." : ""));
    }
    return result;
  }
}
```

即便作者们把这个模块写得已经很棒，但童子军军规[1]却告诉我们，离时要比来时整洁。所以，我们怎样才能改进代码清单 15-2 中的原始代码呢？

我们首先看到的是成员变量 f 前缀[N6]。在现今的运行环境中，这类范围性编码纯属多余。所以，先删除所有的 f 前缀。

```
private int contextLength;
private String expected;
private String actual;
private int prefix;
private int suffix;
```

下一步，在 compact 函数开始处，有一个未封装的条件判断[G28]。

```
public String compact(String message) {
  if (expected == null || actual == null || areStringsEqual())
    return Assert.format(message, expected, actual);

  findCommonPrefix();
  findCommonSuffix();
  String expected = compactString(this.expected);
  String actual = compactString(this.actual);
  return Assert.format(message, expected, actual);
}
```

这个条件判断应当封装起来，从而更清晰地表达代码的意图。我们拆解出一个方法，来解释这个条件判断。

```
public String compact(String message) {
  if (shouldNotCompact())
    return Assert.format(message, expected, actual);

  findCommonPrefix();
  findCommonSuffix();
  String expected = compactString(this.expected);
  String actual = compactString(this.actual);
  return Assert.format(message, expected, actual);
}

private boolean shouldNotCompact() {
  return expected == null || actual == null || areStringsEqual();
}
```

---

[1] 见本书 1.6 节。

我也不太喜欢 compact 函数中的 this.expected 符号和 this.actual 符号。这个是我们把 fExpected 改为 expected 时发生的。为什么函数中的变量会与成员变量同名呢？它们不是该表示其他意思吗[N4]？我们应该区分这些名称。

```
String compactExpected = compactString(expected);
String compactActual = compactString(actual);
```

否定式稍微比肯定式难理解一些[G29]。我们把 if 语句放到上头，调转条件判断。

```
public String compact(String message) {
  if (canBeCompacted()) {
    findCommonPrefix();
    findCommonSuffix();
    String compactExpected = compactString(expected);
    String compactActual = compactString(actual);
    return Assert.format(message, compactExpected, compactActual);
  } else {
    return Assert.format(message, expected, actual);
  }
}

private boolean canBeCompacted() {
  return expected != null && actual != null && !areStringsEqual();
}
```

函数名很奇怪[N7]。尽管它的确会压缩字符串，但如果 canBeCompacted() 为 false，那么它实际上就不会压缩字符串。用 compact 来命名，隐藏了错误检查的副作用。注意，该函数返回一条格式化后的消息，而不仅是压缩后的字符串。所以，函数名其实应该是 formatCompactedComparison。在用以下参数调用时，读起来会好很多：

```
public String formatCompactedComparison(String message) {
```

两个字符串是在 if 语句体中被压缩的。我们应当拆分出一个名为 compactExpectedAndActual 的方法。然而，我们希望 formatCompactComparison 函数完成所有的格式化工作。而 compact... 函数除了压缩之外什么都不做[G30]。所以，做如下拆分：

```
...
  private String compactExpected;
  private String compactActual;
...
  public String formatCompactedComparison(String message) {
    if (canBeCompacted()) {
      compactExpectedAndActual();
      return Assert.format(message, compactExpected, compactActual);
    } else {
      return Assert.format(message, expected, actual);
    }
  }
```

```java
private void compactExpectedAndActual() {
  findCommonPrefix();
  findCommonSuffix();
  compactExpected = compactString(expected);
  compactActual = compactString(actual);
}
```

注意，这要求我们向成员变量举荐 `compactExpected` 和 `compactActual`。我不喜欢新函数最后两行返回变量的方式,但前两个可不是这样。它们没采用一以贯之的约定[G11]。我们应该修改 `findCommonPrefix` 和 `findCommonSuffix`，分别返回前缀和后缀值。

```java
private void compactExpectedAndActual() {
  prefixIndex = findCommonPrefix();
  suffixIndex = findCommonSuffix();
  compactExpected = compactString(expected);
  compactActual = compactString(actual);
}

private int findCommonPrefix() {
  int prefixIndex = 0;
  int end = Math.min(expected.length(), actual.length());
  for (; prefixIndex < end; prefixIndex++) {
    if (expected.charAt(prefixIndex) != actual.charAt(prefixIndex))
      break;
  }
  return prefixIndex;
}

private int findCommonSuffix() {
  int expectedSuffix = expected.length() - 1;
  int actualSuffix = actual.length() - 1;
  for (; actualSuffix >= prefixIndex && expectedSuffix >= prefixIndex;
       actualSuffix--, expectedSuffix--) {
    if (expected.charAt(expectedSuffix) != actual.charAt(actualSuffix))
      break;
  }
  return expected.length() - expectedSuffix;
}
```

我们还应该修改成员变量的名称，使之更准确一点[N1]，毕竟它们都是索引。

仔细检查 `findCommonSuffix`，其中隐藏了一个时序性耦合[G31]，该函数它依赖 `prefixIndex` 是由 `findCommonPrefix` 计算得来的事实。如果这两个方法不按这样的顺序调用, 调试就会变得困难。为了曝露这个时序性耦合，我们将 `prefixIndex` 作为 `find` 的参数。

```java
private void compactExpectedAndActual() {
  prefixIndex = findCommonPrefix();
  suffixIndex = findCommonSuffix(prefixIndex);
  compactExpected = compactString(expected);
  compactActual = compactString(actual);
}
```

```
private int findCommonSuffix(int prefixIndex) {
  int expectedSuffix = expected.length() - 1;
  int actualSuffix = actual.length() - 1;
  for (; actualSuffix >= prefixIndex && expectedSuffix >= prefixIndex;
       actualSuffix--, expectedSuffix--) {
    if (expected.charAt(expectedSuffix) != actual.charAt(actualSuffix))
      break;
  }
  return expected.length() - expectedSuffix;
}
```

我对这样的方式不太满意，因为传递 prefixIndex 参数有些随意[G32]，该参数成功维持了执行次序，但对于解释排序的需要却毫无作用。其他程序员可能会抹杀我们刚完成的工作，因为并没有迹象说明该参数确属必要。所以还是采取别的做法吧。

```
private void compactExpectedAndActual() {
  findCommonPrefixAndSuffix();
  compactExpected = compactString(expected);
  compactActual = compactString(actual);
}

private void findCommonPrefixAndSuffix() {
  findCommonPrefix();
  int expectedSuffix = expected.length() - 1;
  int actualSuffix = actual.length() - 1;
  for (
    ;
       actualSuffix >= prefixIndex && expectedSuffix >= prefixIndex;
       actualSuffix--, expectedSuffix--
  ) {
    if (expected.charAt(expectedSuffix) != actual.charAt(actualSuffix))
      break;
  }
  suffixIndex = expected.length() - expectedSuffix;
}

private void findCommonPrefix() {
  prefixIndex = 0;
  int end = Math.min(expected.length(), actual.length());
  for (; prefixIndex < end; prefixIndex++)
    if (expected.charAt(prefixIndex) != actual.charAt(prefixIndex))
      break;
}
```

我们恢复 findCommonPrefix 和 findCommonSuffix 的原样，把 findCommonSuffix 的名称改为 findCommon**PrefixAnd**Suffix，让它在执行其他操作之前，先调用 findCommonPrefix。这样一来，就以一种比前种手段有效的方式建立了两个函数之间的时序关系。

```
private void findCommonPrefixAndSuffix() {
  findCommonPrefix();
  int suffixLength = 1;
```

```
      for (; !suffixOverlapsPrefix(suffixLength); suffixLength++) {
        if (charFromEnd(expected, suffixLength) !=
            charFromEnd(actual, suffixLength))
          break;
      }
      suffixIndex = suffixLength;
    }

    private char charFromEnd(String s, int i) {
      return s.charAt(s.length()-i);
    }

    private boolean suffixOverlapsPrefix(int suffixLength) {
      return actual.length() - suffixLength < prefixLength ||
        expected.length() - suffixLength < prefixLength;
    }
```

这样就好多了。它曝露出 `suffixIndex` 其实是后缀的长度，而且名字没起好，对于 `prefix` 也是如此。虽然在那种情形下 `index` 和 `length` 是同义的，但使用 `length` 一词更有一贯性。问题在于，`suffixIndex` 变量并不是从 0 开始，而是从 1 开始的，所以并非真正的长度。这也是 `computeCommonSuffix` 中那些 +1 存在的原因[G33]。来修正它们吧，修正结果就是代码清单 15-4。

代码清单 15-4　ComparisonCompactor.java（过渡版本）

```
public class ComparisonCompactor {
...
  private int suffixLength;
...
  private void findCommonPrefixAndSuffix() {
    findCommonPrefix();
    suffixLength = 0;
    for (; !suffixOverlapsPrefix(suffixLength); suffixLength++) {
      if (charFromEnd(expected, suffixLength) !=
          charFromEnd(actual, suffixLength))
        break;
    }
  }

  private char charFromEnd(String s, int i) {
    return s.charAt(s.length() - i - 1);
  }

  private boolean suffixOverlapsPrefix(int suffixLength) {
    return actual.length() - suffixLength <= prefixLength ||
      expected.length() - suffixLength <= prefixLength;
  }
...
  private String compactString(String source) {
    String result =
      DELTA_START +
      source.substring(prefixLength, source.length() - suffixLength) +
```

```
        DELTA_END;
    if (prefixLength > 0)
      result = computeCommonPrefix() + result;
    if (suffixLength > 0)
      result = result + computeCommonSuffix();
    return result;
  }

  ...
  private String computeCommonSuffix() {
    int end = Math.min(expected.length() - suffixLength +
      contextLength, expected.length()
    );
    return
      expected.substring(expected.length() - suffixLength, end) +
      (expected.length() - suffixLength <
        expected.length() - contextLength ?
        ELLIPSIS : "");
  }
```

我们用 charFromEnd 中的那个-1 替代了 computeCommonSuffix 中的一堆+1，前者更为合理，suffixOverlapsPrefix 中的两个 "<=" 操作符也同理。这样我们就能修改 suffixIndex 和 suffixLength 的名称，极大地提升了代码的可读性。

不过还有一个问题。在消灭那些+1 时，我注意到 compactString 中的以下代码：

```
if (suffixLength > 0)
```

看看代码清单 15-4 中的这行代码。因为 suffixLength 现在要比原本少 1，所以我们应该把 ">" 操作符改为 ">=" 操作符。那本无道理，不过现在却有意义！这表示这么做没道理，而且可能是个缺陷。嗯，也不算是个缺陷。从之前的分析中我们可以看到，if 语句现在会放置添加长度为零的后缀。在作出修改之前，if 语句没有作用，因为 suffixIndex 永不会小于 1。

这说明 compactString 中的两个 if 语句都有问题！看起来它们都该被删除。所以，我们将其注释掉，运行测试。测试通过了！那就重新构建 compactString，删除没用的 if 语句，将函数改得更加简洁[G9]。

```
  private String compactString(String source) {
    return
      computeCommonPrefix() +
      DELTA_START +
      source.substring(prefixLength, source.length() - suffixLength) +
      DELTA_END +
      computeCommonSuffix();
  }
```

这样就好多了！现在我们看到，compactString 函数只是把片段组合起来。我们甚至可以让它更清晰，还有许多细微的整理工作可做。与其拖着你遍历剩下的那些修改，我更愿意直接展示代码清单 15-5 中的结果。

代码清单 15-5　ComparisonCompactor.java（最终版）

```java
package junit.framework;

public class ComparisonCompactor {

  private static final String ELLIPSIS = "...";
  private static final String DELTA_END = "]";
  private static final String DELTA_START = "[";

  private int contextLength;
  private String expected;
  private String actual;
  private int prefixLength;
  private int suffixLength;

  public ComparisonCompactor(
      int contextLength, String expected, String actual
  ) {
    this.contextLength = contextLength;
    this.expected = expected;
    this.actual = actual;
  }

  public String formatCompactedComparison(String message) {
    String compactExpected = expected;
    String compactActual = actual;
    if (shouldBeCompacted()) {
      findCommonPrefixAndSuffix();
      compactExpected = compact(expected);
      compactActual = compact(actual);
    }
    return Assert.format(message, compactExpected, compactActual);
  }

  private boolean shouldBeCompacted() {
    return !shouldNotBeCompacted();
  }

  private boolean shouldNotBeCompacted() {
    return expected == null ||
           actual == null ||
           expected.equals(actual);
  }

  private void findCommonPrefixAndSuffix() {
    findCommonPrefix();
    suffixLength = 0;
    for (; !suffixOverlapsPrefix(); suffixLength++) {
      if (charFromEnd(expected, suffixLength) !=
          charFromEnd(actual, suffixLength)
      )
```

```java
      break;
    }
  }

  private char charFromEnd(String s, int i) {
    return s.charAt(s.length() - i - 1);
  }

  private boolean suffixOverlapsPrefix() {
    return actual.length() - suffixLength <= prefixLength ||
      expected.length() - suffixLength <= prefixLength;
  }

  private void findCommonPrefix() {
    prefixLength = 0;
    int end = Math.min(expected.length(), actual.length());
    for (; prefixLength < end; prefixLength++)
      if (expected.charAt(prefixLength) != actual.charAt (prefixLength))
        break;
  }

  private String compact(String s) {
    return new StringBuilder()
      .append(startingEllipsis())
      .append(startingContext())
      .append(DELTA_START)
      .append(delta(s))
      .append(DELTA_END)
      .append(endingContext())
      .append(endingEllipsis())
      .toString();
  }

  private String startingEllipsis() {
    return prefixLength > contextLength ? ELLIPSIS : "";
  }

  private String startingContext() {
    int contextStart = Math.max(0, prefixLength - contextLength);
    int contextEnd = prefixLength;
    return expected.substring(contextStart, contextEnd);
  }

  private String delta(String s) {
    int deltaStart = prefixLength;
    int deltaEnd = s.length() - suffixLength;
    return s.substring(deltaStart, deltaEnd);
  }

  private String endingContext() {
    int contextStart = expected.length() - suffixLength;
    int contextEnd =
      Math.min(contextStart + contextLength, expected.length());
    return expected.substring(contextStart, contextEnd);
```

```
  }

  private String endingEllipsis() {
    return (suffixLength > contextLength ? ELLIPSIS : "");
  }
}
```

这的确很漂亮。模块分解成了一组分析函数和一组合成函数。它们以一种拓扑方式排序，每个函数的定义都正好在其被调用的位置后面。所有的分析函数都先出现，而所有的合成函数都最后出现。

仔细阅读，你会发现我推翻了在本章较前位置做出的几个决定。例如，我将几个分解出来的方法重新内联为 `formatCompactComparison`，修改了 `souldNotBeCompacted` 表达式的意思，这种做法很常见。重构常会导致另一次推翻此次重构的重构。重构是一种不停试错的迭代过程，不可避免地集中于我们认为是专业人员该做的事。

15.2 小结

如此我们遵循了童子军军规。模块比我们发现它时更整洁了，不是说它原本不整洁，作者们做了卓越的工作，但模块都能再改进，我们每个人都有责任把模块改进得比发现它时更整洁。

第 16 章

重构 SerialDate

如果你找到 JCommon 类库，深入该类库，其中有个名为 `org.jfree.date` 的程序包。在该程序包中，有个名为 `SerialDate` 的类，我们即将剖析这个类。

`SerialDate` 的作者是 David Gilbert。David 显然是一位经验丰富、能力很强的程序员。如我们将看到的，他在代码中展示了极高的专业性和原则性。无论怎么说，`SerialDate` 都是"好代码"，而我将把它撕成碎片。

这并非恶意的行为，我也不认为自己比 David 强许多，有权对他的代码说三道四。其实，如果你看过我的代码，我敢说你也会发现好些该埋怨的东西。

不，这也并非傲慢无礼的行为。我所要做的，只是一种专业眼光的检视，不多也不少，那是我们都该坦然接受的做法。那是我们应该欢迎别人对自己做的事。只有通过这样的批评，我们才能学到东西。医生就是这样做的，飞行员就是这样做的，律师就是这样做的，我们程序员也需要学习如何这样做。

多说一句关于 David Gilbert 的事：David 不仅是一位优秀的程序员，他还有着将代码免费呈献给社区的勇气和好心。他公开代码，让所有人都能看到，邀请大众使用并审查。做得真好！

`SerialDate`（见代码清单 B-1）是一个用 Java 呈现日期的类。为什么在 Java 已经有 `java.util.Date` 和 `java.util.Calendar` 的情况下，还需要一个呈现日期的类呢？作者编写这个类，是为了解除我自己也常感到的痛苦。在开放的 Javadoc（第 67 行）中，他很好地解释了原因。我们可以质疑他的初衷，但我的确有处理这个问题的需要，而且我也欢迎有一个关乎日期甚于关乎时间的类存在。

16.1 首先，让它能工作

在一个名为 `SerialDateTests` 的类（见代码清单 B-2）中，有一些单元测试。测试都通过了，但不幸的是，快览一遍测试，发现它们并没有测试所有东西[T1]。例如，用"查找使用"搜索方法 `MonthCodeToQuarter`（第 356 行），会发现没有被用过[F4]。因此，单元测试并没有测试这个方法。

所以，我用 Clover 来检查单元测试覆盖了哪些代码。Clover 报告说，在 `SerialDate` 的 185 个可执行语句中，单元测试只执行了 91 个（约 50%）[T2]。覆盖图看起来像是一床满是补丁的棉被，整个类上布满大块的未执行代码。

我的目标是完整地理解和重构这个类，如果没有好得多的测试覆盖率，就达不到目标。所以，我完全重起炉灶编写了自己的单元测试（见代码清单 B-4）。

在阅读这些测试时，你可以看到，其中许多注释掉了。这些测试不能通过。它们代表了我以为 `SerialDate` 应该有的行为。在我重构 `SerialDate` 时，也将让这些测试通过。

即便有些测试被注释掉，Clover 也还是会报告新的单元测试执行了 185 个可执行语句中的 170 个（92%）。这样就好多了，而且我想我们可以把这个覆盖率提高些。

前几个注释掉的测试（第 23~63 行）是我一厢情愿。程序并没有设计为通过这些测试，但对我来说它们代表的行为显而易见[G2]。我不太确定 `StringToWeekdayCode` 方法为何要写成那样，不过既然它已经在那儿，显然不该是区分大小写的。编写这些测试是区区小事[T3]，通过测试更加容易。我只修改了第 259 行和第 263 行，就能使用 `equalsIgnoreCase` 了。

我注释掉了第 32 行和第 45 行的测试，因为我不太明确是否应该支持 tues 和 thurs

缩写。

第 153 行和第 154 行的测试不能通过。显然，它们本该通过[G2]。我们可以轻易地修正，只要对 `stringToMonthCode` 作出以下修改就行，对于第 163 行和第 213 行的测试也一样。

```
457         if ((result < 1) || (result > 12)) {
              result = -1;
458           for (int i = 0; i < monthNames.length; i++) {
459             if (s.equalsIgnoreCase(shortMonthNames[i])) {
460               result = i + 1;
461               break;
462             }
463             if (s.equalsIgnoreCase(monthNames[i])) {
464               result = i + 1;
465               break;
466             }
467           }
468         }
```

第 318 行注释掉的测试暴露了 `getFollowingDayOfWeek` 方法中的一个缺陷（第 672 行）。2004 年 12 月 25 日是周六。下一个周六是 2005 年 1 月 1 日。然而，运行测试时，会看到 `getFollowingDayOfWeek` 返回 12 月 25 日之后的周六还是 12 月 25 日。显然这不对[G3][T1]。我们看到问题在第 685 行。那是个典型的边界条件错误[T5]。应该是这样：

```
685         if (baseDOW >= targetWeekday) {
```

很有意思，这个函数是之前一次修改的结果。修改记录（第 43 行）显示，`getPreviousDayOfWeek`、`getFollowingDayOfWeek` 和 `getNearestDayOfWeek` 中的"缺陷"已被修正[T6]。

测试 `getNearestDayOfWeek`（第 705 行）的单元测试 `testGetNearestDayOfWeek`（第 329 行）之前的版本不像现在一样没有遗漏。我添加了大量测试用例，因为初始的测试用例并没有全部通过[T6]。查看哪些测试用例被注释掉，你可以看到失败的模式，这很有启发。如果最近的日期是在未来，算法就会失败。显然存在某种边界条件错误[T5]。

Clover 汇报的测试覆盖模式也很有趣[T8]。第 719 行根本没有执行！这意味着第 718 行的 `if` 语句总是得到 `false` 的结果。没错，看一眼代码就知道是这样。变量 `adjust` 总是为负，所以不会大于或等于 4。所以，算法错了。

正确的算法如下所示：

```
int delta = targetDOW - base.getDayOfWeek();
int positiveDelta = delta + 7;
int adjust = positiveDelta % 7;
if (adjust > 3)
  adjust -= 7;

return SerialDate.addDays(adjust, base);
```

最后，只要简单地抛出 `IllegalArgumentException` 异常而不是从 `weekInMonth-ToString` 和 `relativeToString` 返回错误字符串，第 417 行和第 429 行的测试就能通过。

做出这些修改后，所有的单元测试都通过了，我确信 `SerialDate` 现在可以工作。是时候让它"做对"了。

16.2 让它做对

我们将从头到尾遍历 `SerialDate`，同时加以改进。尽管在本章的讨论中你看不到这个过程，在每次做修改后，我还是要运行全部 JCommon 单元测试，包括我为 `SerialDate` 改进的那些单元测试。所以，后面你看到的所有修改，对于 JCommon 都是可工作的。

从第 1 行开始，我看到大量有关许可、版权、作者和修改历史的注释。我明白，的确有些法律事宜要说明，所以版权和许可信息应该保留。另外，修改历史是产生于 19 世纪 60 年代的古董，现今源代码控制工具可以帮我们做到这个。应该删掉修改历史[C1]。

从第 61 行开始的导入列表应该通过使用 `java.text.*` 和 `java.util.*` 来缩短。[J1]

Javadoc 的 HTML 格式化工作（第 67 行）令我畏惧。一个源文件里面有多种语言，我有点发怵。这条注释有 4 种语言：Java、英文、Javadoc 和 html[G1]。有那么多语言，注释就很难直截了当。例如，生成 Javadoc 后，第 71 行和第 72 行原本很好的位置就丢失了，而且谁想在源代码中看到 `<ul>` 和 `<li>` 这样的东西呢？更好的策略可能是用 `<pre>` 标签把整个注释部分包围起来，这样，对于源代码的格式化只会限于 Javadoc 之内①。

第 86 行是类声明。这个类为何要命名为 `SerialDate` 呢？Serial 一词有什么妙处吗？是不是因为该类派生自 `Serializable`？看来不是这样的。

别猜了，我知道为什么（或者我认为自己知道）要用 Serial 一词。线索就在第 98 行和第 101 行的常量 `SERIAL_LOWER_BOUND` 和 `SERIAL_UPPER_BOUND`。更好的线索在从第 830 行开始的注释中。该类被命名为 `SerialDate`，是因为它用"序列数"（serial number）来实现，该序列数恰好是从 1899 年 12 月 30 日后的天数。

对此我有两个问题。首先，术语"序列数"并不真对。可能有点诡辩，但其呈现方式却更接近相对偏移。术语"序列数"更多地用于产品版本标识，而非日期标识。我没发现这个名称特别有描述力[N1]。更有描述力的术语大概是"顺序"（ordinal）。

第二个问题更突出。名称 `SerialDate` 暗示了一种实现。该类是一个抽象类，没必要暗示任何有关实现的事。实际上，没理由隐藏实现！我发现这个名称放在了不正确的抽象层级上[N2]。以我之见，该类的名称应该就是简单的 `Date`。

不幸的是，Java 类库里面有太多名为 `Date` 的类了，所以这大概也不是最好的名称。因

---

① 更好的解决方案是让 Javadoc 不对注释做格式化，这样注释在代码和文档中就会是同一种样式。

为这个类关于日期而非时间,所以我想将其命名为 Day,但 Day 这个名字也在多处被滥用。最后,我选了 DayDate 作为最佳折中方案。

从现在起,我将使用术语 DayDate。请记住,你读到的代码清单,还是用的 SerialDate。

我理解为何 DayDate 继承自 Comparable 和 Serializable。不过,为什么它要继承自 MonthConstants 呢?类 MonthConstants(见代码清单 B-3)只是一大堆定义了月份的静态常量。从常量类继承是 Java 程序员用的一种老花招,这样他们就能避免形如 MonthConstants.January 的表达式,不过这是一个坏主意[J2]。MonthConstants 其实应该是一个枚举。

```
public abstract class DayDate implements Comparable,
                                         Serializable {
  public static enum Month {
    JANUARY(1),
    FEBRUARY(2),
    MARCH(3),
    APRIL(4),
    MAY(5),
    JUNE(6),
    JULY(7),
    AUGUST(8),
    SEPTEMBER(9),
    OCTOBER(10),
    NOVEMBER(11),
    DECEMBER(12);

    Month(int index) {
      this.index = index;
    }
    public static Month make(int monthIndex) {
      for (Month m : Month.values()) {
        if (m.index == monthIndex)
          return m;
      }
      throw new IllegalArgumentException("Invalid month index " + monthIndex);
    }
    public final int index;
  }
```

把 MonthConstants 改成枚举,导致对 DayDate 类和所有用到这个类的代码的一些修改。我花了一小时来改代码。不过,原来以 int 为月份类型的函数,现在都用上 Month 枚举元素了。这意味着我们可以去除 isValidMonthCode 方法(第 326 行),以及 monthCodeToQuarter 等位置的月份代码错误检查(第 356 行)了[G5]。

下一步,我们看到第 91 行的 serialVersionUID。该变量用于控制序列号。如果我们修改了它,那么用这个软件编写的旧版本 DayDate 都将不再可用,而是返回一个 InvalidClassException 异常。如果你没有声明 serialVersionUID 变量,则编译器会自动生成一个，每次修改模块时都会得到不一样的值。我知道，所有的文档都建议手工控制这个变量，但对我来说自动控制序列号安全得多[G4]。我宁可调试 `InvalidClassException`，也不愿意面对因忘记修改 `serialVersionUID` 引起的后续工作。所以，我要删除这个变量——至少暂时这么做①。

我发现第 93 行的注释是多余的。这正是谎言和误导信息所在之地[C2]。所以我要删掉它和它的同类。

第 97 行和第 100 行的注释有关序列数，我之前已经讨论过这个问题[C1]，它们描述的变量是 DayDate 能够描述的最早和最晚的日期。这可以搞得更清楚些[N1]。

```
public static final int EARLIEST_DATE_ORDINAL = 2;        // 1/1/1900
public static final int LATEST_DATE_ORDINAL  = 2958465;   // 12/31/9999
```

我不太清楚为什么 `EARLIEST_DATE_ORDINAL` 的值是 2 而不是 0。在第 829 行的注释中有提示，说明这与用 Microsoft Excel 展示日期的方式有关。在 DayDate 的派生类 SpreadsheetDate 中能看得更深入（见代码清单 B-5）。第 71 行的注释很好地描述了这个问题。

我的问题是，这看来应该与 SpreadsheetDate 有关，而与 DayDate 无关才对。所以，`EARLIEST_DATE_ORDINAL` 和 `LATEST_DATE_ORDINAL` 实在不该属于 DayDate，而应该移到 SpreadsheetDate 中[G6]。

的确，搜索一下代码就知道，这些变量值仅在 SpreadsheetDate 中用到，在 DayDate 中没用到，在 JCommon 框架的其他类中也没用到。所以，我将把这些变量值向下移到 SpreadsheetDate 中。

下面的两个变量，`MINIMUN_YEAR_SUPPORTED` 和 `MAXIMUM_YEAR_SUPPORTED`（第 104 行和第 107 行）地位尴尬。显然，如果 DayDate 是一个没有提供实现铺垫的抽象类，它就不该告知我们有关最小年份和最大年份的信息。同样，我很想把这些变量向下移到 SpreadsheetDate 中[G6]。然而，如果快速查找这些变量的使用情况，会发现另一个类也在用：RelativeDayOfWeekRule（见代码清单 B-6）。在第 177 行和第 178 行的 getDate 函数中，它们被用来检查 getDate 的年份参数是否有效，而抽象类的用户需要得知其实现信息，这是一个矛盾。

我们要做的是既提供信息，又不污染 DayDate。通常，我们会从派生类实体中获取实现信息。不过，我们并未向 getDate 函数传入 DayDate 的实体，反而返回了一个 DayDate 的实体，这意味着必须在某处创建这个实体。第 187~205 行提供了线索。DayDate 实体是在 getPreviousDayOfWeek、getNearestDayOfWeek 或 getFollowingDayOfWeek 这 3

---

① 本章的好几个审读者都不这么认为。他们主张，在开源框架中，手工控制序列号会比较好，因为较小的修改不会导致序列化后的日期无效。这是种中肯的观点。然而，尽管会不方便，但失败就会有清晰的原因。另外，如果该类的作者忘记更新序列号，则失败模式就会不可预期，而且可能会隐藏得很深。我认为，这个故事的精髓在于，不应该跨版本做反序列化处理。

个函数其中之一里面创建的。看一下 DayDate 代码清单，我们看到，这些函数（第 638～724 行）全都返回了由 addDays（第 571 行）创建的日期实体，addDays 调用 CreateInstance（第 808 行），创建出一个 SpreadsheetDate！[G7]。

通常来说，基类不宜了解其派生类的情况。为了修正这个毛病，我们应该利用抽象工厂模式（ABSTRACT FACTORY）[①]，创建一个 DayDateFactory，该工厂类将创建我们所需要的 DayDate 的实体，并回答有关实现的问题，例如最大和最小日期之类。

```java
public abstract class DayDateFactory {
  private static DayDateFactory factory = new SpreadsheetDateFactory();
  public static void setInstance(DayDateFactory factory) {
    DayDateFactory.factory = factory;
  }

  protected abstract DayDate _makeDate(int ordinal);
  protected abstract DayDate _makeDate(int day, DayDate.Month month, int year);
  protected abstract DayDate _makeDate(int day, int month, int year);
  protected abstract DayDate _makeDate(java.util.Date date);
  protected abstract int _getMinimumYear();
  protected abstract int _getMaximumYear();

  public static DayDate makeDate(int ordinal) {
    return factory._makeDate(ordinal);
  }

  public static DayDate makeDate(int day, DayDate.Month month, int year) {
    return factory._makeDate(day, month, year);
  }

  public static DayDate makeDate(int day, int month, int year) {
    return factory._makeDate(day, month, year);
  }

  public static DayDate makeDate(java.util.Date date) {
    return factory._makeDate(date);
  }

  public static int getMinimumYear() {
    return factory._getMinimumYear();
  }

  public static int getMaximumYear() {
    return factory._getMaximumYear();
  }
}
```

该工厂类用 makeDate 方法替代了 createInstance 方法，前者的名称稍好一些[N1]。在初始状态下，它使用 SpreadsheetDateFactory，但随时可以使用其他工厂。委托到抽

---

① [GOF]。

象方法的静态方法混合采用了单件模式（SINGLETON）、油漆工模式[①]和抽象工厂模式[②]，我发现这种手段很有用。

SpreadsheetDateFactory 看起来像这个样子：

```
public class SpreadsheetDateFactory extends DayDateFactory {
  public DayDate _makeDate(int ordinal) {
    return new SpreadsheetDate(ordinal);
  }

  public DayDate _makeDate(int day, DayDate.Month month, int year) {
    return new SpreadsheetDate(day, month, year);
  }

  public DayDate _makeDate(int day, int month, int year) {
    return new SpreadsheetDate(day, month, year);
  }

  public DayDate _makeDate(Date date) {
    final GregorianCalendar calendar = new GregorianCalendar();
    calendar.setTime(date);
    return new SpreadsheetDate(
      calendar.get(Calendar.DATE),
      DayDate.Month.make(calendar.get(Calendar.MONTH) + 1),
      calendar.get(Calendar.YEAR));
  }

  protected int _getMinimumYear() {
    return SpreadsheetDate.MINIMUM_YEAR_SUPPORTED;
  }

  protected int _getMaximumYear() {
    return SpreadsheetDate.MAXIMUM_YEAR_SUPPORTED;
  }
}
```

如你所见，我已经把变量 MINIMUM_YEAR_SUPPORTED 和 MAXIMUM_YEAR_SUPPORTED 移到了它们该在的 SpreadsheetDate 中[G6]。

DayDate 的下一个问题是第 109 行的日期常量。这些常量其实应该是枚举类型[J3]。我们之前见过这种模式，不再赘述。你可以在最终的代码清单中看到。

接着，我们看到第 140 行中一系列以 LAST_DAY_OF_MONTH 开头的数组。首先，描述这些数组的注释全属多余[C3]，光看名称就足够了，所以我要删除这些注释。

这个数组没理由不是私有的[G8]，因为有一个静态函数 lastDayOfMonth 提供同样的数据。

下一个数组 AGGREGATE_DAYS_TO_END_OF_MONTH 更神秘一些，在 JCommon 框架中

[①] Ibid.
[②] Ibid.

根本没用到它[G9]。所以我直接删除了。

对 LEAP_YEAR_AGGREGATE_DAYS_TO_END_OF_MONTH 也一样。

AGGREGATE_DAYS_TO_END_OF_PRECEDING_MONTH 只在 SpreadsheetDate 中用到（第 434 行和第 473 行）。是否把它移到 SpreadsheetDate 中是一个问题。不转移的理由是，该数组并不专属于任何特定的实现[G6]，另外，实际上并不存在 SpreadsheetDate 之外的实现，所以，数组应该移到靠近其使用位置的地方[G10]。

说服我的理由是保持一致[G11]，数组应该私有，并通过类似于 julianDateOfLastDayOfMonth 这样的函数来暴露。看来没人需要那样的函数。而且，如果有新的 DayDate 实现需要该数组，可以轻易地把它移回到 DayDate 中去。所以我就把它移到 SpreadsheetDate 里面了。

对 LEAP_YEAR_AGGREGATE_DAYS_TO_END_OF_MONTH 也一样。

接着，我们看到 3 组可以转换为枚举的常量（第 162～205 行）。第一个用来选择月份中的一周，我将其转换为名为 WeekInMonth 的枚举。

```
public enum WeekInMonth {
  FIRST(1), SECOND(2), THIRD(3), FOURTH(4), LAST(0);
  public final int index;

  WeekInMonth(int index) {
    this.index = index;
  }
}
```

第二组常量（第 177～187 行）有点麻烦。常量 INCLUDE_NONE、INCLUDE_FIRST、INCLUDE_SECOND 和 INCLUDE_BOTH 用于描述某个范围的终止日是否包含在该范围之内。数学上，用术语"开放区间""半开放区间"和"闭合区间"来表示。我想，用数学术语来命名会更清晰[N3]，所以就将其转换为枚举 DateInterval，其中包括 CLOSED、CLOSED_LEFT、CLOSED_RIGHT 和 OPEN 枚举元素。

第三组常量（第 189～205 行）描述了是否该在最后、下一个或最近的日期实体中呈现对某个星期中的特定一天的查找结果。怎么命名是一个难题。最终，我给 WeekdayRange 设定了 LAST、NEXT 和 NEAREST 枚举元素。

你也许不会同意我起的名字。对我而言这些名字有意义，但对你可能不然。要点是它们眼下变成了易于修改的形式[J3]，不再以整数形式传递，而是作为符号传递。我可以用 IDE 的"修改名称"功能来改动名称或类型，无须担忧漏掉代码中某处 -1 或 2 之类的数字，也不必担忧某些 int 参数声明处于描述不佳的状态。

第 208 行的描述字段看来没有任何地方用到，我把它及其取值器和赋值器都删掉了。

我还删除了第 213 行的默认构造器[G12]，编译器会为我们自动生成的。

略过 isValidWeekdayCode 方法（第 216～238 行），在创建 Day 枚举时已经把它删掉了。

于是来到 `stringToWeekdayCode` 方法（第 242~270 行）。没有方法签名增添价值的 Javadoc 都是废话[C3]、[G12]，唯一的价值是对返回值-1 的描述。然而，我们改用了 `Day` 枚举，所以这条注释完全错误了[C2]。该方法现在抛出一个 `IllegalArgumentException` 异常，所以我删除了 Javadoc。

我还删除了参数和变量声明中的全部 `final` 关键字，我敢说，它们毫无价值，只会混淆视听[G12]。删除这些 `final`，不合乎某些成例。例如，Robert Simmons[①]就强烈建议我们"……在代码中遍布 `final`。"我不能苟同。我认为，`final` 有少数的好用法，例如，偶尔使用的 `final` 常量，但除此之外该关键字利小于弊。我之所以这么认为，或许是因为 `final` 可能捕获到的那些错误类型，早已被我编写的单元测试捕获了。

我不喜欢 `for` 循环（第 259 行和第 263 行）中的那些 `if` 语句[G5]，所以我利用"||"操作符把它们连接为单个 `if` 语句。我还使用 `Day` 枚举整理 `for` 循环，做了一些装饰性的修改。

我认为，这个方法并不真属于 `DayDate` 类，它其实是 `Day` 的一个解析函数。所以，我将它移到 `Day` 枚举中。不过，那样 `Day` 枚举就会变得太大。因为 `Day` 的概念并不依赖 `DayDate`，所以我把 `Day` 枚举移到 `DayDate` 类之外，放到它自己的源代码文件中。

我还把下一个函数 `weekdayCodeToString`（第 272~286 行），移植到 `Day` 枚举中，称其为 `toString`。

```
public enum Day {
  MONDAY(Calendar.MONDAY),
  TUESDAY(Calendar.TUESDAY),
  WEDNESDAY(Calendar.WEDNESDAY),
  THURSDAY(Calendar.THURSDAY),
  FRIDAY(Calendar.FRIDAY),
  SATURDAY(Calendar.SATURDAY),
  SUNDAY(Calendar.SUNDAY);

  public final int index;
  private static DateFormatSymbols dateSymbols = new DateFormatSymbols();

  Day(int day) {
    index = day;
  }

  public static Day make(int index) throws IllegalArgumentException {
    for (Day d : Day.values())
      if (d.index == index)
        return d;
    throw new IllegalArgumentException(
      String.format("Illegal day index: %d.", index));
  }

  public static Day parse(String s) throws IllegalArgumentException {
```

---

① [Simmons04], p. 73。

```
    String[] shortWeekdayNames =
      dateSymbols.getShortWeekdays();
    String[] weekDayNames =
      dateSymbols.getWeekdays();

    s = s.trim();
    for (Day day : Day.values()) {
      if (s.equalsIgnoreCase(shortWeekdayNames[day.index]) ||
          s.equalsIgnoreCase(weekDayNames[day.index])) {
        return day;
      }
    }
    throw new IllegalArgumentException(
      String.format("%s is not a valid weekday string", s));
  }

  public String toString() {
    return dateSymbols.getWeekdays()[index];
  }
}
```

有两个 getMonth 函数（第 288～316 行）。第一个函数调用第二个函数。第二个函数只被第一个函数调用。所以，我把这两个函数合二为一，而且极大地简化之[G9][G12][F4]。最后，我把名称修改得更具描述力[N1]。

```
public static String[] getMonthNames() {
  return dateFormatSymbols.getMonths();
}
```

由于有了 Month 枚举，函数 isValidMonthCode（第 326～346 行）就变得没什么用，因此我把它删除了[G9]。

函数 monthCodeToQuarter（第 356～375 行）有特性依恋（FEATURE ENVY）[①]的味道，可以是 Month 枚举中的一个名为 quarter 的方法，我就这么办了。

```
public int quarter() {
  return 1 + (index-1)/3;
}
```

这样一来，Month 枚举就大到需要放到自己的类中了。我把它从 DayDate 中移出来，与 Day 枚举保持一致[G11][G13]。

后面两个方法被命名为 monthCodeToString（第 377～426 行）。我们再次看到其中一个方法使用标识调用其兄弟方法的模式。将标识作为参数传递给函数的做法通常不太好，尤其是当该标识只是有关其输出格式时[G15]。我重命名、简化、重新构架了这些函数，并把它们移到 Month 枚举中[N1][N3][G14]。

```
public String toString() {
  return dateFormatSymbols.getMonths()[index - 1];
```

---

① [Refactoring]。

```
}
public String toShortString() {
  return dateFormatSymbols.getShortMonths()[index - 1];
}
```

下一个方法是 stringToMonthCode（第 428～472 行）。我重新为它命名，转移到 Month 枚举中，并且简化之[N1][N3][C3][G14][G12]。

```
public static Month parse(String s) {
  s = s.trim();
  for (Month m : Month.values())
    if (m.matches(s))
      return m;

  try {
    return make(Integer.parseInt(s));
  }
  catch (NumberFormatException e) {}
  throw new IllegalArgumentException("Invalid month " + s);
}

private boolean matches(String s) {
  return s.equalsIgnoreCase(toString()) ||
         s.equalsIgnoreCase(toShortString());
}
```

方法 isLeapYear（第 495～517 行）可以写得更具表达力一些[G16]。

```
public static boolean isLeapYear(int year) {
  boolean fourth = year % 4 == 0;
  boolean hundredth = year % 100 == 0;
  boolean fourHundredth = year % 400 == 0;
  return fourth && (!hundredth || fourHundredth);
}
```

下一个函数 leapYearCount（第 519～536 行）并不真属于 DayDate。除了 SpreadsheetDate 中的两个方法，没有其他调用者，所以我将它往下放。

函数 lastDayOfMonth（第 538～560 行）使用了 LAST_DAY_OF_MONTH 数组，该数组应该隶属于 Month 枚举[G17]，所以我就把它移到那儿去了。我还简化了这个函数，使其更具表达力[G16]。

```
public static int lastDayOfMonth(Month month, int year) {
  if (month == Month.FEBRUARY && isLeapYear(year))
    return month.lastDay() + 1;
  else
    return month.lastDay();
}
```

现在，事情变得比较有趣了。下一个函数是 addDays（第 562～576 行）。首先，由于该函数对 DayDate 的变量进行操作，它就不该是静态的[G18]，因此，我把它修改为实体方法。其次，它调用了函数 `toSerial`，这个函数应该重新命名为 `toOrdinal`[N1]。最后，该方法可以简化。

```
public DayDate addDays(int days) {
  return DayDateFactory.makeDate(toOrdinal() + days);
}
```

对于 addMonth（第 578～602 行）也一样。它应该是一个实体方法[G18]，算法过于复杂，所以我利用解释临时变量模式（EXPLAINING TEMPORARY VARIABLES）[1]来使其更为透明。我还将方法 `getYYY` 重命名为 `getYear` [N1]。

```
public DayDate addMonths(int months) {
  int thisMonthAsOrdinal = 12 * getYear() + getMonth().index - 1;
  int resultMonthAsOrdinal = thisMonthAsOrdinal + months;
  int resultYear = resultMonthAsOrdinal / 12;
  Month resultMonth = Month.make(resultMonthAsOrdinal % 12 + 1);
  int lastDayOfResultMonth = lastDayOfMonth(resultMonth, resultYear);
  int resultDay = Math.min(getDayOfMonth(), lastDayOfResultMonth);
  return DayDateFactory.makeDate(resultDay, resultMonth, resultYear);
}
```

对函数 addYear（第 604～626 行）也照方办理。

```
public DayDate plusYears(int years) {
  int resultYear = getYear() + years;
  int lastDayOfMonthInResultYear = lastDayOfMonth(getMonth(), resultYear);
  int resultDay = Math.min(getDayOfMonth(), lastDayOfMonthInResultYear);
  return DayDateFactory.makeDate(resultDay, getMonth(), resultYear);
}
```

把这些方法从静态方法变为实体方法，让我有点心头发痒。用 `date.addDays(5)` 这样的表达方法，是不是明确地表示 date 对象并没变动，以及返回了一个 DayDate 的新实体呢？或者，它只是错误地暗示我们往 date 对象添加了 5 天呢？你可能不会认为这是一个大问题，但下列代码可能会有欺骗性。

```
DayDate date = DateFactory.makeDate(5, Month.DECEMBER, 1952);
date.addDays(7); // bump date by one week
```

有些读到这段代码的人会认为 `addDays` 在修改 date 对象。所以，我们需要消除这种歧义[N4]。我把名称改为 `plusDays` 和 `plusMonths`。我认为，方法的初衷很清楚地被

```
DayDate date = oldDate.plusDays(5);
```

所体现，不过下列代码对认为 date 对象被修改的读者来说，看起来并不那么顺畅：

```
date.plusDays(5);
```

算法越来越有趣，`getPreviousDayOfWeek`（第 628～660 行）可以工作，不过有点复

---

[1] [Beck97]。

杂了。经过一番思考，了解到它的功能后[G21]，我就能够使用解释临时变量模式来简化它[G19]，使其更为清晰。我还将它从静态方法改为实体方法[G18]，并删除了重复的实体方法[G5]（第 997～1008 行）。

```
public DayDate getPreviousDayOfWeek(Day targetDayOfWeek) {
  int offsetToTarget = targetDayOfWeek.index - getDayOfWeek().index;
  if (offsetToTarget >= 0)
    offsetToTarget -= 7;
  return plusDays(offsetToTarget);
}
```

对 getFollowingDayOfWeek（第 662～693 行）也如法炮制：

```
public DayDate getFollowingDayOfWeek(Day targetDayOfWeek) {
  int offsetToTarget = targetDayOfWeek.index - getDayOfWeek().index;
  if (offsetToTarget <= 0)
    offsetToTarget += 7;
  return plusDays(offsetToTarget);
}
```

下一个函数是我们之前修改过的 getNearestDayOfWeek（第 695～726 行）。我之前所做的修改和前两个函数没有保持一致[G11]，所以我将它改为和这两个函数保持一致，并且使用解释临时变量模式[G19]来阐明算法。

```
public DayDate getNearestDayOfWeek(final Day targetDay) {
  int offsetToThisWeeksTarget = targetDay.index - getDayOfWeek().index;
  int offsetToFutureTarget = (offsetToThisWeeksTarget + 7) % 7;
  int offsetToPreviousTarget = offsetToFutureTarget - 7;

  if (offsetToFutureTarget > 3)
    return plusDays(offsetToPreviousTarget);
  else
    return plusDays(offsetToFutureTarget);
}
```

方法 getEndOfCurrentMonth（第 728～740 行）有点奇怪，因为它获取了 DayDate 参数，从而成为一个依恋[G14]其自身类的实体方法。我将其改为真正的实体方法，并修改了几个名称。

```
public DayDate getEndOfMonth() {
  Month month = getMonth();
  int year = getYear();
  int lastDay = lastDayOfMonth(month, year);
  return DayDateFactory.makeDate(lastDay, month, year);
}
```

重构 weekInMonthToString（第 742～761 行）的过程非常有趣。利用 IDE 的重构工具，我先将其移到我之前创建的 WeekInMonth 枚举中，再将其重命名为 toString。接着，我把它从静态方法改为实体方法。所有的测试都通过了。（你能猜出来我打算做什么吗？）

接下来，我删掉了整个方法！有 5 个断言失败了（第 411~415 行，见代码清单 B-4）。我改动了这些代码行，让它们使用枚举元素的名称（FIRST、SECOND……）。全部测试都通过了。你知道为什么吗？你是否知道为什么这些步骤都是必要的吗？重构工具确保之前对 weekInMonthToString 方法的调用现在都调用 weekInMonth 枚举元素的 toString 方法，全部枚举元素都以返回其名称的形式实现了 toString 方法……

我不幸有点聪明过头了。经过这一套美妙的重构，我终于意识到，这个函数的唯一调用者，就是我刚修改的测试，所以我删除了这些测试。

愚我一次，是你之耻。愚我两次，是我之耻！所以，在判定除测试之外没有人调用过 relativeToString（第 765~781 行）后，我就删除了该函数及其测试。

我最后将其改为这个抽象类的抽象方法。第一个函数保持了原样：toSerial（第 838~844 行），前文我曾把其名称改为 toOrdinal，以现在的情形看，我决定把名称改为 getOrdinalDay。

下一个抽象方法是 toDate（第 838~844 行）。它将 DayDate 转换为 java.util.Date。这个方法为何是抽象的？查看其在 SpreadsheetDate 中的实现（第 198~207 行，见代码清单 B-5），可以看到它并不依赖该类的实现[G6]，所以，我把它往上推了。

方法 getYYYY、getMonth 和 getDayOfMonth 已经是抽象方法。不过，getDayOfWeek 方法是另一个应该从 SpreadsheetDate 中提出来的方法，因为它不依赖 DayDate 之外的东西[G6]。是这样吗？

仔细阅读代码清单 B-5 第 247 行，可以发现该算法暗中依赖顺序日期的起点（换言之，第 0 天的星期日数）。所以，即便该方法没有物理上的依赖，也不能移到 DayDate 中，因为它的确有逻辑上的依赖。

这样的逻辑依赖困扰了我[G22]。如果有什么东西在逻辑上依赖实现的话，也该有物理上的依赖存在。我也认为，算法本身也该有一小部分依赖实现。

所以我在 DayDate 中创建了一个名为 getDayOfWekForOrdinalZero 的抽象方法，并在 SpreadsheetDate 中实现它，返回 Day.SATURDAY。然后我把 getDayOfWeek 上移到 DayDate 中，并调用 getOrdinalDay 和 getDayOfWeekForOrdinalZero。

```
public Day getDayOfWeek() {
  Day startingDay = getDayOfWeekForOrdinalZero();
  int startingOffset = startingDay.index - Day.SUNDAY.index;
  return Day.make((getOrdinalDay() + startingOffset) % 7 + 1);
}
```

顺便说一句，请仔细阅读第 895~899 行的注释。这样的重复有必要吗？通常，我会删除这类注释。

下一个方法是 compare（第 902~913 行）。同样，该抽象方法是不恰当的[G6]，我将其实现上移到 DayDate，其名称也不足够有沟通意义[N1]。方法实际上返回的是自参数日期以来的天数，所以我把名称改为 daysSince。我还注意到该方法没有测试，就为它编写了测试。

下面 6 个函数（第 915～980 行）全都是应该在 DayDate 中实现的抽象方法。我把它们全都从 SpreadsheetDate 中抽出来了。

最后一个函数 isInRange（第 982～995 行）也需要推到上一层并重构之。那个 switch 语句有点儿丑陋[G23]，可以把那些条件判断移到 DateInterval 枚举中去。

```
public enum DateInterval {
  OPEN {
    public boolean isIn(int d, int left, int right) {
      return d > left && d < right;
    }
  },
  CLOSED_LEFT {
    public boolean isIn(int d, int left, int right) {
      return d >= left && d < right;
    }
  },
  CLOSED_RIGHT {
    public boolean isIn(int d, int left, int right) {
      return d > left && d <= right;
    }
  },
  CLOSED {
    public boolean isIn(int d, int left, int right) {
      return d >= left && d <= right;
    }
  };

  public abstract boolean isIn(int d, int left, int right);
}

public boolean isInRange(DayDate d1, DayDate d2, DateInterval interval) {
  int left = Math.min(d1.getOrdinalDay(), d2.getOrdinalDay());
  int right = Math.max(d1.getOrdinalDay(), d2.getOrdinalDay());
  return interval.isIn(getOrdinalDay(), left, right);
}
```

我们来到了 DayDate 的末尾。现在我们要从头到尾再过一次，看看整个重构过程是怎样良好执行的。

首先，开端注释过时已久，我缩短并改进了它[C2]。

然后，我把全部枚举移到它们自己的文件中[G12]。

接着，我把静态变量（dateFormatSymbols）和 3 个静态方法（getMonthNames、isLeapYear 和 lastDayOfMonth）移到名为 DateUtil 的新类中[G6]。

我把那些抽象方法上移到它们该在的顶层类中[G24]。

我把 Month.make 改为 Month.fromInt [N1]，并如法炮制所有其他枚举。我还为全

部枚举创建了 `toInt()` 访问器，把 `index` 字段改为私有。

在 `plusYears` 和 `plusMonths` 中存在一些有趣的重复[G5]，我通过抽离出名为 `correctLastDayOfMonth` 的新方法消解了重复，使这 3 个方法清晰多了。

我消除了魔术数 1 [G25]，用 `Month.JANUARY.toInt()` 或 `Day.SUNDAY.toInt()` 做了恰当的替换。我在 `SpreadsheetDate` 上花了点儿时间，清理了一下算法。最终结果在代码清单 B-7～代码清单 B-16 中。

有趣的是，`DayDate` 的代码覆盖率降低到了 84.9%！这并不是因为测试到的功能减少了，而是因为该类缩减得太多，导致少量未覆盖到的代码行拥有了更大权重。`DayDate` 的 53 个可执行语句中有 45 个得到测试覆盖。未覆盖的代码行微细到不值得测试。

16.3 小结

我们再一次遵从了童子军军规。我们签入的代码，要比签出时整洁了一点。虽然花了点儿时间，不过很值得。测试覆盖率提升了，修改了一些缺陷，代码清晰并缩短了。后来者有望比我们更容易地应对这些代码，他们也有可能把代码整理得更干净些。

16.4 文献

[GOF]：*Design Patterns: Elements of Reusable Object Oriented Software,* Gamma et al., Addison-Wesley, 1996.

[Simmons04]：*Hardcore Java,* Robert Simmons, Jr., O'Reilly, 2004.

[Refactoring]：*Refactoring: Improving the Design of Existing Code,* Martin Fowler et al., Addison-Wesley, 1999.

[Beck97]：*Smalltalk Best Practice Patterns,* Kent Beck, Prentice Hall, 1997.

第 17 章

味道与启发

Martin Fowler 在其大作《重构：改善既有代码的设计》（*Refactoring: Improving the Design of Existing Code*）[①]中指出了许多不同的"代码的坏味道"。下面的清单包括很多 Martin 指出的"坏味道"，还添加了更多我自己提出的"坏味道"，也包括我借以历练本业的其他珍宝与启发。

我借由遍览和重构几个不同的程序总结出这个清单。每次修改，我都问自己为什么要这样改，把修改的原因写下来，结果就得到相当长的清单，该清单列出了在读代码时让我闻起

---

① [Refactoring]。

来不舒服的味道。

清单应按顺序阅读，并作为一种参考来使用。

17.1 注释

C1：不恰当的信息

让注释传达本该更好地在源代码控制系统、问题追踪系统或任何其他记录系统中保存的信息，是不恰当的。例如，修改历史记录只会用大量过时而无趣的文本搞乱源代码文件。通常，作者、最后修改时间、SPR 数等元数据不该在注释中出现。注释只应该描述有关代码和设计的技术性信息。

C2：废弃的注释

过时、无关或不正确的注释就是废弃的注释。注释会很快过时。最好别编写将被废弃的注释。如果发现废弃的注释，最好尽快更新或删除。废弃的注释会远离它们曾经描述的代码，变成代码中无关和误导阅读者的浮岛。

C3：冗余注释

如果注释描述的是某种充分自我描述了的东西，那么注释就是多余的。例如：

```
i++; // increment i
```

另一个例子是除函数签名之外什么也没多说（或少说）的 Javadoc：

```
/**
 * @param sellRequest
 * @return
 * @throws ManagedComponentException
 */
public SellResponse beginSellItem(SellRequest sellRequest)
  throws ManagedComponentException
```

注释应该谈及代码自身没提到的东西。

C4：糟糕的注释

值得编写的注释，也值得好好写。如果要编写一条注释，就花时间保证写出最好的注释。字斟句酌，使用正确的语法和拼写，别闲扯，别画蛇添足，要保持简洁。

C5：注释掉的代码

看到被注释掉的代码会令我抓狂。谁知道它有多旧？谁知道它有没有意义？没人会删除它，因为大家都假设别人需要它或是有进一步计划。

那样的代码就这样腐烂掉，随着时间推移，它与系统越来越没关系。它调用不复存在的函数，它使用已改名的变量，它遵循已被废弃的约定，它污染了所属的模块，分散了想要读它的人的注意力。注释掉的代码纯属厌物。

看到注释掉的代码，就删除它！别担心，源代码控制系统还会记得它。如果有人真的需要，可以签出较旧的版本。别被它搞得死去活来。

17.2 环境

E1：需要多步才能实现的构建

构建系统应该是单步的小操作。不应该从源代码控制系统中一小点一小点签出代码。不应该需要一系列神秘指令或环境依赖脚本来构建单个元素。不应该四处寻找额外的小JAR、XML 文件和其他系统所需的杂物。你应当能够用单个命令签出系统，并用单个指令构建它。

```
svn get mySystem
cd mySystem
ant all
```

E2：需要多步才能做到的测试

你应当能够发出单个指令就可以运行全部单元测试。能够运行全部测试是如此基础和重要，应该快速、轻易和直截了当地做到。

17.3 函数

F1：过多的参数

函数的参数量应该少。没参数最好，一个次之，两个、三个再次之。三个以上的参数非常值得质疑，应坚决避免。（参见 3.6 节。）

F2：输出参数

输出参数违反直觉，因为读者期望参数用于输入而非输出。如果函数非要修改什么东西的状态，就修改它所在对象的状态好了。（参见 3.7 节。）

F3：标识参数

布尔值参数大声宣告函数做了不止一件事。它们令人迷惑，应该被消灭掉。（参见 3.6.2 节。）

F4：死函数

永不被调用的函数应该被丢弃。保留死函数纯属浪费，别害怕删除死函数，记住，源代码控制系统还会记得它。

17.4　一般性问题

G1：一个源文件中存在多种语言

当今的现代编程环境允许在单个源文件中存在多种不同语言。例如，Java 源文件可能还包括 XML、HTML、YAML、Javadoc、英文、JavaScript 等语言。再如，JSP 文件可能还包括 HTML、Java、标签库语法、英文注释、Javadoc、XML、JavaScript 等。往好处说是令人迷惑，往坏处说就是粗心大意、驳杂不精。

理想的源文件包括且只包括一种语言。现实中，我们可能会不得不使用多于一种语言，但应该尽力缩小源文件中额外语言的数量和范围。

G2：明显的行为未被实现

遵循"最小惊异原则"[①]，函数或类应该实现其他程序员有理由期待的行为。例如，考虑一个将日期名称翻译为表示该日期的枚举的函数。

```
Day day = DayDate.StringToDay(String dayName);
```

我们期望字符串 Monday 翻译为 Day.MONDAY，也期望常用缩写形式能被翻译出来，还期望函数忽略大小写。

如果明显的行为未被实现，读者和用户就不能再依靠他们对函数名称的直觉。他们不再信任原作者，不得不阅读代码细节。

G3：不正确的边界行为

代码应该有正确行为，这话看似明白。问题是我们很难明白正确行为有多复杂。开发者常常写出他们以为能工作的函数，信赖自己的直觉，而不是努力去证明代码在所有的角落和边界情形下都真正能工作。

没什么可以替代谨小慎微。每种边界条件、每种极端情形、每个异常都代表了某种可能

[①] 或称"最少惊诧原则"（The Principle of Least Astonishment）。

搞乱优雅而直白的算法的东西。别依赖直觉。追索每种边界条件，并编写测试。

G4：忽视安全

切尔诺贝利核电站崩塌了，因为电厂经理一再忽视安全机制。遵守安全规则就不便于做试验，结果就造成试验未能运行，全世界都目睹了首个民用核电站的大灾难。

忽视安全相当危险。手工控制 `serialVersionUID` 可能有必要，但总会有风险。关闭某些编译器警告（或者全部警告！）可能有助于构建成功，但可能会陷于无穷无尽的调试中。关闭失败测试、告诉自己过后再处理，这和假装刷信用卡不用还钱一样存在安全隐患。

G5：重复

有一条本书提到过的最重要的规则之一，你应该非常严肃地对待它，实际上，每位编写有关软件设计的作者都提到过这条规则，Dave Thomas 和 Andy Hunt 称之为 DRY 原则（Don't Repeat Yourself，别重复自己）[1]。Kent Beck 将它列为极限编程核心原则之一，并称之为 "一次，也只一次"。Ron Jeffries 将这条规则列在第二位，地位仅次于通过所有测试。

每次看到重复代码，都代表遗漏了抽象。重复的代码可能成为子程序或干脆是另一个类。将重复代码叠放进类似的抽象，增加了你的设计语言的词汇量。其他程序员可以用到你创建的抽象设施。编码变得越来越快，错误越来越少，因为你提升了抽象层级。

重复最明显的形态是你不断看到明显一样的代码，就像是某位程序员疯狂地用鼠标不断复制粘贴代码。可以用单一方法来替代之。

较隐蔽的形态是在不同模块中不断重复出现、检测同一组条件的 `switch/case` 或 `if/else` 链。可以用多态来替代之。

更隐蔽的形态是采用类似算法但具体代码行不同的模块。这也是一种重复，可以使用模板方法模式[2]或策略模式[3]来修正。

的确，过去 15 年内出现的多数设计模式都是消除重复的有名手段。Codd 范式（Codd Normal Forms）是消除数据库规划中的重复的策略。面向对象自身也是组织模块和消除重复的策略。毫不出奇，结构化编程也是。

重点已经在那里了。尽可能找到并消除重复。

G6：在错误的抽象层级上的代码

创建分离较高层级一般性概念与较低层级细节概念的抽象模型，这很重要。有时，我们创建抽象类来容纳较高层级概念，创建派生类来容纳较低层级概念。这样做的时候，需要确保分离完整。所有较低层级概念放在派生类中，所有较高层级概念放在基类中。

---

[1] [FRAG]。
[2] [GOF]。
[3] [GOF]。

例如，只与细节实现有关的常量、变量或工具函数不应该在基类中出现，基类应该对这些东西一无所知。

这条规则对于源文件、组件和模块也适用。良好的软件设计要求分离位于不同层级的概念，将它们放到不同容器中。这些容器有时是基类或派生类，有时是源文件、模块或组件。无论哪种情况，分离都要完整。较低层级概念和较高层级概念不应混杂在一起。看看下面的代码：

```
public interface Stack {
  Object pop() throws EmptyException;
  void push(Object o) throws FullException;
  double percentFull();
  class EmptyException extends Exception {}
  class FullException extends Exception {}
}
```

函数 `percentFull` 位于错误的抽象层级。尽管存在许多在其中"充满"（fullness）概念有意义的 `Stack` 的实现，但也有其他不能知道自己有多满的实现存在。所以，最好将该函数放在类似于 `BoundedStack` 之类的派生接口中。

你或许会认为，如果栈无边界，实现可以返回 0，但问题是，不存在真的无边界的栈。你不能真的避免在做以下检查时出现 `OutOfMemoryException` 异常：

```
stack.percentFull() < 50.0.
```

实现返回 0 的函数可能是在撒谎。

要点是你不能就错误放置的抽象模型撒谎。孤立抽象是软件开发者最难做到的事之一，而且一旦做错就没有快捷的修复手段。

G7：基类依赖派生类

将概念分解到基类和派生类的最普遍的原因是，较高层级基类概念可以不依赖较低层级派生类概念。这样，如果看到基类而提到派生类名称，就可能发现了问题。通常来说，基类对派生类应该一无所知。

当然也有例外。有时，派生类数量严格固定，而基类中拥有在派生类之间选择的代码。在有限状态机的实现中这种情形很多见。然而，在那种情况下，派生类和基类紧密耦合，总是在同一个 jar 文件中部署。一般情况下，我们会想要把派生类和基类部署到不同的 jar 文件中。

将派生类和基类部署到不同的 jar 文件中，确保基类 jar 文件对派生类 jar 文件的内容一无所知，我们就能把系统部署为分散和独立的组件。修改了这些组件时，不必重新部署基组件就能部署它们。这意味着修改产生的影响极大地降低了，而维护系统也变得更加简单。

G8：信息过多

设计良好的模块有着非常小的接口，让你事半功倍。设计低劣的模块有着广阔、深入的接口，你不得不事倍功半。设计良好的接口并不提供许多需要依赖的函数，所以耦合度也较低。设计低劣的接口提供大量你必须调用的函数，耦合度较高。

优秀的软件开发人员要学会限制类或模块中暴露的接口数量。类中的方法越少越好，函数知道的变量越少越好，类拥有的实体变量越少越好。

隐藏你的数据，隐藏你的工具函数，隐藏你的常量和你的临时变量。不要创建拥有大量方法或大量实体变量的类，不要为子类创建大量受保护变量和函数。尽力保持接口紧凑。通过限制信息来控制耦合度。

G9：死代码

死代码就是不执行的代码。可以在检查不会发生的条件的 `if` 语句体中找到，可以在从不抛出异常的 `try` 语句的 `catch` 块中找到，可以在从不被调用的小工具方法中找到，也可以在永不会发生的 `switch/case` 条件中找到。

死代码的问题是过不久它就会发出"坏味道"。时间越久，味道就越酸臭。这是因为，在设计改变时，死代码不会随之更新。它还能通过编译，但并不会遵循较新的约定或规则。编写它的时候，系统是另一番模样。如果你找到死代码，就体面地埋葬它，将它从系统中删除掉。

G10：垂直分隔

变量和函数应该在靠近被使用的地方定义。本地变量应该正好在其首次被使用的位置上面声明，垂直距离要短。本地变量不该在距离其被使用之处几百行以外的位置声明。

私有函数应该刚好在其首次被使用的位置下面定义。私有函数属于整个类，但我们还是要限制调用和定义之间的垂直距离。找一个私有函数，应该只是从其首次被使用处往下一点儿的位置那么简单。

G11：前后不一致

从一而终。这可以追溯到最小惊异原则。小心选择约定，一旦选中，就小心持续遵循。

如果在特定函数中用名为 `response` 的变量来持有 `HttpServletResponse` 对象，则在其他用到 `HttpServletResponse` 对象的函数中也用同样的变量名。如果将某个方法命名为 `processVerificationRequest`，则给处理其他请求类型的方法起类似的名字，例如 `processDeletionRequest`。

如此简单的前后一致，一旦坚决贯彻，就能让代码更加易于阅读和修改。

G12：混淆视听

没有实现的默认构造器有何用处呢？它只会用无意义的杂碎搞乱对代码的理解。没有用到的变量，从不调用的函数，没有信息量的注释，等等，这些都是应该移除的废物。保持源文件整洁，组织良好，不被搞乱。

G13：人为耦合

不互相依赖的东西不该耦合。例如，普通的 enum 不应被包括在特殊类中，因为这样一来应用程序就要了解这些更为特殊的类。对于在特殊类中声明一般目的的 static 函数也是如此。

一般来说，人为耦合是指两个没有直接目的的模块之间的耦合。其根源是将变量、常量或函数不恰当地放在临时方便的位置。这是一种漫不经心的偷懒行为。

花点儿时间研究应该在什么地方声明函数、常量和变量。不要为了方便随手放置，然后置之不理。

G14：特性依恋

这是 Martin Fowler 提出的代码的"坏味道"之一[①]。类的方法只应对其所属类中的变量和函数感兴趣，不该垂青其他类中的变量和函数。当方法通过某个其他对象的访问器和修改器来操作该对象内部数据时，它就依恋该对象所属类的范围。它期望自己在那个类里面，这样就能直接访问它操作的变量。例如：

```
public class HourlyPayCalculator {
  public Money calculateWeeklyPay(HourlyEmployee e) {
    int tenthRate = e.getTenthRate().getPennies();
    int tenthsWorked = e.getTenthsWorked();
    int straightTime = Math.min(400, tenthsWorked);
    int overTime = Math.max(0, tenthsWorked - straightTime);
    int straightPay = straightTime * tenthRate;
    int overtimePay = (int)Math.round(overTime*tenthRate*1.5);
    return new Money(straightPay + overtimePay);
  }
}
```

方法 calculateWeeklyPay 深入到 HourlyEmployee 对象，获取要操作的数据。方法 calculateWeeklyPay 依恋 HourlyEmployee 的作用范围，它"期望"自己能够在 HourlyEmployee 中。

同样情况下，我们要消除特性依恋，因为它将一个类的内部情形暴露给了另外一个类。不过，有时特性依恋是一种有必要的恶行。看一下下面的代码：

```
public class HourlyEmployeeReport {
  private HourlyEmployee employee ;

  public HourlyEmployeeReport(HourlyEmployee e) {
    this.employee = e;
  }

  String reportHours() {
```

---

① [Refactoring]。

```
    return String.format(
      "Name: %s\tHours:%d.%1d\n",
      employee.getName(),
      employee.getTenthsWorked()/10,
      employee.getTenthsWorked()%10);
  }
}
```

显然，reportHours 方法依恋 HourlyEmployee 类。另外，我们并不想要 HourlyEmployee 得知报告的格式。把格式化字符串移到 HourlyEmployee 会破坏好几种面向对象设计原则[①]。它将把 HourlyEmployee 与报告的格式耦合起来，向该格式的修改暴露这个类。

G15：选择算子参数

没有什么比在函数调用末尾遇到一个 false 参数更为可憎的事情了，这个 false 是什么意思呢？如果它是 true，会有什么变化吗？不仅一个选择算子（selector）参数的目的令人难以记住，而且每个选择算子参数将多个函数绑到了一起。选择算子参数只是一种避免把大函数切分为多个小函数的偷懒做法。考虑下面这段代码：

```java
public int calculateWeeklyPay(boolean overtime) {
  int tenthRate = getTenthRate();
  int tenthsWorked = getTenthsWorked();
  int straightTime = Math.min(400, tenthsWorked);
  int overTime = Math.max(0, tenthsWorked - straightTime);
  int straightPay = straightTime * tenthRate;
  double overtimeRate = overtime ? 1.5 : 1.0 * tenthRate;
  int overtimePay = (int)Math.round(overTime*overtimeRate);
  return straightPay + overtimePay;
}
```

当加班时间以一倍半计算薪资时，用 true 调用这个函数，false 则表示直接计算。每次用到这个函数，你都得记住 calculateWeeklyPay(false) 表示什么，这已经足够糟糕了，但这种函数真正的坏处在于作者错过了这样写的机会：

```java
public int straightPay() {
  return getTenthsWorked() * getTenthRate();
}

public int overTimePay() {
  int overTimeTenths = Math.max(0, getTenthsWorked() - 400);
  int overTimePay = overTimeBonus(overTimeTenths);
  return straightPay() + overTimePay;
}

private int overTimeBonus(int overTimeTenths) {
  double bonus = 0.5 * getTenthRate() * overTimeTenths;
```

[①] 具体是单一权责原则、开放闭合原则和公共关闭原则。参见[PPP]。

```
    return (int) Math.round(bonus);
}
```

当然，选择算子不一定是 `boolean` 类型，而可能是枚举元素、整数或任何一种用于选择函数行为的参数。使用多个函数，通常优于向单个函数传递某些代码来选择函数行为。

G16：晦涩的意图

代码要尽可能具有表达力。联排表达式、匈牙利语标记法和魔术数都遮蔽了作者的意图。例如，下面是 overTimePay 函数可能的一种表现形式：

```
public int m_otCalc() {
  return iThsWkd * iThsRte +
    (int) Math.round(0.5 * iThsRte *
      Math.max(0, iThsWkd - 400)
    );
}
```

它既短小又紧凑，但实际上不可捉摸。值得花时间将代码的意图呈现给读者。

G17：位置错误的权责

软件开发者做出的最重要决定之一就是在哪里放代码。例如，`PI` 常量放在何处？是应该放在 `Math` 类中吗？或者应该属于 `Trigonometry` 类？还是属于 `Circle` 类？

最小惊异原则在这里起作用了。代码应该放在读者自然而然期待它所在的地方。`PI` 常量应该出现在声明三角函数的地方。`OVERTIME_RATE` 常量应该在 `HourlyPayCalculator` 类中声明。

有时，我们"聪明"地知道在何处放置功能代码，我们会将其放在自己方便而读者不能随直觉找到的地方。例如，也许我们需要打印出某个雇员的总工作时间的报表。我们可以在打印报表的代码中做工作时间统计，或者我们可以在接受工作时间卡的代码中保留一份工作时间记录。

做这个决定的途径之一是看函数名称。例如，报表模块有一个名为 `getTotalHours` 的函数。接受时间卡的模块有一个 `saveTimeCard` 函数。顾名思义，哪个名称暗示了函数会计算总时间呢？答案显而易见。

显然，对于总时间应该在接受时间卡的时候计算，而不是在打印报表时计算，这里面有些性能上的考量。没问题，但函数名称应该反映这种考虑。例如，应该在时间卡模块中有个 `computeRunningTotalOfHours` 函数。

G18：不恰当的静态方法

`Math.max(double a, double b)` 是一个良好的静态方法。它并不在单个实体上操作。的确，不得不写 `new Math().max(a,b)` 甚至 `a.max(b)` 实在愚蠢。那个 `max` 用到的全部数据来自其两个参数，而不是来自"所属"对象，而且，我们也没机会用到 `Math.max` 的多态特征。

不过，我们有时也编写不该是静态的静态方法。例如：

`HourlyPayCalculator.calculatePay(employee, overtimeRate)`。

这看起来像是一个有道理的 `static` 函数。它并不在任何特定对象上操作，而且从参数中获得全部数据。然而，我们却有理由希望这个函数是多态的。我们可能希望为计算每小时支付的工资实现几种不同算法，如 `OvertimeHourlyPayCalculator` 和 `StraightTimeHourlyPayCalculator`。所以，在这种情况下，该函数就不该是静态的。它应该是 `Employee` 的非静态成员函数。

通常应该倾向于选用非静态方法。如果有疑问，就用非静态函数。如果的确需要静态函数，确保没机会打算让它有多态行为。

G19：使用解释性变量

Kent Beck 在其巨著 *Smalltalk Best Practice Patterns*[1] 和 *Implementation Patterns*[2] 中都写到这个。让程序可读的最有力方法之一就是将计算过程打散成在用有意义的单词命名的变量中放置的中间值。

看看来自 FitNesse 的这个例子：

```
Matcher match = headerPattern.matcher(line);
if(match.find())
{
  String key = match.group(1);
  String value = match.group(2);
  headers.put(key.toLowerCase(), value);
}
```

解释性变量的这种简单用法，说明了第一个匹配组是 `key`，而第二个匹配组是 `value`。这事很难做过火。解释性变量多比少好。只要把计算过程打散成一系列命名良好的中间值，不透明的模块就会突然变得透明，这很值得注意。

G20：函数名称应该表达其行为

看看这行代码：

`Date newDate = date.add(5);`

你会期望它向日期添加 5 天吗？或者是 5 星期？5 小时？该 `date` 实体会变化吗？或者该函数只返回一个新的 `Date` 实体，并不改动旧的？从函数调用中看不出函数的行为。

如果函数向日期添加 5 天并且修改该日期，就该命名为 `addDaysTo` 或 `increaseByDays`。如果函数返回一个表示 5 天后的日期，而不修改日期实体，就该叫作

---

[1] [Beck97]。
[2] [Beck07]。

daysLater 或 daysSince。

如果你必须查看函数的实现（或文档）才知道它是做什么的，就该换个更好的函数名，或者重新安排功能代码，放到有较好名称的函数中。

G21：理解算法

好多可笑代码的出现，是因为人们没花时间去理解算法。他们硬塞进足够多的 if 语句和标识，从不真正停下来考虑发生了什么，勉强让系统能工作。

编程常常是一种探险。你以为自己知道某事的正确算法，然后就卷起袖子瞎干一气，搞到"可以工作"为止。你怎么知道它"可以工作"？因为它通过了你能想到的单元测试。这种做法没错。实际上，这也是让函数按你设想的方式执行的唯一途径。不过，"可以工作"周围的引号可不能一直保留。

在你认为自己完成某个函数之前，确认自己理解了它是怎么工作的。通过全部测试还不够好。你必须知道[1]解决方案是正确的。

获得这种知识和理解的最好途径，往往是重构函数，得到某种整洁而足具表达力、清楚呈示如何工作的东西。

G22：把逻辑依赖改为物理依赖

如果某个模块依赖另一个模块，依赖就该是物理上的而不是逻辑上的。依赖者模块不应对被依赖者模块有假定（换言之，逻辑依赖），它应当明确地询问后者全部信息。

例如，想象你在编写一个打印出雇员工作时长的纯文本报表的函数。有一个名为 HourlyReporter 的类把数据收集为某种方便的形式，传递到 HourlyReportFormatter 中，再打印出来。（如代码清单 17-1 所示。）

代码清单 17-1　HourlyReporter.java

```java
public class HourlyReporter {
  private HourlyReportFormatter formatter;
  private List<LineItem> page;
  private final int PAGE_SIZE = 55;

  public HourlyReporter(HourlyReportFormatter formatter) {
    this.formatter = formatter;
    page = new ArrayList<LineItem>();
  }

  public void generateReport(List<HourlyEmployee> employees) {
    for (HourlyEmployee e : employees) {
      addLineItemToPage(e);
      if (page.size() == PAGE_SIZE)
```

---

[1] 了解代码如何工作与了解算法是否按需要执行是不一样的。不确定算法是否恰当司空见惯，而不确定代码做什么却是一种懒惰行为。

```
      printAndClearItemList();
    }
    if (page.size() > 0)
      printAndClearItemList();
  }

  private void printAndClearItemList() {
    formatter.format(page);
    page.clear();
  }

  private void addLineItemToPage(HourlyEmployee e) {
    LineItem item = new LineItem();
    item.name = e.getName();
    item.hours = e.getTenthsWorked() / 10;
    item.tenths = e.getTenthsWorked() % 10;
    page.add(item);
  }

  public class LineItem {
    public String name;
    public int hours;
    public int tenths;
  }
}
```

这段代码有尚未物理化的逻辑依赖。你能指出来吗？那就是常量 PAGE_SIZE。Hourly-Reporter 为什么要知道页面尺寸？页面尺寸只该是 HourlyReportFormatter 的权责。

PAGE_SIZE 在 HourlyReporter 中声明，代表了一种位置错误的权责[G17]，导致 HourlyReporter 假定它知道页面尺寸。这类假设是一种逻辑依赖。HourlyReporter 依赖 HourlyReportFormatter 能应对 55 的页面尺寸。如果 HourlyReportFormatter 的某些实现不能处理这样的尺寸，就会出错。

可以通过创建 HourlyReport 中名为 getMaxPageSize() 的新方法来物理化这种依赖。HourlyReporter 将调用这个方法，而不是使用 PAGE_SIZE 常量。

G23：用多态替代 If/Else 或 Switch/Case

有了第 6 章谈及的主题，这条建议看似奇怪。在第 6 章中，我提出在添加新函数甚于添加新类型的系统中，switch 语句是恰当的。

首先，多数人使用 switch 语句，因为它是最直截了当又有力的方案，而不是因为它适合当前情形。这给我们的启发是在使用 switch 之前，先考虑使用多态。

其次，函数变化甚于类型变化的情形相对罕见。每个 switch 语句都值得怀疑。

我使用所谓"单个 switch"规则：对于给定的选择类型，不应有多于一个的 switch 语句。在那个 switch 语句中的多个 case，必须创建多态对象，取代系统中其他类似的 switch 语句。

G24：遵循标准约定

每个团队都应遵循基于通用行业规范的一套编码标准。编码标准应指定诸如在何处声明实体变量，如何命名类、方法和变量，在何处放置括号，等等。团队不应用文档描述这些约定，因为代码本身提供了范例。

团队中的每个成员都应遵循这些约定。这意味着每个团队成员必须成熟到能了解只要全体同意在何处放置括号，那么在哪里放置都无关紧要。

如果你想知道我遵循哪些约定，可以查看代码清单 B-7 至代码清单 B-14 中重构之后的代码。

G25：用命名常量替代魔术数

这大概是软件开发中最古老的规则之一了。我记得，在 20 世纪 60 年代介绍 COBOL、FORTRAN 和 PL/1 的手册中就读到过。在代码中出现原始形态数字通常来说是坏现象。应该用命名良好的常量来隐藏它。

例如，数字 86400 应当藏在常量 SECONDS_PER_DAY 后面。如果每页打印 55 行，则常数 55 应该藏在常量 LINES_PER_PAGE 后面。

有些常量与非常具有自我解释能力的代码协同工作时，如此易于识别，也就不必总是需要命名常量来隐藏了。例如：

```
double milesWalked = feetWalked/5280.0;
int dailyPay = hourlyRate * 8;
double circumference = radius * Math.PI * 2;
```

在上例中，我们真需要常量 FEET_PER_MILE、WORK_HOURS_PER_DAY 和 TWO 吗？显然，最后那个很可笑。有些情况下，常量直接写作原始形态数字会更好。你可能会质疑 WORK_HOURS_PER_DAY，因为约定规则可能会改变。另外，在这里直接用数字 8 读起来很舒服，也就没必要非用 17 个额外的字母来加重读者负担。对于 FEET_PER_MILE，数字 5280 众人皆知，意义独特，即便没有上下文环境，读者也能识别它。

3.141 592 653 589 793 之类的常数也众所周知，很容易识别。不过，如果直接使用原始形式，则很有可能出错。每次有人看到 3.141 592 653 589 793，都会知道那是 π 值，从而不会去仔细查看。（你发现那个错误的数字了吗？）我们不想要人们使用 3.14、3.14 159 或 3.142 等。所以，为我们定义好 Math.PI 是一件好事。

术语"魔术数"不仅指数字，它还泛指任何不能自我描述的符号。例如：

```
assertEquals(7777, Employee.find("John Doe").employeeNumber());
```

上述断言中有两个魔术数。第一个魔术数显然是 7777，它的意义并不明确。第二个魔术数是"John Doe"，因为其意图不明显。

"John Doe"是开发团队创建的测试数据中编号为#7777 的雇员。团队中每个成员都知道，当连接到数据库时，里面已经有数个雇员信息，其值和属性都是大家熟知的。所以，这个测试应该读作：

```
assertEquals(
  HOURLY_EMPLOYEE_ID,
  Employee.find(HOURLY_EMPLOYEE_NAME).employeeNumber());
```

G26：准确

期望某个查询的第一次匹配就是唯一匹配可能过于天真。用浮点数表示货币几近于犯罪。因为你不想做并发更新就避免使用锁和/或事务管理往好处说也是一种懒惰行为。在可以用 `List` 的时候非要把变量声明为 `ArrayList` 就过于拘束了。把所有变量设置为 `protected` 却不够自律。

在代码中做决定时，确认自己足够准确。明确自己为何要这么做，如果遇到异常情况如何处理。别懒得理会决定的准确性。如果你打算调用可能返回 `null` 的函数，确认自己检查了 `null` 值。如果查询你认为是数据库中唯一的记录，确保代码检查不存在其他记录。如果要处理货币数据，使用整数[①]，并恰当地处理四舍五入。如果可能有并发更新，确认你实现了某种锁定机制。

代码中的含糊和不准确要么是意见不同的结果，要么源于懒惰。无论原因是什么，都要消除。

G27：结构甚于约定

坚守结构甚于约定的设计原则。命名约定很好，但却次于强制性的结构。例如，用到命名良好的枚举的 `switch/case` 要弱于拥有抽象方法的基类。没人会被强迫每次都以同样方式实现 `switch/case` 语句，但基类却让具体类必须实现所有抽象方法。

G28：封装条件

如果没有 `if` 或 `while` 语句的上下文，布尔逻辑就难以理解。应该把解释了条件意图的函数抽离出来。

例如：

```
if (shouldBeDeleted(timer))
```

要好于

```
if (timer.hasExpired() && !timer.isRecurrent())
```

G29：避免否定性条件

否定式要比肯定式难明白一些。所以，尽可能将条件表示为肯定形式。例如：

```
if (buffer.shouldCompact())
```

[①] 或者用更好的使用整数的 `Money` 类。

要好于

```
if (!buffer.shouldNotCompact())
```

G30：函数只该做一件事

编写执行一系列操作的包括多段代码的函数常常是诱人的。这类函数做了不只一件事，应该转换为多个更小的函数，每个小函数只做一件事。

例如：

```
public void pay() {
  for (Employee e : employees) {
    if (e.isPayday()) {
      Money pay = e.calculatePay();
      e.deliverPay(pay);
    }
  }
}
```

这段代码做了 3 件事。它遍历所有雇员，检查是否该给雇员付工资，然后支付薪水。代码可以写得更好，如：

```
public void pay() {
  for (Employee e : employees)
    payIfNecessary(e);
}

private void payIfNecessary(Employee e) {
  if (e.isPayday())
    calculateAndDeliverPay(e);
}

private void calculateAndDeliverPay(Employee e) {
  Money pay = e.calculatePay();
  e.deliverPay(pay);
}
```

上列每个函数都只做一件事。（见 3.2 节。）

G31：掩蔽时序耦合

常常有必要使用时序耦合，但你不应该掩蔽它。排列函数参数，好让它们被调用的次序显而易见。看下列代码：

```
public class MoogDiver {
  Gradient gradient;
  List<Spline> splines;

  public void dive(String reason) {
    saturateGradient();
    reticulateSplines();
```

```
    diveForMoog(reason);
  }
  ...
}
```

3 个函数的次序很重要。捕鱼之前先织网，织网之前先编绳。不幸的是，代码并没有强制这种时序耦合。其他程序员可以在调用 `saturateGradient` 之前调用 `reticulateSplines`，从而导致抛出 `UnsaturatedGradientException` 异常。更好的方式是：

```
public class MoogDiver {
  Gradient gradient;
  List<Spline> splines;

  public void dive(String reason) {
    Gradient gradient = saturateGradient();
    List<Spline> splines = reticulateSplines(gradient);
    diveForMoog(splines, reason);
  }
  ...
}
```

这样就通过创建顺序队列暴露了时序耦合。每个函数都产生出下一个函数所需的结果，这样一来就没理由不按顺序调用了。

你可能会抱怨这增加了函数的复杂度，没错，不过这点额外的复杂度却曝露了该种情况真正的时序复杂性。

注意，我保留了那些实体变量。我假设类中的私有方法可能会用到它们。即便如此，我还是希望参数能让时序耦合变得可见。

G32：别随意

构建代码需要理由，而且理由应与代码结构相契合。如果结构显得太随意，其他人就会想修改它。如果结构自始至终保持一致，其他人就会使用它，并且遵循其约定。例如，我最近对 FitNesse 做合并修改，发现有位贡献者这么做：

```
public class AliasLinkWidget extends ParentWidget
{
  public static class VariableExpandingWidgetRoot {
    ...

    ...
  }
```

问题在于，`VariableExpandingWidgetRoot` 没必要在 `AliasLinkWidget` 作用范围之内。而且，其他无关的类也用到 `AliasLinkWidget.VariableExpandingWidgetRoot`，这些类没必要了解 `AliasLinkWidget`。

或许那位程序员只是循例把 `VariableExpandingWidgetRoot` 放到 `AliasWidget` 里面，或者他真认为这么做是对的。不管是什么原因，结果都显得随意。不作为类工具的公共类，不应该放到其他类里面，惯例是将它置为 `public`，并且放在代码包的顶部。

G33：封装边界条件

边界条件难以追踪。把处理边界条件的代码集中到一处，不要散落于代码中。我们不想见到四处散见的 +1 和 -1 字样。看看这个来自 FIT 的简单例子：

```
if(level + 1 < tags.length)
{
  parts = new Parse(body, tags, level + 1, offset + endTag);
  body = null;
}
```

注意，`level+1` 出现了两次。这是一个应该封装到名为 `nextLevel` 之类的变量中的边界条件。

```
int nextLevel = level + 1;
if(nextLevel < tags.length)
{
  parts = new Parse(body, tags, nextLevel, offset + endTag);
  body = null;
}
```

G34：函数应该只在一个抽象层级上

函数中的语句应该在同一抽象层级上，该层级应该是函数名所示操作的下一层。这可能是最难理解和遵循的启发。尽管概念足够直白，但是人们还是很容易混淆抽象层级。例如，请看下面来自 FitNesse 的例子：

```
public String render() throws Exception
{
  StringBuffer html = new StringBuffer("<hr");
  if(size > 0)
    html.append(" size=\"").append(size + 1).append("\"");
  html.append(">");

  return html.toString();
}
```

稍微研究一下，你就会看到发生了什么。该函数构建了绘制横贯页面线条的 HTML 标记。线条高度在 `size` 变量中指定。

再看一遍。方法混杂了至少两个抽象层级，第一个是横线有尺寸这个概念，第二个是 `hr` 标记自身的语法。这段代码来自 FitNesse 的 `HruleWidget` 模块。该模块检测一行 4 个或更多个破折号，并将其转换为恰当的 `hr` 标记。破折号越多，尺寸越大。

我重构了这段代码。注意，我修改了 `size` 字段的名称，新名称反映其真正目的，表示额外的破折号的数量。

```
public String render() throws Exception
{
  HtmlTag hr = new HtmlTag("hr");
  if (extraDashes > 0)
    hr.addAttribute("size", hrSize(extraDashes));
  return hr.html();
}

private String hrSize(int height)
{
  int hrSize = height + 1;
  return String.format("%d", hrSize);
}
```

这次修改很好地拆开了两个抽象层级。函数 render 只构造一个 hr 标记，不去管该标记的 HTML 语法。而 HtmlTag 模块则照管所有这些肮脏的语法问题。

做出修改时，我发现了一处微小的错误。原始代码没有加上 hr 标记的结束斜线符，而 XHTML 标准要求这样做。（换言之，代码使用了\<hr>而不是\<hr />。）HtmlTag 模块很早就改造成符合 XHTML 标准了。

拆分不同抽象层级是重构的最重要功能之一，也是最难实现的一个功能。以下面的代码为例。这是我第一次尝试拆分 `HruleWidget.rendermethod` 中的抽象层级的结果。

```
public String render() throws Exception
{
  HtmlTag hr = new HtmlTag("hr");
  if (size > 0) {
    hr.addAttribute("size", ""+(size+1));
  }
  return hr.html();
}
```

此时，我的目的是做必要的拆分，并让测试通过。我轻易达到了这一目的，但结果是该函数仍然混杂了多个抽象层级。此时，混杂的层级是 hr 标记的构建，以及 size 变量的翻译和格式化。这说明当你循抽象界线拆解函数时，经常会挖出原本被之前的结构所掩蔽的新抽象界线。

G35：在较高层级放置可配置数据

如果你有一个已知并该在较高抽象层级的默认常量或配置值，不要将它埋藏到较低层级的函数中。把它作为较高层级函数调用较低层级函数时的一个参数。看看以下来自 FitNesse 的代码：

```
public static void main(String[] args) throws Exception
{
  Arguments arguments = parseCommandLine(args);
  ...
}
```

```java
public class Arguments
{
  public static final String DEFAULT_PATH = ".";
  public static final String DEFAULT_ROOT = "FitNesseRoot";
  public static final int DEFAULT_PORT = 80;
  public static final int DEFAULT_VERSION_DAYS = 14;
  ...
}
```

命令行参数在 FitNesse 中的第一行可执行代码得到解析。这些参数的默认值在 Argument 类的顶部指定。你不必到系统的较低层级去查看类似的语句：

```java
if (arguments.port == 0) // use 80 by default
```

位于较高层级的配置性常量易于修改，它们向下贯穿应用程序。应用程序的较低层级并不拥有这些常量的值。

G36：避免传递浏览

通常我们不想让某个模块了解太多其协作者的信息。更具体地说，如果 A 与 B 协作，B 与 C 协作，我们不想让使用 A 的模块了解 C 的信息。（例如，我们不想写类似于 a.getB().getC().doSomething() 的代码。）

这就是所谓的得墨式耳律。《程序员修炼之道》(*The Pragmatic Programmers*) 称之为 "编写害羞代码"[1]。两者都归结为确保模块只了解其直接协作者，而不了解整个系统的游览图。

如果有多个模块使用类似 a.getB().getC() 这样的语句形式，就难以修改设计和架构，要在 B 和 C 之间插进一个 Q，就得找到 a.getB().getC() 出现的所有地方，并将其改为 a.getB().getQ().getC()。系统就会变得缺乏柔韧性。太多的模块了解了太多有关架构的信息。

正确的做法是让直接协作者提供所需的全部服务，而不必逛遍系统的对象全图，搜寻我们要调用的方法。只要简单地写：

```
myCollaborator.doSomething().
```

17.5 Java

J1：通过使用通配符避免过长的导入清单

如果使用了来自同一程序包的两个或多个类，用以下语句导入整个包：

```
import package.*;
```

---

[1] [PRAG]，P. 138。

过长的导入清单令读者望而却步。我们不想用 80 行导入语句搞乱模块顶部位置。我们想要导入语句简约地列出我们要使用的包。

指定导入包是一种硬依赖，而通配符导入则不是。如果你具体指定导入某个类，那么该类必须存在。但如果你用通配符导入某个包，则不需要存在具体的类。导入语句只在搜寻名称时把这个包列入查找路径。所以，这种导入并未构成真正的依赖，也就让我们的模块较少耦合。

有时，长长的具体导入清单也会有用。例如，如果你在处理遗留下来的代码，想要找出需要为哪些类构造替身类和占位代码，就可以遍历导入清单，找出这些类的真名，再恰当地放置占位代码。不过，这种用法很罕见。而且，多数现代 IDE 允许你用一个命令就把通配符导入语句转换为指定导入清单。所以，即便在处理遗留代码时，最好也用通配符导入。

通配符导入有时会导致名称冲突和歧义。两个同名但位于不同包中的类需要指名导入，或至少在使用时指定名称。这种情形的确讨厌，不过很罕见，所以使用通配符导入通常仍优于指定名称导入。

J2：不要继承常量

我见过这种情况好几次，它总是让我面露苦笑。某个程序在接口中放了一些常量，再通过继承结构来访问这些常量。看看以下代码：

```java
public class HourlyEmployee extends Employee {
  private int tenthsWorked;
  private double hourlyRate;

  public Money calculatePay() {
    int straightTime = Math.min(tenthsWorked, TENTHS_PER_WEEK);
    int overTime = tenthsWorked - straightTime;
    return new Money(
      hourlyRate * (tenthsWorked + OVERTIME_RATE * overTime)
    );
  }
  ...
}
```

常量 `TENTHS_PER_WEEK` 和 `OVERTIME_RATE` 来自何方？它们可能来自 `Employee` 类。来看看：

```java
public abstract class Employee implements PayrollConstants {
  public abstract boolean isPayday();
  public abstract Money calculatePay();
  public abstract void deliverPay(Money pay);
}
```

不，不在那儿，但在哪儿呢？再仔细看 `Employee` 类，它实现了 `PayrollConstants`

接口。

```java
public interface PayrollConstants {
  public static final int TENTHS_PER_WEEK = 400;
  public static final double OVERTIME_RATE = 1.5;
}
```

真是丑陋不堪！常量躲在了继承结构的最顶端。呸！别利用继承欺骗编程语言的作用范围规则，而应该用静态导入。

```java
import static PayrollConstants.*;

public class HourlyEmployee extends Employee {
  private int tenthsWorked;
  private double hourlyRate;

  public Money calculatePay() {
    int straightTime = Math.min(tenthsWorked, TENTHS_PER_WEEK);
    int overTime = tenthsWorked - straightTime;
    return new Money(
      hourlyRate * (tenthsWorked + OVERTIME_RATE * overTime)
    );
  }
  ...
}
```

J3：常量和枚举

现在 enum 已经加入 Java 语言（Java 5），放心用吧！别再用那个 `public static final int` 老花招，那样做 int 的意义就丧失了，而用 enum 则不然，因为它隶属于有名称的枚举。

而且，仔细研究 enum 的语法，它可以拥有方法和字段，从而成为能比 int 提供更多表达力和灵活性的强有力工具。看看以下发薪代码中的不同做法：

```java
public class HourlyEmployee extends Employee {
  private int tenthsWorked;
  HourlyPayGrade grade;

  public Money calculatePay() {
    int straightTime = Math.min(tenthsWorked, TENTHS_PER_WEEK);
    int overTime = tenthsWorked - straightTime;
    return new Money(
      grade.rate() * (tenthsWorked + OVERTIME_RATE * overTime)
    );
  }
  ...
}

public enum HourlyPayGrade {
  APPRENTICE {
    public double rate() {
      return 1.0;
```

```
    }
  },
  LIEUTENANT_JOURNEYMAN {
    public double rate() {
      return 1.2;
    }
  },
  JOURNEYMAN {
    public double rate() {
      return 1.5;
    }
  },
  MASTER {
    public double rate() {
      return 2.0;
    }
  };

  public abstract double rate();
}
```

17.6 名称

N1：采用描述性名称

不要太快起名。确认名称具有描述性。记住，事物的意义随着软件的演化而变化，所以，要经常性地重新估量名称是否恰当。

这不仅是一条"感觉良好式"建议。软件中的名称对于软件可读性有 90% 的作用。你要花时间明智地起名，保持名称有关。名称太重要了，不可随意对待。

看看以下代码。这段代码是做什么的？用了好名称的代码一目了然，而这样的代码却是符号和魔术数的大杂烩。

```
public int x() {
  int q = 0;
  int z = 0;
  for (int kk = 0; kk < 10; kk++) {
    if (l[z] == 10)
    {
      q += 10 + (l[z + 1] + l[z + 2]);
      z += 1;
    }
    else if (l[z] + l[z + 1] == 10)
    {
      q += 10 + l[z + 2];
      z += 2;
    } else {
      q += l[z] + l[z + 1];
```

```
      z += 2;
    }
  }
  return q;
}
```

下面是这段代码应该写成的样子。代码片段实际上不如上段完整，但你还是能马上推断出它要做什么，而且很有可能依据推断出的意思写出遗漏的函数。魔术数不复神秘，算法的结构也足具描述性。

```
public int score() {
  int score = 0;
  int frame = 0;
  for (int frameNumber = 0; frameNumber < 10; frameNumber++) {
    if (isStrike(frame)) {
      score += 10 + nextTwoBallsForStrike(frame);
      frame += 1;
    } else if (isSpare(frame)) {
      score += 10 + nextBallForSpare(frame);
      frame += 2;
    } else {
      score += twoBallsInFrame(frame);
      frame += 2;
    }
  }
  return score;
}
```

仔细起好的名称的威力在于，它用描述性信息覆盖了代码。这种信息覆盖设定了读者对于模块中其他函数行为的期待。看看上面的代码，你就能推断出 isStrike() 的实现。读到 isStrick 方法时，它"深合你意"[①]。

```
private boolean isStrike(int frame) {
  return rolls[frame] == 10;
}
```

N2：名称应与抽象层级相符

不要起沟通实现的名称，而起反映类或函数抽象层级的名称。这样做不容易。人们擅长于混杂抽象层级。每次浏览代码，你总会发现有些变量的名称层级太低。你应当趁机为之改名。要让代码可读，需要持续不断的改进。看看下面的 Modem 接口：

```
public interface Modem {
  boolean dial(String phoneNumber);
  boolean disconnect();
  boolean send(char c);
```

---

① 见 1.3.5 节中 Ward Cunningham 的引语。

```
  char recv();
  String getConnectedPhoneNumber();
}
```

粗看还行。函数看来都很合适，对多数应用程序来说是这样。不过，想想看某个应用中有些调制解调器并不用拨号连接的情形，有些用线缆直连，有些通过向 USB 口发送端口信息连接。显然，有关电话号码的信息就是位于错误的抽象层级了。对于这种情形，更好的命名策略可能是：

```
public interface Modem {
  boolean connect(String connectionLocator);
  boolean disconnect();
  boolean send(char c);
  char recv();
  String getConnectedLocator();
}
```

现在名称不再与电话号码有关系，并且还是可以用于用电话号码的情形，也可以用于其他连接策略。

N3：尽可能使用标准命名法

如果名称基于既存约定或用法，就比较易于理解。例如，如果你采用油漆工模式，就该在给油漆类命名时用上 Decorator 字样。例如，AutoHangupModemDecorator 可能是某个给 Modem 类刷上在会话结束时自动挂机的能力的类的名称。

模式只是标准的一种。例如，在 Java 中，将对象转换为字符串的函数通常命名为 toString。最好是遵循这些约定，而不是自己创造命名法。

对于特定项目，开发团队常常发明自己的命名标准系统。Eric Evans 称之为项目的共同语言[①]。代码应该使用来自这种语言的术语。简言之，具有与项目有关的特定意义的名称用得越多，读者就越容易明白你的代码是做什么的。

N4：无歧义的名称

选用不会混淆函数或变量意义的名称。看看来自 FitNesse 的这个例子：

```
private String doRename() throws Exception
{
  if(refactorReferences)
    renameReferences();
  renamePage();

  pathToRename.removeNameFromEnd();
  pathToRename.addNameToEnd(newName);
  return PathParser.render(pathToRename);
}
```

---

① [DDD].

该函数的名称含混不清，没有说明函数的作用。由于在 doRename 函数里面还有一个名为 renamePage 的函数，这就更不明白了！这些名称有没有说明两个函数之间的区别呢？没有。

更好的函数名称应该是 renamePageAndOptionallyAllReferences。看似太长，的确是很长，不过它只在模块中的一处被调用，所以其解释性的好处大过了长度的坏处。

N5：为较大作用范围选用较长名称

名称的长度应与作用范围的广泛度相关。对于较小的作用范围，可以用很短的名称，而对于较大的作用范围，就该用较长的名称。

类似 i 和 j 之类的变量名对于作用范围在 5 行之内的情形没问题。看看以下来自老"标准保龄球游戏"的代码片段：

```
private void rollMany(int n, int pins)
{
  for (int i=0; i<n; i++)
    g.roll(pins);
}
```

这段代码很明白，如果用 rollCount 之类烦人的名称代替变量 i，反而是徒增混乱。另外，在较长距离上，使用短名称的变量和函数会丧失其含义。名称的作用范围越大，名称就该越长、越准确。

N6：避免编码

不应在名称中包括类型或作用范围信息。在如今的开发环境中，m_ 或 f 之类的前缀完全无用。类似 vis_（表示图形系统）之类的项目或子系统名称也属多余。当今的开发环境不用纠缠于名称也能提供这些信息。不要用匈牙利语命名法污染你的名称。

N7：名称应该说明副作用

名称应该说明函数、变量或类的一切信息。不要用名称掩蔽副作用。不要用简单的动词来描述做了不止一个简单动作的函数。例如，请看以下来自 TestNG 的代码：

```
public ObjectOutputStream getOos() throws IOException {
  if (m_oos == null) {
    m_oos = new ObjectOutputStream(m_socket.getOutputStream());
  }
  return m_oos;
}
```

该函数不只是获取一个"oos"，如果"oos"不存在，还会创建一个"oos"。所以，更好的名称大概是 createOrReturnOos。

17.7 测试

T1：测试不足

一套测试中应该有多少个测试？不幸的是，许多程序员的衡量标准是"看起来够了"。一套测试应该测到所有可能失败的东西。只要还有没被测试探测过的条件，或是还有没被验证过的计算，测试就还不够。

T2：使用覆盖率工具

覆盖率工具能汇报你测试策略中的缺口。使用覆盖率工具能更容易地找到测试不足的模块、类和函数。多数 IDE 都能给出直观的指示，用绿色标记测试覆盖了的代码行，而未覆盖的代码行则是红色。这样就能又快又容易地找到尚未检测过的 if 或 catch 语句。

T3：别略过小测试

小测试易于编写，其文档上的价值高于编写成本。

T4：被忽略的测试就是对不确定事物的疑问

有时，我们会因为需求不明而不能确定某个行为细节。可以用注释掉的测试或者用 @Ignore 注解的测试来表达我们对于需求的疑问。使用哪种方式，取决于该不确定性所涉及的代码是否要编译。

T5：测试边界条件

特别注意测试边界条件。算法的中间部分正确但边界判断错误的情形很常见。

T6：全面测试相近的缺陷

缺陷趋向于扎堆。在某个函数中发现一个缺陷时，最好全面测试那个函数。你可能会发现缺陷不止一个。

T7：测试失败的模式有启发性

有时，你可以通过找到测试用例失败的模式来诊断问题所在。这也是尽可能编写足够完整的测试用例的理由之一。完整的测试用例，按合理的顺序排列，能暴露出模式。

简单举例，假设你注意到所有长于 5 个字符的输入，或者向函数的第二个参数传入负数都会导致测试失败。有时，只要看看测试报告的红绿模式，就足以绽放出那句带来解决方法的"啊哈！"回头看看第 16 章中的有趣例子吧。

T8：测试覆盖率的模式有启发性

查看被或未被已通过的测试执行的代码，往往能发现失败的测试为何失败的线索。

T9：测试应该快速

慢速的测试是不会被运行的测试。时间一紧，较慢的测试就会被摘掉。所以，竭尽所能让测试够快。

17.8 小结

这份关于启发与坏味道的清单很难说已完备无缺。我不能确定这样一份清单会不会完备无缺。但或许完整性不该是目标，因为该清单确实给出了一套价值体系。

这套价值体系才该是目标，也是本书的主题所在。整洁代码并非遵循一套规则写就。学习一系列启发并不足以让你成为软件匠人。专业性和技艺来自驱动规程的价值观。

17.9 文献

[Refactoring]：*Refactoring: Improving the Design of Existing Code,* Martin Fowler et al., Addison-Wesley, 1999.
[PRAG]：*The Pragmatic Programmer,* Andrew Hunt, Dave Thomas, Addison-Wesley, 2000.
[GOF]：*Design Patterns: Elements of Reusable Object Oriented Software,* Gamma et al., Addison-Wesley, 1996.
[Beck97]：*Smalltalk Best Practice Patterns,* Kent Beck, Prentice Hall, 1997.
[Beck07]：*Implementation Patterns,* Kent Beck, Addison-Wesley, 2008.
[PPP]：*Agile Software Development: Principles, Patterns, and Practices,* Robert C. Martin, Prentice Hall, 2002.
[DDD]：*Domain Driven Design,* Eric Evans, Addison-Wesley, 2003.

附录 A

并发编程 II

Brett L. Schuchert

本附录扩充了第 13 章的内容，由一组相互独立的主题组成，你可以按随意顺序阅读。为了实现这样的阅读方式，节与节之间存在一些重复内容。

A.1 客户端/服务器的例子

想象一个简单的客户端/服务器应用程序。服务器在一个套接字上等待接受来自客户端的连接请求。客户端连接到服务器并发送请求。

A.1.1 服务器

下面是服务器应用程序的简化版本代码。在 A.10 节中有完整的代码。

```
ServerSocket serverSocket = new ServerSocket(8009);

while (keepProcessing) {
  try {
    Socket socket = serverSocket.accept();
    process(socket);
  } catch (Exception e) {
    handle(e);
  }
}
```

这个简单的应用等待连接请求，处理接收到的新消息，再等待下一个客户端请求。下面是连接到服务器的客户端代码：

```
private void connectSendReceive(int i) {
  try {
    Socket socket = new Socket("localhost", PORT);
    MessageUtils.sendMessage(socket, Integer.toString(i));
    MessageUtils.getMessage(socket);
    socket.close();
  } catch (Exception e) {
    e.printStackTrace();
  }
}
```

这对客户端/服务器程序运行得如何呢？怎样才能正式地描述其性能？下面是断言其性能"可接受"的测试：

```
@Test(timeout = 10000)
public void shouldRunInUnder10Seconds() throws Exception {
  Thread[] threads = createThreads();
  startAllThreads(threads);
  waitForAllThreadsToFinish(threads);
}
```

为了让例子够简单，设置过程被忽略了（见代码清单 A-4）。测试断言程序应该在 10 000 毫秒内完成。

这是个验证系统吞吐量的典型例子。系统应该在 10 秒内完成一组客户端请求。只要服务器能在时限内处理每个客户端请求，测试就通过了。

如果测试失败会怎样？缺少了某些事件轮询机制，在单个线程上也没什么可让代码更快的手段。使用多线程能解决问题吗？可能会，我们先得了解什么地方耗费时间。下面是两种可能：

- I/O——使用套接字、连接到数据库、等待虚拟内存交换等；
- 处理器——数值计算、正则表达式处理、垃圾回收等。

以上在系统中都会部分存在，但对于特定的操作，其中之一会起主导作用。如果代码运行速度主要与处理器有关，增加处理器硬件就能提升吞吐量，从而通过测试。但 CPU 运算周期是有上限的，因此，只是增加线程的话并不会提升受处理器限制的代码的速度。

另外，如果吞吐量与 I/O 有关，则并发编程能提升运行效率。当系统的某个部分在等待 I/O，另一部分就可以利用等待的时间处理其他事务，从而更有效地利用 CPU 能力。

A.1.2 添加线程代码

假定性能测试失败了。如何才能提高吞吐量、通过性能测试呢？如果服务器的 `process` 方法与 I/O 有关，就有一个办法让服务器利用线程（只需要修改 `process` 方法）：

```
void process(final Socket socket) {
  if (socket == null)
    return;

  Runnable clientHandler = new Runnable() {
    public void run() {
      try {
        String message = MessageUtils.getMessage(socket);
        MessageUtils.sendMessage(socket, "Processed: " + message);
        closeIgnoringException(socket);
      } catch (Exception e) {
        e.printStackTrace();
      }
    }
  };

  Thread clientConnection = new Thread(clientHandler);
  clientConnection.start();
}
```

假设修改后测试通过了[①]。代码是否完整、正确了呢？

A.1.3 观察服务器端

修改了的服务器成功通过测试，只花费了一秒多时间。不幸的是，这种解决手段有点一厢情愿，而且导致了新问题产生。

服务器应该创建多少个线程？代码没有设置上限，所以我们很有可能达到 Java 虚拟机（JVM）的限制。对许多简单系统来说这无所谓。但如果系统要支持公众网络上的众多用户呢？如果有太多用户同时连接，系统就有可能挂掉。

不过先把性能问题放到一边吧。这种手段还有整洁性和结构上的问题。服务器代码有多少种权责呢？如下所列：

- 套接字连接管理；
- 客户端处理；
- 线程策略；
- 服务器关闭策略。

这些权责不幸全在 `process` 函数中。而且，代码跨越多个抽象层级。所以，即便 `process` 函数这么短小，也还是需要再加以切分。

服务器有多个修改的原因，所以它违反了单一权责原则。要保持并发系统整洁，应该将线程管理代码约束于少数几处控制良好的地方。而且，管理线程的代码只应该做管理线程的事。为什么？即便无须同时考虑其他非多线程代码，跟踪并发问题也已经足够困难了。

[①] 你可以自行验证修改之前和之后的代码。复查前文的非多线程代码。复查之后的多线程代码。

如果为上述每个权责（包括线程管理权责在内）创建单独的类，当改动线程管理策略时，就会对整个代码产生较小影响，不至于污染其他权责。这样一来，也能在不担心线程问题的前提下测试所有其他权责。下面是修改过的版本：

```
public void run() {
  while (keepProcessing) {
    try {
      ClientConnection clientConnection = connectionManager.awaitClient();
      ClientRequestProcessor requestProcessor
        = new ClientRequestProcessor(clientConnection);
      clientScheduler.schedule(requestProcessor);
    } catch (Exception e) {
      e.printStackTrace();
    }
  }
  connectionManager.shutdown();
}
```

所有与线程相关的东西都放到了 `clientScheduler` 里面。如果出现并发问题，只要看这个地方就可以了：

```
public interface ClientScheduler {
  void schedule(ClientRequestProcessor requestProcessor);
}
```

并发策略易于实现：

```
public class ThreadPerRequestScheduler implements ClientScheduler {
  public void schedule(final ClientRequestProcessor requestProcessor) {
    Runnable runnable = new Runnable() {
      public void run() {
        requestProcessor.process();
      }
    };

    Thread thread = new Thread(runnable);
    thread.start();
  }
}
```

把所有线程管理隔离到一个位置，修改控制线程的方式就容易多了。例如，移植到 Java 5 Executor 框架就只需要编写一个新类并插进来即可（如代码清单 A-1 所示）。

代码清单 A-1　ExecutorClientScheduler.java

```
import java.util.concurrent.Executor;
import java.util.concurrent.Executors;

public class ExecutorClientScheduler implements ClientScheduler {
  Executor executor;

  public ExecutorClientScheduler(int availableThreads) {
    executor = Executors.newFixedThreadPool(availableThreads);
```

```
    }

    public void schedule(final ClientRequestProcessor requestProcessor) {
        Runnable runnable = new Runnable() {
            public void run() {
                requestProcessor.process();
            }
        };
        executor.execute(runnable);
    }
}
```

A.1.4 小结

本例介绍的并发编程,演示了一种提高系统吞吐量的方法,以及一种通过测试框架验证吞吐量的方法。将全部并发代码放到少数类中,是应用单一权责原则的范例。对于并发编程,因其复杂性,这一点尤其重要。

A.2 执行的可能路径

复查没有循环或条件分支的单行 Java 方法 incrementValue:

```
public class IdGenerator {
  int lastIdUsed;

  public int incrementValue() {
    return ++lastIdUsed;
  }
}
```

忽略整数溢出的情形,假定只有单个线程能访问 `IdGenerator` 的单个实体。这种情况下,只有一种执行路径和一个确定的结果:

- 返回值等于 `lastIdUsed` 的值,两者都比调用方法前大 1。

如果使用两个线程、不修改方法的话会发生什么?如果每个线程都调用一次 incrementValue,可能得到什么结果呢?有多少种可能执行路径?首先来看结果(假定 `lastIdUsed` 初始值为 93):

- 线程 1 得到 94,线程 2 得到 95,`lastIdUsed` 为 95;
- 线程 1 得到 95,线程 2 得到 94,`lastIdUsed` 为 95;
- 线程 1 得到 94,线程 2 得到 94,`lastIdUsed` 为 94。

最后一个结果尽管令人吃惊,但还是有可能出现的。要想明白为何可能出现这些结果,就需要理解可能执行路径的数量以及 Java 虚拟机是如何执行这些路径的。

A.2.1　路径数量

为了算出可能执行路径的数量,我们从生成的字节码开始研究。那行 Java 代码(`return ++lastIdUsed;`)变成了 8 个字节码指令。两个线程有可能交错执行这 8 个指令,就像庄家在洗牌时交错牌张一样[①]。即便每只手上只有 8 张牌,洗牌得到的结果数量也很可观。

对于指令系列中有 N 个指令和 T 个线程、没有循环或条件分支的简单情况,总的可能执行路径数量等于

$$\frac{(NT)!}{N!^T}$$

计算可能执行次序

以下摘自鲍勃大叔给 Brett 的一封电子邮件。

对于 N 步指令和 T 个线程,总共有 $T*N$ 个步骤。在执行每步指令之前,会有在 T 个线程中选择其一的环境开关。因而每条路径都能以一个数字字符串的形式来表示该环境开关。对于步骤 A、B 及线程 1 和 2,可能有 6 条可能路径:1122、1212、1221、2112、2121 和 2211。或者以指令步骤表示为 A1B1A2B2、A1A2B1B2、A1A2B2B1、A2A1B1B2、A2A1B2B1 及 A2B2A1B1。对于 3 个线程,执行序列就是 112233、112323、113223、113232、112233、121233、121323、121332、123132、123123……

这些字符串的特征之一是每个 T 总会出现 N 次。所以字符串 111111 是无效的,因为里面有 6 个 1,而 2 和 3 则未出现过。

所以要排列组合 N1、N2……直至 NT。这其实就是 $N*T$ 对应 $N*T$ 的排列,即 $(N*T)!$,但要剔除重复的情形。所以,巧妙之处就在于计算重复次数并从 $(N*T)!$ 中剔除掉。

对于两步指令和两个线程,有多少重复呢?每个四位数字符串中都有两个 1 和两个 2。每个这种配对都可以在不影响字符串意义的前提下调换。可以同时调换全部 1 和 2,也可以都不调换。所以每个字符串就有四种同构形态,即存在 3 次重复。所以四分之三的路径是重复的,而四分之一的排列则不重复。4!×0.25=6。这样计算看来可行。

有多少重复呢?对 $N=1$ 且 $T=2$ 的情形,我可以调换 1,调换 2,或两者都调换。对 $N=2$ 且 $T=3$ 的情形,我可以调换 1、2、3,1 和 2,1 和 3,或 2 和 3。调换只是 N 的排列组合罢了。设有 N 的 P 种排列组合。排列组合的方式总共有 $P**T$ 种。

所以可能的同构形态数量为 $N!**T$。路径的数量就是 $(T*N)!/(N!**T)$。对 $T=2$ 且 $N=2$ 的情况,结果就是 6 (即 24/4)。

对 $N=2$ 且 $T=3$ 的情况,结果是 720/8=90。

对 $N=3$ 且 $T=3$ 的情况,结果是 $9!/6^3=1680$。

对于一行 Java 代码(等同于 8 行字节码)和两个线程的简单情况,可能执行路径的总数

[①] 这说得有点儿简单了。鉴于讨论的目的,我们就用这个简化模型了。

量就是 12 870。如果 `lastIdUsed` 的类型为 `long`，每次读/写操作都变成了两次操作，而可能的次序高达 2 704 156 种。

如果改动一下该方法会怎样？

```
public synchronized void incrementValue() {
  ++lastIdUsed;
}
```

这样一来，对于两个线程的情况，可能执行路径的数量就是 2，即 $N!$。

A.2.2 深入挖掘

两个线程都调用方法一次（在添加 `synchronized` 之前），得到同一结果数字的惊异结果又怎样呢？怎么可能出现这种情况？一样一样来。

什么是原子操作？可以把原子操作定义为不可中断的操作。例如，在下列代码的第 5 行，0 被赋值给 `lastId`，就是一个原子操作。因为依据 Java 内存模型，32 位值的赋值操作是不可中断的。

```
01:    public class Example {
02:      int lastId;
03:
04:      public void resetId()  {
05:        lastId = 0;
06:      }
07:
08:      public int getNextId()  {
09:        ++lastId;
10:      }
11:    }
```

如果把 `lastId` 的类型从 `int` 改为 `long` 会怎样？第 5 行还是原子操作吗？如果不考虑 JVM 规约，则有可能根据处理器不同而不同。不过，根据 JVM 规约，64 位值的赋值需要两次 32 位赋值。这意味着在第一次和第二次 32 位赋值之间，其他线程可能插进来，修改其中一个值。

第 9 行的前递增操作符++又怎样呢？前递增操作符可以被中断，所以它不是原子的。为了理解这点，仔细复查一下这些方法的字节码吧。

在更进一步说明之前，有以下 3 个重要的定义。

- 框架——每个方法调用都需要一个框架。该框架包括返回地址、传入方法的参数，以及方法中定义的本地变量。框架是定义一个调用栈的标准技术，现代编程语言用框架来实现基本函数/方法调用和递归调用。
- 本地变量——方法作用范围内定义的每个变量。所有非静态方法至少有一个变量 `this`，代表当前对象，即接收导致方法调用的（当前线程内）大多数最新消息的对象。

- 运算对象栈——Java 虚拟机中的许多指令都有参数。运算对象栈是放置参数的地方。栈是一个标准的后入先出（LIFO）数据结构。

下面是 restId() 的字节码，如表 A-1 所示。

表 A-1　　　　　　　　　　　　restId() 的字节码

指令	描述	操作对象栈
ALOAD 0	将第 0 个变量放到操作对象栈中。什么是第 0 个变量？就是 this，即当前对象。当方法被调用，消息接收者，即 Example 的一个实体，被推到为方法调用创建的框架的本地变量数组中。这总是放进每个实体方法的第一个变量	this
ICONST_0	将常量值 0 放到操作对象栈中	this、0
PUTFIELD lastId	将栈中的第一个值（即 0）存储到引用对象的字段值，距栈顶部 this 一个对象引用的距离	<空>

这 3 条指令确保是原子的，因为尽管执行它们的线程可能在其中任何一个指令后被打断，但 PUTFIELD 指令（栈顶部的常量值 0 和顶端之下的 this 引用及其字段值）的信息并不能为其他线程所触及。所以，当赋值操作发生时，值 0 一定将存储到字段值中。该操作是原子的。操作对象都处理对方法而言是本地的信息，故在多个线程之间并无冲突。

所以，如果这 3 条指令由 10 个线程执行，就会有 4.38679733629e+24 种可能的执行次序。不过，只会有一种可能的结果，所以执行次序不同无关紧要。对于本例中的 long 常量，总是有同一种运算结果。为什么？因为 10 个线程的赋值操作都是针对一个常量的。即便它们互相干涉，结果也是一样。

方法 getNextId 中的 ++ 操作就会有问题了。假定 lastId 在方法开始时的值为 42。下面是新方法的字节码，如表 A-2 所示。

表 A-2　　　　　　　　　　　　新方法的字节码

指令	描述	操作对象栈
ALOAD 0	将 this 装载到操作对象栈	this
DUP	复制栈顶部内容。在对象栈中有两个 this 的副本	this、this
GETFIELD lastId	从指向栈顶部（this）的对象中取得字段 lastId 的值，并存储回栈中	this、42
ICONST_1	将整数常量 1 推入栈	this、42、1
IADD	对栈顶部的两个值做整数加操作，将结果存储回栈	this、43
DUP_X1	复制值 43，放到 this 之前	43、this、43
PUTFIELD lastId	将栈顶部的值 43 放到当前对象的字段值中，表现为对象栈中的下一个值 this	43
IRETURN	返回栈顶部（而且只是顶部）的值	<空>

设想第一个线程完成了前 3 个操作，直到执行完 GETFIELD，然后被打断。第二个线程接手并完成整个方法调用，lastId 的值递增 1，得到的值为 43。第一个线程再从中断处继续执行，操作对象栈中的值还是 42，因为那就是该线程执行 GETFIELD 时的 lastId 值。线程给 lastId 加 1，得到 43，存储这个结果。第一个线程也得到了值 43。结果就是其中一个递增操作丢失了，因为第一个线程在被第二个线程打断后又踏入了第二个线程中。

将 getNextId() 方法修改为同步方法就能修正这个问题。

A.2.3 小结

理解线程之间如何互相干涉，并不一定要精通字节码。如果你能看明白这个例子，它应该已经展示了多个线程之间互相干涉的可能性，这已经足够了。

这个小例子说明，有必要尽量理解内存模型，明白什么是安全的，什么是不安全的。有一种普遍的误解，认为 ++（前递增或后递增）操作符是原子的，其实并非如此。你必须知道：

- 什么地方有共享对象/值；
- 哪些代码会导致并发读/写问题；
- 如何防止这种并发问题发生。

A.3 了解类库

A.3.1 Executor 框架

如前文 ExecutorClientScheduler.java 所演示的，Java 5 中引入的 Executor 框架支持利用线程池进行复杂的执行。那就是 java.util.concurrent 包中的一个类。

如果在创建线程时没有使用线程池或自行编写线程池，可以考虑使用 Executor。它能让代码更整洁，易于理解，且更加短小。

Executor 框架将把线程放到池中，自动调整其大小，并在必要时重建线程。它还支持 future（一种通用的并发编程结构）。Executor 能与实现了 Runnable 的类协同工作，也能与实现了 Callable 接口的类协同工作。Callback 看来就像是 Runnable，但它能返回一个结果，这在多线程解决方案中是普遍的需求。

当代码需要执行多个相互独立的操作并等待这些操作结束时，future 刚好就手：

```
public String processRequest(String message) throws Exception {
    Callable<String> makeExternalCall = new Callable<String>() {
```

```java
    public String call() throws Exception {
      String result = "";
      // make external request
      return result;
    }
  };

  Future<String> result = executorService.submit(makeExternalCall);
  String partialResult = doSomeLocalProcessing();
  return result.get() + partialResult;
}
```

在本例中，方法开始执行 `makeExternalCall` 对象。然后该方法继续其他操作。最后一行代码调用 `result.get()`，在 future 代码执行完成前，这个操作是锁定的。

A.3.2 非锁定的解决方案

Java 5 虚拟机利用了现代处理器支持可靠、非锁定更新的设计优点。例如，考虑某个使用同步（从而也是锁定的）来提供线程安全地更新一个值的类：

```java
public class ObjectWithValue {
  private int value;
  public void synchronized incrementValue() { ++value; }
  public int getValue() { return value; }
}
```

Java 5 有一系列用于此类情况的新类，其中包括 `AtomicBoolean`、`AtomicInteger` 和 `AtomicReference` 等，还有另外一些类。我们可以重写上面的代码，使用非锁定的手段，如下所示：

```java
public class ObjectWithValue {
  private AtomicInteger value = new AtomicInteger(0);

  public void incrementValue() {
    value.incrementAndGet();
  }
  public int getValue() {
    return value.get();
  }
}
```

即便使用了对象而非直接操作，若使用了 `incrementAndGet()` 这样的信息发送方式而非 ++ 操作，则这个类的性能也还是几乎总能胜过上一版本。在某些情况下只会快一点点，但较慢的情形几乎不存在。

怎么会这样？现代处理器拥有一种通常称为比较和交换（Compare and Swap，CAS）的操作。这种操作类似于数据库中的乐观锁定，而其同步版本则类似于保守锁定。

关键字 `synchronized` 总是要求上锁，即便第二个线程并不更新同一值时也如此。尽管这种固有锁的性能一直在提升，但仍然代价高昂。

非上锁的版本假定多个线程通常并不频繁修改同一个值，导致问题产生。它高效地监测这种情形是否发生，并不断尝试，直至更新成功。这种监测行为几乎总是比上锁来得划算，在争用激烈的情况下也是如此。

虚拟机如何实现这种机制？CAS 的操作是原子的。逻辑上，CAS 操作看起来像这样：

```
int variableBeingSet;

void simulateNonBlockingSet(int newValue) {
  int currentValue;
  do {
    currentValue = variableBeingSet
  } while(currentValue != compareAndSwap(currentValue, newValue));
}

int synchronized compareAndSwap(int currentValue, int newValue) {
  if(variableBeingSet == currentValue) {
    variableBeingSet = newValue;
    return currentValue;
  }
  return variableBeingSet;
}
```

当某个方法试图更新一个共享变量，CAS 操作就会验证要赋值的变量是否保存有上一次的已知值。如果是，就修改变量值。如果不是，则不会碰变量，因为另一个线程正在试图更新变量值。要更新数据的方法（通过 CAS 操作）查看是否修改并持续尝试。

A.3.3 非线程安全类

有些类天生不是线程安全的。下面是几个例子：
- `SimpleDateFormat`；
- 数据库连接；
- `java.util` 中的容器；
- **Servlet**。

注意，有些群集类拥有一些线程安全的方法。不过，涉及调用多个方法的操作都不是线程安全的。例如，如果因为 `Hashtable` 中已经有某物而不打算替换它，可能会写出以下代码：

```
if(!hashTable.containsKey(someKey)) {
  hashTable.put(someKey, new SomeValue());
}
```

单个方法是线程安全的。不过，另一个线程却可能在 `containsKey` 和 `put` 调用之间塞进一个值。有几种修正这个问题的手段。

- 先锁定 Hashtable，确定其他使用者都做了基于客户端的锁定：

  ```
  synchronized(map) {
  if(!map.containsKey(key))
    map.put(key,value);
  }
  ```

- 用其对象包装 Hashtable，并使用不同的 API——利用 ADAPTER 模式做基于服务端的锁定：

  ```
  public class WrappedHashtable<K, V> {
    private Map<K, V> map = new Hashtable<K, V>();

    public synchronized void putIfAbsent(K key, V value) {
      if (map.containsKey(key))
        map.put(key, value);
    }
  }
  ```

- 采用线程安全的群集：

  ```
  ConcurrentHashMap<Integer, String> map = new ConcurrentHashMap<Integer, String>();
  map.putIfAbsent(key, value);
  ```

在 `java.util.concurrent` 中的群集都有 `putIfAbsent()` 之类提供这种操作的方法。

A.4 方法之间的依赖可能破坏并发代码

下面是一个有关在方法间引入依赖的小例子：

```
public class IntegerIterator implements Iterator<Integer>
  private Integer nextValue = 0;

  public synchronized boolean hasNext() {
    return nextValue < 100000;
  }

  public synchronized Integer next() {
    if (nextValue == 100000)
      throw new IteratorPastEndException();
    return nextValue++;
  }

  public synchronized Integer getNextValue() {
    return nextValue;
  }
}
```

A.4 方法之间的依赖可能破坏并发代码

下面是使用 `IntegerIterator` 的代码：

```
IntegerIterator iterator = new IntegerIterator();
while(iterator.hasNext()) {
  int nextValue = iterator.next();
  // do something with nextValue
}
```

如果只有一个线程执行这段代码，不会有什么问题。但如果有两个线程，各自抱着每个线程都处理它获得的值但列表中的每个元素都只被处理一次的意图，尝试共享 `IntegerIterator` 的单个实体，会发生什么事？多数时候什么也不会发生，线程开心地共享着列表，处理从迭代器获取的元素，在迭代器完成执行时停下。然而，在迭代的末尾，两个线程也有可能少量互相干涉，导致其中一个超出迭代器末尾，抛出异常。

问题在这里。线程 1 调用 `hasNext()` 方法，该方法返回 `true`。线程 1 占先，然后线程 2 也调用这个方法，同样返回 `true`。线程 2 接着调用 `next()`，该方法如期返回一个值，但副作用是，之后再调用 `hasNext()` 就会返回 `false`。线程 1 继续执行，以为 `hasNext()` 还是 `true`，然后调用 `next()`。即便单个方法是同步的，客户端也还是使用了两个方法。

这的确是个问题，也是并发代码中此类问题的典型例子。在这个特殊例子中，问题尤其隐蔽，因为只有在迭代器最后一次发生迭代时才会导致错误。如果线程刚好在那个点中断，其中一个线程就可能超出迭代器末尾。这类错误往往在系统部署之后很久才发生，而且很难追踪。

出现错误时，有 3 种做法。
- 容忍错误；
- 修改客户代码解决问题：基于客户代码的锁定；
- 修改服务端代码解决问题，同时也修改客户代码：基于服务端的锁定。

A.4.1 容忍错误

有时，可以通过一些设置让错误不会导致损害。例如，上述客户代码可以捕捉并清理异常。坦白地说，这有点草草从事，就像是通过半夜重启解决内存泄漏问题一样。

A.4.2 基于客户代码的锁定

要让 `IntegerIterator` 在多线程情况下正确运行，可以对客户代码做如下修改：

```
IntegerIterator iterator = new IntegerIterator();

while (true) {
  int nextValue;
  synchronized (iterator) {
```

```
    if (!iterator.hasNext())
      break;
    nextValue = iterator.next();
  }
  doSometingWith(nextValue);
}
```

每个客户端都通过 synchronized 关键字引入一个锁。这种重复违反了 DRY 原则，但如果代码使用非线程安全的第三方工具，可能必须这样做。

这种策略有风险，因为使用服务端的程序员都得记住在使用前上锁、用过后解锁。许多（许多！）年前，我遇到过一个在共享资源上应用基于客户代码锁定的系统。代码中有几百处用到这个资源的地方。有位可怜的程序员忘记在其中一处做资源锁定。

该系统是一个多终端分时系统，为 Local 705 卡车司机联盟运行会计软件。计算机放在距 Local 705 总部 50 英里（约 80.47km）以北的一间镶有高于地面的地板、环境可控的机房中。总部有几十位数据录入员，往终端输入记录。终端使用电话专线和 600 bit/s 的半双工调制解调器连接到计算机。（这可是很久很久以前的事了。）

每天大概都会有一台终端毫无理由地"死锁"。死锁也不限定在某些终端或特定时间。就像是有人掷骰子选择死锁的时机和终端一般。有时，会有几台终端死锁。有时，好几天都不出现死锁情况。

刚开始，唯一的解决手段就是重启，但协同起来很不便。我们得打电话给总部，让大家都完成在终端上的工作。然后我们才能关机、重启。如果有人在做要花一两小时才能完成的事，被锁定的终端就只能一直等着。

经过几个星期的调试，我们发现，原因在于一个指针不同步的环形缓冲区计数器。该缓冲区控制向终端的输出。指针值说明缓冲区是空的，但计数器却指出缓冲区是满的。因为指针值说明缓冲区是空的，所以没什么可显示；但因为计数器指出缓冲区是满的，所以无法向其中加入可在屏幕上显示的内容。

我们知道了终端为何会死锁，却不知道为什么环形缓冲区会不同步。我们用了点手段发现问题所在。当时程序能够读取计算机的前面板开关状态（这可是很久很久以前的事了）。我们写了一个陷阱程序，监测这些开关何时被拨动，然后查找既空又满的环形缓冲区。如果找到，就重置该缓冲区为空。乌拉！锁定的终端又重新开始显示了。

这样，在终端锁定时就不必重启系统了。客户只需要打电话告诉我们出现死锁，我们就径直走到机房，拨动一下开关即可。

当然，有时他们会在周末加班，但是我们不加班。所以我们又在计划列表中添加了一个函数，每分钟检查一次全部环形缓冲区，重置既空又满的缓冲区。在客户打电话之前，显示就已经恢复正常了。

在发现问题原因之前，我们花了好几个星期查看一页又一页的单片机汇编语言代码。我们已经完成计算，算出死锁的频率是周期性的，而且其中有一处未受保护的环形缓冲区使用。

所以，剩下的任务就是找出那个错误的用法。不幸这是多年以前的事，那时既没有搜索工具，也没有交叉引用或任何其他自动化帮助手段。我们只能细查代码清单。

在芝加哥 1971 年的寒冬，我学到了重要的一课。基于客户代码的锁定实在不可靠。

A.4.3 基于服务端的锁定

按照以下方式修改 `IntegerIterator` 也能消除重复：

```java
public class IntegerIteratorServerLocked {
  private Integer nextValue = 0;
  public synchronized Integer getNextOrNull() {
    if (nextValue < 100000)
      return nextValue++;
    else
      return null;
  }
}
```

客户代码也要修改：

```java
while (true) {
  Integer nextValue = iterator.getNextOrNull();
  if (next == null)
    break;
  // do something with nextValue
}
```

在这种情形下，我们实际上是修改了类的 API，使其能适应多线程①。客户端需要做 `null` 检查，而不是检查 `hasNext()`。

通常你应该选用基于服务端的锁定，因为：

- 它减少了重复代码——采用基于客户代码的锁定，每个客户端都要正确锁定服务端，把锁定代码放到服务端，客户端就能自由使用对象，不必费心编写额外的锁定代码；
- 它提升了性能——在单线程部署中，可以用非多线程安全服务端代码替代线程安全客户端，从而省去开销；
- 它减少了出错的可能性——只会有一个程序员忘记上锁；
- 它执行了单一策略——该策略只在服务端这一处地方实施，而不是在许多地方（每个客户端）实施；
- 它缩减了共享变量的作用范围——客户端不必关心它们或它们是如何锁定的，一切都隐藏在服务端。如果出错，要监测的范围就小多了。

---

① 实际上，`Iterator` 接口天生不是线程安全的。它并不为多线程而设计，所以出现这种情况也不奇怪。

如果你无法修改服务端代码又该如何？
- 使用 ADAPTER 模式修改 API，添加锁定；

```
public class ThreadSafeIntegerIterator {
  private IntegerIterator iterator = new IntegerIterator();

  public synchronized Integer getNextOrNull() {
    if(iterator.hasNext())
      return iterator.next();
    return null;
  }
}
```

- 更好的方法是使用线程安全的群集和扩展接口。

A.5 提升吞吐量

假设我们打算连接上网，从一个 URL 列表中读取一组页面的内容。读到一个页面时，解析该页面并得到一些统计结果。读完所有页面后，打印出一份提要报表。

下面的类返回给定 URL 的页面内容：

```
public class PageReader {
  //...
  public String getPageFor(String url) {
    HttpMethod method = new GetMethod(url);

    try {
      httpClient.executeMethod(method);
      String response = method.getResponseBodyAsString();
      return response;
    } catch (Exception e) {
      handle(e);
    } finally {
      method.releaseConnection();
    }
  }
}
```

下一个类是给出 URL 迭代器中每个页面的内容的迭代器：

```
public class PageIterator {
  private PageReader reader;
  private URLIterator urls;

  public PageIterator(PageReader reader, URLIterator urls) {
    this.urls = urls;
    this.reader = reader;
  }
```

```
  public synchronized String getNextPageOrNull() {
    if (urls.hasNext())
      getPageFor(urls.next());
    else
      return null;
  }
  public String getPageFor(String url) {
    return reader.getPageFor(url);
  }
}
```

`PageIterator` 的一个实体可为多个不同线程共享，每个线程使用自己的 `PageReader` 实体读取并解析从迭代器中得到的页面。

注意，我们把 `synchronized` 代码块的数量限制在小范围之内，该范围只包括深处于 `PageIterator` 内部的临界区。最好尽可能少地使用同步。

A.5.1 单线程条件下的吞吐量

来做个简单计算。鉴于讨论的目的，假定：
- 获取一个页面的 I/O 时间（平均）是 1 秒；
- 解析一个页面的处理时间（平均）是 0.5 秒；
- I/O 操作不耗费处理器能力，而解析页面耗费 100% 处理器能力。

对于单个线程要处理的 N 个页面，总的执行时间为 $1.5*N$ 秒。图 A-1 显示了 13 个页面或大概 19.5 秒的快照。

图 A-1　单线程

A.5.2 多线程条件下的吞吐量

如果能够以任意次序获得页面并独立处理页面，就有可能利用多线程提升吞吐量。如果我们使用 3 个线程会如何？在同一时间内能获取多少个页面呢？

如你在图 A-2 中所见，多线程方案中与处理器能力有关的页面解析操作可以和与 I/O 有关的页面读取操作叠加进行。在理想状态下，这意味着处理器力尽其用。每个耗时一秒的页面读取操作都与两次解析操作叠加进行。这样，我们就能在每秒内处理两个页面，即 3 倍于单线程方案的吞吐量。

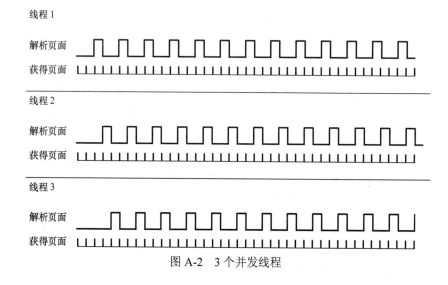

图 A-2　3 个并发线程

A.6　死锁

想象一个拥有两个有限共享资源池的 Web 应用程序：

- 一个用于本地临时工作存储的数据库连接池；
- 一个用于连接到主存储库的 MQ 池。

假定该应用中有两个操作，即创建和更新：

- 创建——获取到主存储库和数据库的连接，与主存储库协调，并把工作保存到本地临时工作数据库；
- 更新——先获取到数据库的连接，再获取到主存储库的连接，从临时工作数据库中读取数据，再发送给主存储库。

如果用户数量多于池的大小会怎样？假设每个池中能容纳 10 个资源：

- 有 10 个用户尝试创建，获取了 10 个数据库连接，每个线程在获取到数据库连接之后、获取到主存储库连接之前都被打断；
- 有 10 个用户尝试更新，获取了 10 个主存储库连接，每个线程在获取到主存储库连接之后、获取到数据库连接之前都会被打断；
- 现在那 10 个"创建"线程必须等待获取主存储库连接，但那 10 个"更新"线程必须等待获取数据库连接；
- 死锁。系统永远无法恢复。

这听起来不太会出现，但谁会想要一个每隔一周就僵在那里不动的系统呢？谁想要调试出现了难以复现的症状的系统呢？这种问题突然产生，然后得花上好几个星期才能解决。

典型的"解决方案"是加入调试语句，发现问题。当然，调试语句对代码的修改足以令死锁在不同情况下发生，而且要几个月后才会再出现[①]。

要真正地解决死锁问题，我们需要理解死锁的原因。死锁的发生需要满足 4 个条件：
- 互斥；
- 上锁及等待；
- 无抢先机制；
- 循环等待。

A.6.1　互斥

当多个线程需要使用同一资源，且这些资源满足下列条件时，互斥就会发生。
- 无法在同一时间为多个线程所用；
- 数量上有限制。

这种资源的常见例子是数据库连接、打开后用于写入的文件、记录锁或信号量。

A.6.2　上锁及等待

当某个线程获取一个资源，在获取到其他全部所需资源并完成其工作之前，不会释放这个资源。

A.6.3　无抢先机制

线程无法从其他线程处夺取资源。一个线程持有资源时，其他线程获得这个资源的唯一手段就是等待该线程释放资源。

[①] 例如，有人添加了一些调试输出，问题"不见了"。调试代码"修正"了问题，其实问题还在系统中存在。

A.6.4 循环等待

这也被称为"死命拥抱"。想象两个线程，T1 和 T2，还有两个资源，R1 和 R2。T1 拥有 R1，T2 拥有 R2。T1 需要 R2，T2 需要 R1。如此就出现了如图 A-3 所示的情形。

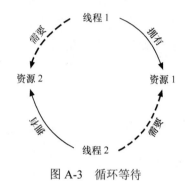

图 A-3　循环等待

这 4 种条件都是死锁所必需的。只要其中一个不满足，死锁就不会发生。

A.6.5 不互斥

避免死锁的一种策略是规避互斥条件。你可以：
- 使用允许同时使用的资源，如 `AtomicInteger`；
- 增加资源数量，使其等于或大于竞争线程的数量；
- 在获取资源之前，检查是否可用。

不幸的是，多数资源都有上限，且不能同时使用，而且第二个资源的标识也常常要依据对第一个资源的操作结果来判断。不过别丧气，还有其他 3 个条件呢。

A.6.6 不上锁及等待

如果拒绝等待，就能消除死锁。在获得资源之前检查资源，如果遇到某个资源繁忙，就释放所有资源，重新来过。

这种手段带来几个潜在问题：
- 线程饥饿——某个线程一直无法获得它所需的资源（它可能需要某种很少能同时获得的资源组合）；
- 活锁——几个线程可能会前后相连地要求获得某个资源，然后再释放一个资源，如此循环。这在单纯的 CPU 任务排列算法中尤其有可能出现（想想嵌入式设备或单纯的手写线程平衡算法）。

上面二者都能导致较差的吞吐量。第一个的结果是 CPU 利用率低，第二个的结果是较高但无用的 CPU 利用率。

尽管这种策略听起来没效率，但也比没有好。至少，如果其他方案不奏效，这种手段几乎总可以用上。

A.6.7 满足抢先机制

避免死锁的另一策略是允许线程从其他线程上夺取资源。这通常利用一种简单的请求机制来实现。当线程发现资源繁忙，就要求其拥有者释放之。如果拥有者还在等待其他资源，就释放全部资源并重新来过。

这和上一种手段相似，但好处是允许线程等待资源，这减少了线程重新启动的次数。不过，管理所有请求可要花点儿心思。

A.6.8 不做循环等待

这是避免死锁的最常用手段。对于多数系统，它只要求一个为各方认同的约定。

在上面的例子中，线程 1 同时需要资源 1 和资源 2，线程 2 同时需要资源 2 和资源 1，只要强制线程 1 和线程 2 以同样次序分配资源，循环等待就不会发生。

更普遍地，如果所有线程都认同一种资源获取次序，并按照这种次序获取资源，死锁就不会发生。就像其他策略一样，这也会有问题：
- 获取资源的次序可能与使用资源的次序不匹配，一开始获取的资源可能在最后才会用到。这可能导致资源不必要地被长时间锁定；
- 有时无法强求资源获取次序，如果第二个资源的 ID 来自对第一个资源操作的结果，获取次序就无从谈起。

有许多避免死锁的方法。有些会导致饥饿，另外一些会导致对 CPU 能力的大量耗费和降低响应率。TANSTAAFL[①]！

将解决方案中与线程相关的部分分隔出来，再加以调整和试验，是获得判断最佳策略所需的洞见的正道。

---

① 世上没有免费的午餐（There ain't no such thing as a free lunch）。

A.7 测试多线程代码

怎么才能编写显示以下代码有错的测试呢？

```
01: public class ClassWithThreadingProblem {
02:   int nextId;
03:
04:   public int takeNextId() {
05:     return nextId++;
06:   }
07: }
```

下面是对能证明上述代码有错的测试的描述：

- 记住 `nextId` 的当前值；
- 创建两个线程，每个都调用 `takeNextId()` 一次；
- 验证 `nextId` 比开始时大 2；
- 持续运行，直至发现 `nextId` 只比开始时大 1 为止。

代码清单 A-2 展示了这样一个测试。表 A-3 是对代码清单 A-2 的注解。

代码清单 A-2　ClassWithThreadingProblemTest.java

```
01: package example;
02:
03: import static org.junit.Assert.fail;
04:
05: import org.junit.Test;
06:
07: public class ClassWithThreadingProblemTest {
08:   @Test
09:   public void twoThreadsShouldFailEventually() throws Exception {
10:     final ClassWithThreadingProblem classWithThreadingProblem
            = new ClassWithThreadingProblem();
11:
12:     Runnable runnable = new Runnable() {
13:       public void run() {
14:         classWithThreadingProblem.takeNextId();
15:       }
16:     };
17:
18:     for (int i = 0; i < 50000; ++i) {
19:       int startingId = classWithThreadingProblem.lastId;
20:       int expectedResult = 2 + startingId;
21:
22:       Thread t1 = new Thread(runnable);
23:       Thread t2 = new Thread(runnable);
24:       t1.start();
25:       t2.start();
```

```
26:        t1.join();
27:        t2.join();
28:
29:        int endingId = classWithThreadingProblem.lastId;
30:
31:        if (endingId != expectedResult)
32:            return;
33:        }
34:
35:        fail("Should have exposed a threading issue but it did not.");
36:    }
37: }
```

表 A-3	代码清单 A-2 的注解
代码行	描述
10	创建 ClassWithThreadingProblem 的单个实体。注意，必须使用 final 关键字，因为要在一个匿名内部类中用到它
12~16	创建一个匿名内部类，该类用到 ClassWithThreadingProblem 的单个实体
18	运行这段代码"足够多"次以展示代码失败，但不要多到"花太长时间"。这是一种平衡行为；我们不想等太久。选择这个数字有点难——尽管我们稍后会看到能够极大地降低这个数字
19	记住开始时的值。这个测试试图证明 ClassWithThreadingProblem 中的代码有错误。如果测试通过，它就证明了这一点。如果测试失败，它就没能证明代码出错
20	我们期望最终值比当前值大 2
22~23	创建两个线程，都使用我们在第 12~16 行创建的对象。这样两个线程就有可能用到 ClassWithThreadingProblem 的单个实体，互相干涉
24~25	开始运行两个线程
26~27	在检查结果之前等待两个线程结束
29	记录真实的最终值
31~32	endingId 是否与期待值不一样？如果是，测试结束——我们已经证明了代码有错误。如果不是，再试一次
35	到达这一步，测试无法证明产品代码在"合理范围"的时间内出错，测试失败了。要么是代码没错，要么是没有运行足够多次，错误条件还没满足

这个测试当然设置了满足并发更新问题发生的条件。不过，问题发生得如此频繁，测试也就极有可能监测不到。

实际上，要真正监测到问题，需要将循环数量设置到 100 万次以上。即便是这样，在 10 个 100 万次循环的执行中，错误也只发生了一次。这意味着我们可能要把循环次数设置为超过亿次才能获得可靠的失败证明。要等多久呢？

即便我们调优测试，在单台机器上得到可靠的失败证明，我们也可能还需要用不同的值来重新设置测试，得到在其他机器、操作系统或不同版本的 JVM 上的失败证明。

而且这只是一个简单的问题。如果连这个简单的问题都无法轻易获得出错证明,我们怎么能真正监测复杂问题呢?

我们能用什么手段来证明这个简单的错误呢?而且,更重要的是,我们如何能写出证明更复杂代码中的错误的测试呢?我们怎样才能在不知道从何处着手时知道代码是否出错了呢?

下面是一些想法。

- **蒙特卡洛测试**。测试要灵活,便于调整。多次运行测试——在一台测试服务器上——随机改变调整值。如果测试失败,代码就有错。确保及早编写这些测试,好让持续集成服务器尽快开始运行测试。另外,确认小心记录了在何种条件下测试失败。
- 在每种目标部署平台上运行测试。重复运行。持续运行。测试在不失败的前提下运行得越久,就越能说明:生产代码正确或测试不足以暴露问题。
- 在另一台有不同负载的机器上运行测试。能模拟生产环境的负载,就模拟之。
- 即便你做了所有这些,也还是不见得有很好的机会发现代码中的线程问题。最阴险的问题拥有很小的截面,在十亿次执行中只会发生一次。这类错误是复杂系统的噩梦。

A.8 测试线程代码的工具支持

IBM 提供了一个名为 ConTest 的工具。它能对类进行装置,令非线程安全代码更有可能失败。

我们与 IBM 或开发 ConTest 的团队没有直接关系。有位同事发现了这个工具。在用了几分钟后,我们发现自己发现线程问题的能力得到了很大提升。

下面是使用 ConTest 的简要步骤:

- 编写测试和生产代码,确保有专门模拟多用户在多种负载情况下操作的测试,如上文所述;
- 用 ConTest 装置测试和生产代码;
- 运行测试。

用 ConTest 装置代码后,原本千万次循环才能暴露一个错误的比率提升到 30 次循环就能找到错误。以下是装置代码后的几次测试运行结果值:13、23、0、54、16、14、6、69、107、49 和 2。显然装置后的类更加容易和可靠地被证明失败。

A.9 小结

本章只是在并发编程广阔而可怕的领地上的短暂逗留罢了。我们只触及了地表。我们在这里强调的,只是保持并发代码整洁的一些规程,如果要编写并发系统,还有许多东西要学。

建议从 Doug Lea 的大作 *Concurrent Programming in Java: Design Principles and Patterns* 开始[1]。

在本章中，我们谈到并发更新，还有清理及避免同步的规程。我们谈到线程如何提升与 I/O 有关的系统的吞吐量，展示了获得这种提升的整洁技术。我们谈到死锁及干净地避免死锁的规程。最后，我们谈到通过装置代码暴露并发问题的策略。

A.10 教程：完整代码范例

A.10.1 客户端/服务器非线程代码

代码清单 A-3　Server.java

```java
package com.objectmentor.clientserver.nonthreaded;

import java.io.IOException;
import java.net.ServerSocket;
import java.net.Socket;
import java.net.SocketException;

import common.MessageUtils;

public class Server implements Runnable {
  ServerSocket serverSocket;
  volatile boolean keepProcessing = true;

  public Server(int port, int millisecondsTimeout) throws IOException {
    serverSocket = new ServerSocket(port);
    serverSocket.setSoTimeout(millisecondsTimeout);
  }

  public void run() {
    System.out.printf("Server Starting\n");

    while (keepProcessing) {
      try {
        System.out.printf("accepting client\n");
        Socket socket = serverSocket.accept();
        System.out.printf("got client\n");
        process(socket);
      } catch (Exception e) {
        handle(e);
      }
    }
  }
```

---

[1] [Lea99]。

```java
    private void handle(Exception e) {
      if (!(e instanceof SocketException)) {
        e.printStackTrace();
      }
    }

    public void stopProcessing() {
      keepProcessing = false;
      closeIgnoringException(serverSocket);
    }

    void process(Socket socket) {
      if (socket == null)
        return;

      try {
        System.out.printf("Server: getting message\n");
        String message = MessageUtils.getMessage(socket);
        System.out.printf("Server: got message: %s\n", message);
        Thread.sleep(1000);
        System.out.printf("Server: sending reply: %s\n", message);
        MessageUtils.sendMessage(socket, "Processed: " + message);
        System.out.printf("Server: sent\n");
        closeIgnoringException(socket);
      } catch (Exception e) {
        e.printStackTrace();
      }
    }

    private void closeIgnoringException(Socket socket) {
      if (socket != null)
        try {
            socket.close();
        } catch (IOException ignore) {
        }
    }

    private void closeIgnoringException(ServerSocket serverSocket) {
      if (serverSocket != null)
        try {
            serverSocket.close();
        } catch (IOException ignore) {
        }
    }
}
```

代码清单 A-4　ClientTest.java

```java
public class ClientTest {
  private static final int PORT = 8009;
  private static final int TIMEOUT = 2000;

  Server server;
```

```java
Thread serverThread;

@Before
public void createServer() throws Exception {
  try {
    server = new Server(PORT, TIMEOUT);
    serverThread = new Thread(server);
    serverThread.start();
  } catch (Exception e){
    e.printStackTrace(System.err);
    throw e;
  }
}

@After
public void shutdownServer() throws InterruptedException {
  if (server != null) {
    server.stopProcessing();
    serverThread.join();
  }
}

class TrivialClient implements Runnable {
  int clientNumber;

  TrivialClient(int clientNumber) {
    this.clientNumber = clientNumber;
  }

  public void run() {
    try {
      connectSendReceive(clientNumber);
    } catch (IOException e) {
      e.printStackTrace();
    }
  }
}

@Test(timeout = 10000)
public void shouldRunInUnder10Seconds() throws Exception {
  Thread[] threads = new Threads[10];

  for (int i = 0; i < threads.length; ++i) {
    threads[i] = new Thread(new TrivialClient(i));
    threads[i].start();
  }

  for (int i = 0; i < threads.length; ++i) {
    threads[i].join();
  }
}

private void connectSendReceive(int i) throws IOException {
  System.out.printf("Client %2d: connecting\n", i);
```

```java
            Socket socket = new Socket("localhost", PORT);
            System.out.printf("Client %2d: sending message\n", i);
            MessageUtils.sendMessage(socket, Integer.toString(i));
            System.out.printf("Client %2d: getting reply\n", i);
            MessageUtils.getMessage(socket);
            System.out.printf("Client %2d: finished\n", i);
            socket.close();
        }

    }
```

代码清单 A-5　MessageUtils.java

```java
package common;

import java.io.IOException;
import java.io.InputStream;
import java.io.ObjectInputStream;
import java.io.ObjectOutputStream;
import java.io.OutputStream;
import java.net.Socket;

public class MessageUtils {
    public static void sendMessage(Socket socket, String message)
        throws IOException {
        OutputStream stream = socket.getOutputStream();
        ObjectOutputStream oos = new ObjectOutputStream(stream);
        oos.writeUTF(message);
        oos.flush();
    }

    public static String getMessage(Socket socket) throws IOException {
        InputStream stream = socket.getInputStream();
        ObjectInputStream ois = new ObjectInputStream(stream);
        return ois.readUTF();
    }
}
```

A.10.2　使用线程的客户端/服务器代码

把服务器修改为使用多线程，只需要对处理方法进行修改即可（新的代码行用粗体标出）：

```java
void process(final Socket socket) {
    if (socket == null)
        return;

    Runnable clientHandler = new Runnable() {
        public void run() {
            try {
                System.out.printf("Server: getting message\n");
                String message = MessageUtils.getMessage(socket);
```

```
            System.out.printf("Server: got message: %s\n", message);
            Thread.sleep(1000);
            System.out.printf("Server: sending reply: %s\n", message);
            MessageUtils.sendMessage(socket, "Processed: " + message);
            System.out.printf("Server: sent\n");
            closeIgnoringException(socket);
        } catch (Exception e) {
            e.printStackTrace();
        }
    }
};

Thread clientConnection = new Thread(clientHandler);
clientConnection.start();
}
```

附录 B

org.jfree.date.SerialDate

代码清单 B-1   SerialDate.Java

```
 1  /* ========================================================================
 2   * JCommon : a free general purpose class library for the Java(tm) platform
 3   * ========================================================================
 4   *
 5   * (C) Copyright 2000-2005, by Object Refinery Limited and Contributors.
 6   *
 7   *
 8   *
 9   * This library is free software; you can redistribute it and/or modify it
10   * under the terms of the GNU Lesser General Public License as published by
11   * the Free Software Foundation; either version 2.1 of the License, or
12   * (at your option) any later version.
13   *
14   * This library is distributed in the hope that it will be useful, but
15   * WITHOUT ANY WARRANTY; without even the implied warranty of MERCHANTABILITY
16   * or FITNESS FOR A PARTICULAR PURPOSE. See the GNU Lesser General Public
17   * License for more details.
18   *
19   * You should have received a copy of the GNU Lesser General Public
20   * License along with this library; if not, write to the Free Software
21   * Foundation, Inc., 51 Franklin Street, Fifth Floor, Boston, MA  02110-1301,
22   * USA.
23   *
24   * [Java is a trademark or registered trademark of Sun Microsystems, Inc.
25   * in the United States and other countries.]
26   *
27   * ---------------
28   * SerialDate.java
29   * ---------------
30   * (C) Copyright 2001-2005, by Object Refinery Limited.
```

```
31  *
32  * Original Author:  David Gilbert (for Object Refinery Limited);
33  * Contributor(s):   -;
34  *
35  * $Id: SerialDate.java,v 1.7 2005/11/03 09:25:17 mungady Exp $
36  *
37  * Changes (from 11-Oct-2001)
38  * --------------------------
39  * 11-Oct-2001 : Re-organised the class and moved it to new package
40  *               com.jrefinery.date (DG);
41  * 05-Nov-2001 : Added a getDescription() method, and eliminated NotableDate
42  *               class (DG);
43  * 12-Nov-2001 : IBD requires setDescription() method, now that NotableDate
44  *               class is gone (DG);  Changed getPreviousDayOfWeek(),
45  *               getFollowingDayOfWeek() and getNearestDayOfWeek() to correct
46  *               bugs (DG);
47  * 05-Dec-2001 : Fixed bug in SpreadsheetDate class (DG);
48  * 29-May-2002 : Moved the month constants into a separate interface
49  *               (MonthConstants) (DG);
50  * 27-Aug-2002 : Fixed bug in addMonths() method, thanks to N???levka Petr (DG);
51  * 03-Oct-2002 : Fixed errors reported by Checkstyle (DG);
52  * 13-Mar-2003 : Implemented Serializable (DG);
53  * 29-May-2003 : Fixed bug in addMonths method (DG);
54  * 04-Sep-2003 : Implemented Comparable.  Updated the isInRange javadocs (DG);
55  * 05-Jan-2005 : Fixed bug in addYears() method (1096282) (DG);
56  *
57  */
58
59  package org.jfree.date;
60
61  import java.io.Serializable;
62  import java.text.DateFormatSymbols;
63  import java.text.SimpleDateFormat;
64  import java.util.Calendar;
65  import java.util.GregorianCalendar;
66
67  /**
68   * An abstract class that defines our requirements for manipulating dates,
69   * without tying down a particular implementation.
70   * <P>
71   * Requirement 1 : match at least what Excel does for dates;
72   * Requirement 2 : class is immutable;
73   * <P>
74   * Why not just use java.util.Date?  We will, when it makes sense.  At times,
75   * java.util.Date can be *too* precise - it represents an instant in time,
76   * accurate to 1/1000th of a second (with the date itself depending on the
77   * time-zone).  Sometimes we just want to represent a particular day (e.g. 21
78   * January 2015) without concerning ourselves about the time of day, or the
79   * time-zone, or anything else.  That's what we've defined SerialDate for.
80   * <P>
81   * You can call getInstance() to get a concrete subclass of SerialDate,
82   * without worrying about the exact implementation.
83   *
84   * @author David Gilbert
```

```java
 85   */
 86  public abstract class SerialDate implements Comparable,
 87                                              Serializable,
 88                                              MonthConstants {
 89
 90      /** For serialization. */
 91      private static final long serialVersionUID = -293716040467423637L;
 92
 93      /** Date format symbols. */
 94      public static final DateFormatSymbols
 95          DATE_FORMAT_SYMBOLS = new SimpleDateFormat().getDateFormatSymbols();
 96
 97      /** The serial number for 1 January 1900. */
 98      public static final int SERIAL_LOWER_BOUND = 2;
 99
100      /** The serial number for 31 December 9999. */
101      public static final int SERIAL_UPPER_BOUND = 2958465;
102
103      /** The lowest year value supported by this date format. */
104      public static final int MINIMUM_YEAR_SUPPORTED = 1900;
105
106      /** The highest year value supported by this date format. */
107      public static final int MAXIMUM_YEAR_SUPPORTED = 9999;
108
109      /** Useful constant for Monday. Equivalent to java.util.Calendar.MONDAY. */
110      public static final int MONDAY = Calendar.MONDAY;
111
112      /**
113       * Useful constant for Tuesday. Equivalent to java.util.Calendar.TUESDAY.
114       */
115      public static final int TUESDAY = Calendar.TUESDAY;
116
117      /**
118       * Useful constant for Wednesday. Equivalent to
119       * java.util.Calendar.WEDNESDAY.
120       */
121      public static final int WEDNESDAY = Calendar.WEDNESDAY;
122
123      /**
124       * Useful constant for Thrusday. Equivalent to java.util.Calendar.THURSDAY.
125       */
126      public static final int THURSDAY = Calendar.THURSDAY;
127
128      /** Useful constant for Friday. Equivalent to java.util.Calendar.FRIDAY. */
129      public static final int FRIDAY = Calendar.FRIDAY;
130
131      /**
132       * Useful constant for Saturday. Equivalent to java.util.Calendar.SATURDAY.
133       */
134      public static final int SATURDAY = Calendar.SATURDAY;
135
136      /** Useful constant for Sunday. Equivalent to java.util.Calendar.SUNDAY. */
137      public static final int SUNDAY = Calendar.SUNDAY;
138
```

```java
139    /** The number of days in each month in non leap years. */
140    static final int[] LAST_DAY_OF_MONTH =
141        {0, 31, 28, 31, 30, 31, 30, 31, 31, 30, 31, 30, 31};
142
143    /** The number of days in a (non-leap) year up to the end of each month. */
144    static final int[] AGGREGATE_DAYS_TO_END_OF_MONTH =
145        {0, 31, 59, 90, 120, 151, 181, 212, 243, 273, 304, 334, 365};
146
147    /** The number of days in a year up to the end of the preceding month. */
148    static final int[] AGGREGATE_DAYS_TO_END_OF_PRECEDING_MONTH =
149        {0, 0, 31, 59, 90, 120, 151, 181, 212, 243, 273, 304, 334, 365};
150
151    /** The number of days in a leap year up to the end of each month. */
152    static final int[] LEAP_YEAR_AGGREGATE_DAYS_TO_END_OF_MONTH =
153        {0, 31, 60, 91, 121, 152, 182, 213, 244, 274, 305, 335, 366};
154
155    /**
156     * The number of days in a leap year up to the end of the preceding month.
157     */
158    static final int[]
159        LEAP_YEAR_AGGREGATE_DAYS_TO_END_OF_PRECEDING_MONTH =
160            {0, 0, 31, 60, 91, 121, 152, 182, 213, 244, 274, 305, 335, 366};
161
162    /** A useful constant for referring to the first week in a month. */
163    public static final int FIRST_WEEK_IN_MONTH = 1;
164
165    /** A useful constant for referring to the second week in a month. */
166    public static final int SECOND_WEEK_IN_MONTH = 2;
167
168    /** A useful constant for referring to the third week in a month. */
169    public static final int THIRD_WEEK_IN_MONTH = 3;
170
171    /** A useful constant for referring to the fourth week in a month. */
172    public static final int FOURTH_WEEK_IN_MONTH = 4;
173
174    /** A useful constant for referring to the last week in a month. */
175    public static final int LAST_WEEK_IN_MONTH = 0;
176
177    /** Useful range constant. */
178    public static final int INCLUDE_NONE = 0;
179
180    /** Useful range constant. */
181    public static final int INCLUDE_FIRST = 1;
182
183    /** Useful range constant. */
184    public static final int INCLUDE_SECOND = 2;
185
186    /** Useful range constant. */
187    public static final int INCLUDE_BOTH = 3;
188
189    /**
190     * Useful constant for specifying a day of the week relative to a fixed
191     * date.
192     */
```

```
193    public static final int PRECEDING = -1;
194
195    /**
196     * Useful constant for specifying a day of the week relative to a fixed
197     * date.
198     */
199    public static final int NEAREST = 0;
200
201    /**
202     * Useful constant for specifying a day of the week relative to a fixed
203     * date.
204     */
205    public static final int FOLLOWING = 1;
206
207    /** A description for the date. */
208    private String description;
209
210    /**
211     * Default constructor.
212     */
213    protected SerialDate() {
214    }
215
216    /**
217     * Returns <code>true</code> if the supplied integer code represents a
218     * valid day-of-the-week, and <code>false</code> otherwise.
219     *
220     * @param code  the code being checked for validity.
221     *
222     * @return <code>true</code> if the supplied integer code represents a
223     *         valid day-of-the-week, and <code>false</code> otherwise.
224     */
225    public static boolean isValidWeekdayCode(final int code) {
226
227        switch(code) {
228            case SUNDAY:
229            case MONDAY:
230            case TUESDAY:
231            case WEDNESDAY:
232            case THURSDAY:
233            case FRIDAY:
234            case SATURDAY:
235                return true;
236            default:
237                return false;
238        }
239
240    }
241
242    /**
243     * Converts the supplied string to a day of the week.
244     *
245     * @param s  a string representing the day of the week.
246     *
```

```java
     * @return <code>-1</code> if the string is not convertable, the day of
     *         the week otherwise.
     */
    public static int stringToWeekdayCode(String s) {

      final String[] shortWeekdayNames
          = DATE_FORMAT_SYMBOLS.getShortWeekdays();
      final String[] weekDayNames = DATE_FORMAT_SYMBOLS.getWeekdays();

      int result = -1;
      s = s.trim();
      for (int i = 0; i < weekDayNames.length; i++) {
        if (s.equals(shortWeekdayNames[i])) {
          result = i;
          break;
        }
        if (s.equals(weekDayNames[i])) {
          result = i;
          break;
        }
      }
      return result;

    }

    /**
     * Returns a string representing the supplied day-of-the-week.
     * <P>
     * Need to find a better approach.
     *
     * @param weekday  the day of the week.
     *
     * @return a string representing the supplied day-of-the-week.
     */
    public static String weekdayCodeToString(final int weekday) {

        final String[] weekdays = DATE_FORMAT_SYMBOLS.getWeekdays();
        return weekdays[weekday];

    }

    /**
     * Returns an array of month names.
     *
     * @return an array of month names.
     */
    public static String[] getMonths() {

      return getMonths(false);

    }

    /**
     * Returns an array of month names.
```

```
301       *
302       * @param shortened  a flag indicating that shortened month names should
303       *                   be returned.
304       *
305       * @return an array of month names.
306       */
307      public static String[] getMonths(final boolean shortened) {
308
309          if (shortened) {
310              return DATE_FORMAT_SYMBOLS.getShortMonths();
311          }
312          else {
313              return DATE_FORMAT_SYMBOLS.getMonths();
314          }
315
316      }
317
318      /**
319       * Returns true if the supplied integer code represents a valid month.
320       *
321       * @param code  the code being checked for validity.
322       *
323       * @return <code>true</code> if the supplied integer code represents a valid month.
324       *
325       */
326      public static boolean isValidMonthCode(final int code) {
327
328          switch(code) {
329              case JANUARY:
330              case FEBRUARY:
331              case MARCH:
332              case APRIL:
333              case MAY:
334              case JUNE:
335              case JULY:
336              case AUGUST:
337              case SEPTEMBER:
338              case OCTOBER:
339              case NOVEMBER:
340              case DECEMBER:
341                  return true;
342              default:
343                  return false;
344          }
345
346      }
347
348      /**
349       * Returns the quarter for the specified month.
350       *
351       * @param code  the month code (1-12).
352       *
353       * @return the quarter that the month belongs to.
354       * @throws java.lang.IllegalArgumentException
```

```
355      */
356     public static int monthCodeToQuarter(final int code) {
357
358       switch(code) {
359         case JANUARY:
360         case FEBRUARY:
361         case MARCH: return 1;
362         case APRIL:
363         case MAY:
364         case JUNE: return 2;
365         case JULY:
366         case AUGUST:
367         case SEPTEMBER: return 3;
368         case OCTOBER:
369         case NOVEMBER:
370         case DECEMBER: return 4;
371         default: throw new IllegalArgumentException(
372             "SerialDate.monthCodeToQuarter: invalid month code.");
373       }
374
375     }
376
377     /**
378      * Returns a string representing the supplied month.
379      * <P>
380      * The string returned is the long form of the month name taken from the
381      * default locale.
382      *
383      * @param month  the month.
384      *
385      * @return a string representing the supplied month.
386      */
387     public static String monthCodeToString(final int month) {
388
389       return monthCodeToString(month, false);
390
391     }
392
393     /**
394      * Returns a string representing the supplied month.
395      * <P>
396      * The string returned is the long or short form of the month name taken
397      * from the default locale.
398      *
399      * @param month  the month.
400      * @param shortened  if <code>true</code> return the abbreviation of the
401      *                   month.
402      *
403      * @return a string representing the supplied month.
404      * @throws java.lang.IllegalArgumentException
405      */
406     public static String monthCodeToString(final int month,
407                                            final boolean shortened) {
408
```

```
409        // check arguments...
410        if (!isValidMonthCode(month)) {
411          throw new IllegalArgumentException(
412            "SerialDate.monthCodeToString: month outside valid range.");
413        }
414
415        final String[] months;
416
417        if (shortened) {
418          months = DATE_FORMAT_SYMBOLS.getShortMonths();
419        }
420        else {
421            months = DATE_FORMAT_SYMBOLS.getMonths();
422        }
423
424        return months[month - 1];
425
426    }
427
428    /**
429     * Converts a string to a month code.
430     * <P>
431     * This method will return one of the constants JANUARY, FEBRUARY, ...,
432     * DECEMBER that corresponds to the string.  If the string is not
433     * recognised, this method returns -1.
434     *
435     * @param s  the string to parse.
436     *
437     * @return <code>-1</code> if the string is not parseable, the month of the
438     *         year otherwise.
439     */
440    public static int stringToMonthCode(String s) {
441
442        final String[] shortMonthNames = DATE_FORMAT_SYMBOLS.getShortMonths();
443        final String[] monthNames = DATE_FORMAT_SYMBOLS.getMonths();
444
445        int result = -1;
446        s = s.trim();
447
448        // first try parsing the string as an integer (1-12)...
449        try {
450          result = Integer.parseInt(s);
451        }
452        catch (NumberFormatException e) {
453          // suppress
454        }
455
456        // now search through the month names...
457        if ((result < 1) || (result > 12)) {
458          for (int i = 0; i < monthNames.length; i++) {
459            if (s.equals(shortMonthNames[i])) {
460              result = i + 1;
461              break;
462            }
```

```
463            if (s.equals(monthNames[i])) {
464                result = i + 1;
465                break;
466            }
467        }
468    }
469
470    return result;
471
472 }
473
474 /**
475  * Returns true if the supplied integer code represents a valid
476  * week-in-the-month, and false otherwise.
477  *
478  * @param code  the code being checked for validity.
479  * @return <code>true</code> if the supplied integer code represents a
480  *         valid week-in-the-month.
481  */
482 public static boolean isValidWeekInMonthCode(final int code) {
483
484    switch(code) {
485        case FIRST_WEEK_IN_MONTH:
486        case SECOND_WEEK_IN_MONTH:
487        case THIRD_WEEK_IN_MONTH:
488        case FOURTH_WEEK_IN_MONTH:
489        case LAST_WEEK_IN_MONTH: return true;
490        default: return false;
491    }
492
493 }
494
495 /**
496  * Determines whether or not the specified year is a leap year.
497  *
498  * @param yyyy  the year (in the range 1900 to 9999).
499  *
500  * @return <code>true</code> if the specified year is a leap year.
501  */
502 public static boolean isLeapYear(final int yyyy) {
503
504    if ((yyyy % 4) != 0) {
505        return false;
506    }
507    else if ((yyyy % 400) == 0) {
508        return true;
509    }
510    else if ((yyyy % 100) == 0) {
511        return false;
512    }
513    else {
514        return true;
515    }
516
```

```
517    }
518
519    /**
520     * Returns the number of leap years from 1900 to the specified year
521     * INCLUSIVE.
522     * <P>
523     * Note that 1900 is not a leap year.
524     *
525     * @param yyyy  the year (in the range 1900 to 9999).
526     *
527     * @return the number of leap years from 1900 to the specified year.
528     */
529    public static int leapYearCount(final int yyyy) {
530
531       final int leap4 = (yyyy - 1896) / 4;
532       final int leap100 = (yyyy - 1800) / 100;
533       final int leap400 = (yyyy - 1600) / 400;
534       return leap4 - leap100 + leap400;
535
536    }
537
538    /**
539     * Returns the number of the last day of the month, taking into account
540     * leap years.
541     *
542     * @param month  the month.
543     * @param yyyy  the year (in the range 1900 to 9999).
544     *
545     * @return the number of the last day of the month.
546     */
547    public static int lastDayOfMonth(final int month, final int yyyy) {
548
549       final int result = LAST_DAY_OF_MONTH[month];
550       if (month != FEBRUARY) {
551          return result;
552       }
553       else if (isLeapYear(yyyy)) {
554          return result + 1;
555       }
556       else {
557          return result;
558       }
559
560    }
561
562    /**
563     * Creates a new date by adding the specified number of days to the base
564     * date.
565     *
566     * @param days  the number of days to add (can be negative).
567     * @param base  the base date.
568     *
569     * @return a new date.
570     */
```

```java
571    public static SerialDate addDays(final int days, final SerialDate base) {
572
573        final int serialDayNumber = base.toSerial() + days;
574        return SerialDate.createInstance(serialDayNumber);
575
576    }
577
578    /**
579     * Creates a new date by adding the specified number of months to the base
580     * date.
581     * <P>
582     * If the base date is close to the end of the month, the day on the result
583     * may be adjusted slightly:  31 May + 1 month = 30 June.
584     *
585     * @param months  the number of months to add (can be negative).
586     * @param base  the base date.
587     *
588     * @return a new date.
589     */
590    public static SerialDate addMonths(final int months,
591                                       final SerialDate base) {
592
593        final int yy = (12 * base.getYYYY() + base.getMonth() + months - 1) / 12;
594
595        final int mm = (12 * base.getYYYY() + base.getMonth() + months - 1) % 12 + 1;
596
597        final int dd = Math.min(
598            base.getDayOfMonth(), SerialDate.lastDayOfMonth(mm, yy)
599        );
600        return SerialDate.createInstance(dd, mm, yy);
601
602    }
603
604    /**
605     * Creates a new date by adding the specified number of years to the base
606     * date.
607     *
608     * @param years  the number of years to add (can be negative).
609     * @param base  the base date.
610     *
611     * @return A new date.
612     */
613    public static SerialDate addYears(final int years, final SerialDate base) {
614
615        final int baseY = base.getYYYY();
616        final int baseM = base.getMonth();
617        final int baseD = base.getDayOfMonth();
618
619        final int targetY = baseY + years;
620        final int targetD = Math.min(
621          baseD, SerialDate.lastDayOfMonth(baseM, targetY)
622        );
623
624        return SerialDate.createInstance(targetD, baseM, targetY);
```

```
625
626    }
627
628    /**
629     * Returns the latest date that falls on the specified day-of-the-week and
630     * is BEFORE the base date.
631     *
632     * @param targetWeekday  a code for the target day-of-the-week.
633     * @param base  the base date.
634     *
635     * @return the latest date that falls on the specified day-of-the-week and
636     *         is BEFORE the base date.
637     */
638    public static SerialDate getPreviousDayOfWeek(final int targetWeekday,
639                                                  final SerialDate base) {
640
641        // check arguments...
642        if (!SerialDate.isValidWeekdayCode(targetWeekday)) {
643            throw new IllegalArgumentException(
644                "Invalid day-of-the-week code."
645            );
646        }
647
648        // find the date...
649        final int adjust;
650        final int baseDOW = base.getDayOfWeek();
651        if (baseDOW > targetWeekday) {
652            adjust = Math.min(0, targetWeekday - baseDOW);
653        }
654        else {
655            adjust = -7 + Math.max(0, targetWeekday - baseDOW);
656        }
657
658        return SerialDate.addDays(adjust, base);
659
660    }
661
662    /**
663     * Returns the earliest date that falls on the specified day-of-the-week
664     * and is AFTER the base date.
665     *
666     * @param targetWeekday  a code for the target day-of-the-week.
667     * @param base  the base date.
668     *
669     * @return the earliest date that falls on the specified day-of-the-week
670     *         and is AFTER the base date.
671     */
672    public static SerialDate getFollowingDayOfWeek(final int targetWeekday,
673                                                   final SerialDate base) {
674
675        // check arguments...
676        if (!SerialDate.isValidWeekdayCode(targetWeekday)) {
677            throw new IllegalArgumentException(
678                "Invalid day-of-the-week code."
```

```
            );
        }

        // find the date...
        final int adjust;
        final int baseDOW = base.getDayOfWeek();
        if (baseDOW > targetWeekday) {
            adjust = 7 + Math.min(0, targetWeekday - baseDOW);
        }
        else {
            adjust = Math.max(0, targetWeekday - baseDOW);
        }

        return SerialDate.addDays(adjust, base);
    }

    /**
     * Returns the date that falls on the specified day-of-the-week and is
     * CLOSEST to the base date.
     *
     * @param targetDOW  a code for the target day-of-the-week.
     * @param base  the base date.
     *
     * @return the date that falls on the specified day-of-the-week and is
     *         CLOSEST to the base date.
     */
    public static SerialDate getNearestDayOfWeek(final int targetDOW,
                                                 final SerialDate base) {

        // check arguments...
        if (!SerialDate.isValidWeekdayCode(targetDOW)) {
            throw new IllegalArgumentException(
                "Invalid day-of-the-week code."
            );
        }

        // find the date...
        final int baseDOW = base.getDayOfWeek();
        int adjust = -Math.abs(targetDOW - baseDOW);
        if (adjust >= 4) {
            adjust = 7 - adjust;
        }
        if (adjust <= -4) {
            adjust = 7 + adjust;
        }
        return SerialDate.addDays(adjust, base);

    }

    /**
     * Rolls the date forward to the last day of the month.
     *
     * @param base  the base date.
     *
```

```
733         * @return a new serial date.
734         */
735        public SerialDate getEndOfCurrentMonth(final SerialDate base) {
736            final int last = SerialDate.lastDayOfMonth(
737                base.getMonth(), base.getYYYY()
738            );
739            return SerialDate.createInstance(last, base.getMonth(), base.getYYYY());
740        }
741
742        /**
743         * Returns a string corresponding to the week-in-the-month code.
744         * <P>
745         * Need to find a better approach.
746         *
747         * @param count  an integer code representing the week-in-the-month.
748         *
749         * @return a string corresponding to the week-in-the-month code.
750         */
751        public static String weekInMonthToString(final int count) {
752
753            switch (count) {
754                case SerialDate.FIRST_WEEK_IN_MONTH : return "First";
755                case SerialDate.SECOND_WEEK_IN_MONTH : return "Second";
756                case SerialDate.THIRD_WEEK_IN_MONTH : return "Third";
757                case SerialDate.FOURTH_WEEK_IN_MONTH : return "Fourth";
758                case SerialDate.LAST_WEEK_IN_MONTH : return "Last";
759                default :
760                    return "SerialDate.weekInMonthToString(): invalid code.";
761            }
762
763        }
764
765        /**
766         * Returns a string representing the supplied 'relative'.
767         * <P>
768         * Need to find a better approach.
769         *
770         * @param relative  a constant representing the 'relative'.
771         *
772         * @return a string representing the supplied 'relative'.
773         */
774        public static String relativeToString(final int relative) {
775
776            switch (relative) {
777                case SerialDate.PRECEDING : return "Preceding";
778                case SerialDate.NEAREST : return "Nearest";
779                case SerialDate.FOLLOWING : return "Following";
780                default : return "ERROR : Relative To String";
781            }
782
783        }
784
785        /**
786         * Factory method that returns an instance of some concrete subclass of
```

```
787      * {@link SerialDate}.
788      *
789      * @param day   the day (1-31).
790      * @param month   the month (1-12).
791      * @param yyyy   the year (in the range 1900 to 9999).
792      *
793      * @return An instance of {@link SerialDate}.
794      */
795     public static SerialDate createInstance(final int day, final int month,
796                                             final int yyyy) {
797         return new SpreadsheetDate(day, month, yyyy);
798     }
799
800     /**
801      * Factory method that returns an instance of some concrete subclass of
802      * {@link SerialDate}.
803      *
804      * @param serial   the serial number for the day (1 January 1900 = 2).
805      *
806      * @return a instance of SerialDate.
807      */
808     public static SerialDate createInstance(final int serial) {
809         return new SpreadsheetDate(serial);
810     }
811
812     /**
813      * Factory method that returns an instance of a subclass of SerialDate.
814      *
815      * @param date   A Java date object.
816      *
817      * @return a instance of SerialDate.
818      */
819     public static SerialDate createInstance(final java.util.Date date) {
820
821         final GregorianCalendar calendar = new GregorianCalendar();
822         calendar.setTime(date);
823         return new SpreadsheetDate(calendar.get(Calendar.DATE),
824                                    calendar.get(Calendar.MONTH) + 1,
825                                    calendar.get(Calendar.YEAR));
826
827     }
828
829     /**
830      * Returns the serial number for the date, where 1 January 1900 = 2 (this
831      * corresponds, almost, to the numbering system used in Microsoft Excel for
832      * Windows and Lotus 1-2-3).
833      *
834      * @return the serial number for the date.
835      */
836     public abstract int toSerial();
837
838     /**
839      * Returns a java.util.Date.  Since java.util.Date has more precision than
840      * SerialDate, we need to define a convention for the 'time of day'.
```

```
841       *
842       * @return this as <code>java.util.Date</code>.
843       */
844      public abstract java.util.Date toDate();
845
846      /**
847       * Returns a description of the date.
848       *
849       * @return a description of the date.
850       */
851      public String getDescription() {
852          return this.description;
853      }
854
855      /**
856       * Sets the description for the date.
857       *
858       * @param description  the new description for the date.
859       */
860      public void setDescription(final String description) {
861          this.description = description;
862      }
863
864      /**
865       * Converts the date to a string.
866       *
867       * @return  a string representation of the date.
868       */
869      public String toString() {
870          return getDayOfMonth() + "-" + SerialDate.monthCodeToString(getMonth())
871                             + "-" + getYYYY();
872      }
873
874      /**
875       * Returns the year (assume a valid range of 1900 to 9999).
876       *
877       * @return the year.
878       */
879      public abstract int getYYYY();
880
881      /**
882       * Returns the month (January = 1, February = 2, March = 3).
883       *
884       * @return the month of the year.
885       */
886      public abstract int getMonth();
887
888      /**
889       * Returns the day of the month.
890       *
891       * @return the day of the month.
892       */
893      public abstract int getDayOfMonth();
894
```

```
895    /**
896     * Returns the day of the week.
897     *
898     * @return the day of the week.
899     */
900    public abstract int getDayOfWeek();
901
902    /**
903     * Returns the difference (in days) between this date and the specified
904     * 'other' date.
905     * <P>
906     * The result is positive if this date is after the 'other' date and
907     * negative if it is before the 'other' date.
908     *
909     * @param other  the date being compared to.
910     *
911     * @return the difference between this and the other date.
912     */
913    public abstract int compare(SerialDate other);
914
915    /**
916     * Returns true if this SerialDate represents the same date as the
917     * specified SerialDate.
918     *
919     * @param other  the date being compared to.
920     *
921     * @return <code>true</code> if this SerialDate represents the same date as
922     *         the specified SerialDate.
923     */
924    public abstract boolean isOn(SerialDate other);
925
926    /**
927     * Returns true if this SerialDate represents an earlier date compared to
928     * the specified SerialDate.
929     *
930     * @param other  The date being compared to.
931     *
932     * @return <code>true</code> if this SerialDate represents an earlier date
933     *         compared to the specified SerialDate.
934     */
935    public abstract boolean isBefore(SerialDate other);
936
937    /**
938     * Returns true if this SerialDate represents the same date as the
939     * specified SerialDate.
940     *
941     * @param other  the date being compared to.
942     *
943     * @return <code>true<code> if this SerialDate represents the same date
944     *         as the specified SerialDate.
945     */
946    public abstract boolean isOnOrBefore(SerialDate other);
947
948    /**
```

```
949      * Returns true if this SerialDate represents the same date as the
950      * specified SerialDate.
951      *
952      * @param other  the date being compared to.
953      *
954      * @return <code>true</code> if this SerialDate represents the same date
955      *         as the specified SerialDate.
956      */
957     public abstract boolean isAfter(SerialDate other);
958
959     /**
960      * Returns true if this SerialDate represents the same date as the
961      * specified SerialDate.
962      *
963      * @param other  the date being compared to.
964      *
965      * @return <code>true</code> if this SerialDate represents the same date
966      *         as the specified SerialDate.
967      */
968     public abstract boolean isOnOrAfter(SerialDate other);
969
970     /**
971      * Returns <code>true</code> if this {@link SerialDate} is within the
972      * specified range (INCLUSIVE).  The date order of d1 and d2 is not
973      * important.
974      *
975      * @param d1  a boundary date for the range.
976      * @param d2  the other boundary date for the range.
977      *
978      * @return A boolean.
979      */
980     public abstract boolean isInRange(SerialDate d1, SerialDate d2);
981
982     /**
983      * Returns <code>true</code> if this {@link SerialDate} is within the
984      * specified range (caller specifies whether or not the end-points are
985      * included).  The date order of d1 and d2 is not important.
986      *
987      * @param d1  a boundary date for the range.
988      * @param d2  the other boundary date for the range.
989      * @param include  a code that controls whether or not the start and end
990      *                 dates are included in the range.
991      *
992      * @return A boolean.
993      */
994     public abstract boolean isInRange(SerialDate d1, SerialDate d2,
995                                       int include);
996
997     /**
998      * Returns the latest date that falls on the specified day-of-the-week and
999      * is BEFORE this date.
1000     *
1001     * @param targetDOW  a code for the target day-of-the-week.
1002     *
```

```java
1003      * @return the latest date that falls on the specified day-of-the-week and
1004      *         is BEFORE this date.
1005      */
1006     public SerialDate getPreviousDayOfWeek(final int targetDOW) {
1007         return getPreviousDayOfWeek(targetDOW, this);
1008     }
1009
1010     /**
1011      * Returns the earliest date that falls on the specified day-of-the-week
1012      * and is AFTER this date.
1013      *
1014      * @param targetDOW  a code for the target day-of-the-week.
1015      *
1016      * @return the earliest date that falls on the specified day-of-the-week
1017      *         and is AFTER this date.
1018      */
1019     public SerialDate getFollowingDayOfWeek(final int targetDOW) {
1020         return getFollowingDayOfWeek(targetDOW, this);
1021     }
1022
1023     /**
1024      * Returns the nearest date that falls on the specified day-of-the-week.
1025      *
1026      * @param targetDOW  a code for the target day-of-the-week.
1027      *
1028      * @return the nearest date that falls on the specified day-of-the-week.
1029      */
1030     public SerialDate getNearestDayOfWeek(final int targetDOW) {
1031         return getNearestDayOfWeek(targetDOW, this);
1032     }
1033
1034 }
```

代码清单 B-2　SerialDateTest.java

```java
 1 /* ========================================================================
 2  * JCommon : a free general purpose class library for the Java(tm) platform
 3  * ========================================================================
 4  *
 5  * (C) Copyright 2000-2005, by Object Refinery Limited and Contributors.
 6  *
 7  *
 8  *
 9  * This library is free software; you can redistribute it and/or modify it
10  * under the terms of the GNU Lesser General Public License as published by
11  * the Free Software Foundation; either version 2.1 of the License, or
12  * (at your option) any later version.
13  *
14  * This library is distributed in the hope that it will be useful, but
15  * WITHOUT ANY WARRANTY; without even the implied warranty of MERCHANTABILITY
16  * or FITNESS FOR A PARTICULAR PURPOSE. See the GNU Lesser General Public
17  * License for more details.
18  *
19  * You should have received a copy of the GNU Lesser General Public
```

```
20   * License along with this library; if not, write to the Free Software
21   * Foundation, Inc., 51 Franklin Street, Fifth Floor, Boston, MA  02110-1301,
22   * USA.
23   *
24   * [Java is a trademark or registered trademark of Sun Microsystems, Inc.
25   * in the United States and other countries.]
26   *
27   * --------------------
28   * SerialDateTests.java
29   * --------------------
30   * (C) Copyright 2001-2005, by Object Refinery Limited.
31   *
32   * Original Author:  David Gilbert (for Object Refinery Limited);
33   * Contributor(s):   -;
34   *
35   * $Id: SerialDateTests.java,v 1.6 2005/11/16 15:58:40 taqua Exp $
36   *
37   * Changes
38   * -------
39   * 15-Nov-2001 : Version 1 (DG);
40   * 25-Jun-2002 : Removed unnecessary import (DG);
41   * 24-Oct-2002 : Fixed errors reported by Checkstyle (DG);
42   * 13-Mar-2003 : Added serialization test (DG);
43   * 05-Jan-2005 : Added test for bug report 1096282 (DG);
44   *
45   */
46
47  package org.jfree.date.junit;
48
49  import java.io.ByteArrayInputStream;
50  import java.io.ByteArrayOutputStream;
51  import java.io.ObjectInput;
52  import java.io.ObjectInputStream;
53  import java.io.ObjectOutput;
54  import java.io.ObjectOutputStream;
55
56  import junit.framework.Test;
57  import junit.framework.TestCase;
58  import junit.framework.TestSuite;
59
60  import org.jfree.date.MonthConstants;
61  import org.jfree.date.SerialDate;
62
63  /**
64   * Some JUnit tests for the {@link SerialDate} class.
65   */
66  public class SerialDateTests extends TestCase {
67
68      /** Date representing November 9. */
69      private SerialDate nov9Y2001;
70
71      /**
72       * Creates a new test case.
73       *
```

```
 74      * @param name  the name.
 75      */
 76     public SerialDateTests(final String name) {
 77         super(name);
 78     }
 79
 80     /**
 81      * Returns a test suite for the JUnit test runner.
 82      *
 83      * @return The test suite.
 84      */
 85     public static Test suite() {
 86         return new TestSuite(SerialDateTests.class);
 87     }
 88
 89     /**
 90      * Problem set up.
 91      */
 92     protected void setUp() {
 93         this.nov9Y2001 = SerialDate.createInstance(9, MonthConstants.NOVEMBER, 2001);
 94     }
 95
 96     /**
 97      * 9 Nov 2001 plus two months should be 9 Jan 2002.
 98      */
 99     public void testAddMonthsTo9Nov2001() {
100         final SerialDate jan9Y2002 = SerialDate.addMonths(2, this.nov9Y2001);
101         final SerialDate answer = SerialDate.createInstance(9, 1, 2002);
102         assertEquals(answer, jan9Y2002);
103     }
104
105     /**
106      * A test case for a reported bug, now fixed.
107      */
108     public void testAddMonthsTo5Oct2003() {
109         final SerialDate d1 = SerialDate.createInstance(5, MonthConstants.OCTOBER,2003);
110         final SerialDate d2 = SerialDate.addMonths(2, d1);
111         assertEquals(d2, SerialDate.createInstance(5, MonthConstants.DECEMBER, 2003));
112     }
113
114     /**
115      * A test case for a reported bug, now fixed.
116      */
117     public void testAddMonthsTo1Jan2003() {
118         final SerialDate d1 = SerialDate.createInstance(1, MonthConstants.JANUARY, 2003);
119         final SerialDate d2 = SerialDate.addMonths(0, d1);
120         assertEquals(d2, d1);
121     }
122
123     /**
124      * Monday preceding Friday 9 November 2001 should be 5 November.
125      */
126     public void testMondayPrecedingFriday9Nov2001() {
127         SerialDate mondayBefore = SerialDate.getPreviousDayOfWeek(
```

```
128        SerialDate.MONDAY, this.nov9Y2001
129    );
130    assertEquals(5, mondayBefore.getDayOfMonth());
131  }
132
133  /**
134   * Monday following Friday 9 November 2001 should be 12 November.
135   */
136  public void testMondayFollowingFriday9Nov2001() {
137    SerialDate mondayAfter = SerialDate.getFollowingDayOfWeek(
138      SerialDate.MONDAY, this.nov9Y2001
139    );
140    assertEquals(12, mondayAfter.getDayOfMonth());
141  }
142
143  /**
144   * Monday nearest Friday 9 November 2001 should be 12 November.
145   */
146  public void testMondayNearestFriday9Nov2001() {
147    SerialDate mondayNearest = SerialDate.getNearestDayOfWeek(
148      SerialDate.MONDAY, this.nov9Y2001
149    );
150    assertEquals(12, mondayNearest.getDayOfMonth());
151  }
152
153  /**
154   * The Monday nearest to 22nd January 1970 falls on the 19th.
155   */
156  public void testMondayNearest22Jan1970() {
157    SerialDate jan22Y1970 = SerialDate.createInstance(22, MonthConstants.JANUARY, 1970);
158    SerialDate mondayNearest=SerialDate.getNearestDayOfWeek(SerialDate.MONDAY,
          jan22Y1970);
159    assertEquals(19, mondayNearest.getDayOfMonth());
160  }
161
162  /**
163   * Problem that the conversion of days to strings returns the right result. Actually,
164   * this result depends on the Locale so this test needs to be modified.
165   */
166  public void testWeekdayCodeToString() {
167
168    final String test = SerialDate.weekdayCodeToString(SerialDate.SATURDAY);
169    assertEquals("Saturday", test);
170
171  }
172
173  /**
174   * Test the conversion of a string to a weekday. Note that this test will fail if
175   * the default locale doesn't use English weekday names...devise a better test!
176   */
177  public void testStringToWeekday() {
178
179    int weekday = SerialDate.stringToWeekdayCode("Wednesday");
180    assertEquals(SerialDate.WEDNESDAY, weekday);
```

```
181
182        weekday = SerialDate.stringToWeekdayCode(" Wednesday ");
183        assertEquals(SerialDate.WEDNESDAY, weekday);
184
185        weekday = SerialDate.stringToWeekdayCode("Wed");
186        assertEquals(SerialDate.WEDNESDAY, weekday);
187
188    }
189
190    /**
191     * Test the conversion of a string to a month.  Note that this test will fail if
192     * the default locale doesn't use English month names...devise a better test!
193     */
194    public void testStringToMonthCode() {
195
196        int m = SerialDate.stringToMonthCode("January");
197        assertEquals(MonthConstants.JANUARY, m);
198
199        m = SerialDate.stringToMonthCode(" January ");
200        assertEquals(MonthConstants.JANUARY, m);
201
202        m = SerialDate.stringToMonthCode("Jan");
203        assertEquals(MonthConstants.JANUARY, m);
204
205    }
206
207    /**
208     * Tests the conversion of a month code to a string.
209     */
210    public void testMonthCodeToStringCode() {
211
212        final String test = SerialDate.monthCodeToString(MonthConstants.DECEMBER);
213        assertEquals("December", test);
214
215    }
216
217    /**
218     * 1900 is not a leap year.
219     */
220    public void testIsNotLeapYear1900() {
221        assertTrue(!SerialDate.isLeapYear(1900));
222    }
223
224    /**
225     * 2000 is a leap year.
226     */
227    public void testIsLeapYear2000() {
228        assertTrue(SerialDate.isLeapYear(2000));
229    }
230
231    /**
232     * The number of leap years from 1900 up-to-and-including 1899 is 0.
233     */
234    public void testLeapYearCount1899() {
```

```
235      assertEquals(SerialDate.leapYearCount(1899), 0);
236    }
237
238    /**
239     * The number of leap years from 1900 up-to-and-including 1903 is 0.
240     */
241    public void testLeapYearCount1903() {
242      assertEquals(SerialDate.leapYearCount(1903), 0);
243    }
244
245    /**
246     * The number of leap years from 1900 up-to-and-including 1904 is 1.
247     */
248    public void testLeapYearCount1904() {
249      assertEquals(SerialDate.leapYearCount(1904), 1);
250    }
251
252    /**
253     * The number of leap years from 1900 up-to-and-including 1999 is 24.
254     */
255    public void testLeapYearCount1999() {
256      assertEquals(SerialDate.leapYearCount(1999), 24);
257    }
258
259    /**
260     * The number of leap years from 1900 up-to-and-including 2000 is 25.
261     */
262    public void testLeapYearCount2000() {
263      assertEquals(SerialDate.leapYearCount(2000), 25);
264    }
265
266    /**
267     * Serialize an instance, restore it, and check for equality.
268     */
269    public void testSerialization() {
270
271      SerialDate d1 = SerialDate.createInstance(15, 4, 2000);
272      SerialDate d2 = null;
273
274      try {
275        ByteArrayOutputStream buffer = new ByteArrayOutputStream();
276        ObjectOutput out = new ObjectOutputStream(buffer);
277        out.writeObject(d1);
278        out.close();
279
280        ObjectInput in = new ObjectInputStream(
                              new ByteArrayInputStream(buffer.toByteArray()));
281        d2 = (SerialDate) in.readObject();
282        in.close();
283      }
284      catch (Exception e) {
285        System.out.println(e.toString());
286      }
287      assertEquals(d1, d2);
```

```
288
289    }
290
291    /**
292     * A test for bug report 1096282 (now fixed).
293     */
294    public void test1096282() {
295        SerialDate d = SerialDate.createInstance(29, 2, 2004);
296        d = SerialDate.addYears(1, d);
297        SerialDate expected = SerialDate.createInstance(28, 2, 2005);
298        assertTrue(d.isOn(expected));
299    }
300
301    /**
302     * Miscellaneous tests for the addMonths() method.
303     */
304    public void testAddMonths() {
305        SerialDate d1 = SerialDate.createInstance(31, 5, 2004);
306
307        SerialDate d2 = SerialDate.addMonths(1, d1);
308        assertEquals(30, d2.getDayOfMonth());
309        assertEquals(6, d2.getMonth());
310        assertEquals(2004, d2.getYYYY());
311
312        SerialDate d3 = SerialDate.addMonths(2, d1);
313        assertEquals(31, d3.getDayOfMonth());
314        assertEquals(7, d3.getMonth());
315        assertEquals(2004, d3.getYYYY());
316
317        SerialDate d4 = SerialDate.addMonths(1, SerialDate.addMonths(1, d1));
318        assertEquals(30, d4.getDayOfMonth());
319        assertEquals(7, d4.getMonth());
320        assertEquals(2004, d4.getYYYY());
321    }
322 }
```

代码清单 B-3   MonthConstants.java

```
 1  /* ========================================================================
 2   * JCommon : a free general purpose class library for the Java(tm) platform
 3   * ========================================================================
 4   *
 5   * (C) Copyright 2000-2005, by Object Refinery Limited and Contributors.
 6   *
 7   *
 8   *
 9   * This library is free software; you can redistribute it and/or modify it
10   * under the terms of the GNU Lesser General Public License as published by
11   * the Free Software Foundation; either version 2.1 of the License, or
12   * (at your option) any later version.
13   *
14   * This library is distributed in the hope that it will be useful, but
15   * WITHOUT ANY WARRANTY; without even the implied warranty of MERCHANTABILITY
16   * or FITNESS FOR A PARTICULAR PURPOSE. See the GNU Lesser General Public
```

```
17   * License for more details.
18   *
19   * You should have received a copy of the GNU Lesser General Public
20   * License along with this library; if not, write to the Free Software
21   * Foundation, Inc., 51 Franklin Street, Fifth Floor, Boston, MA  02110-1301,
22   * USA.
23   *
24   * [Java is a trademark or registered trademark of Sun Microsystems, Inc.
25   * in the United States and other countries.]
26   *
27   * -------------------
28   * MonthConstants.java
29   * -------------------
30   * (C) Copyright 2002, 2003, by Object Refinery Limited.
31   *
32   * Original Author:  David Gilbert (for Object Refinery Limited);
33   * Contributor(s):   -;
34   *
35   * $Id: MonthConstants.java,v 1.4 2005/11/16 15:58:40 taqua Exp $
36   *
37   * Changes
38   * -------
39   * 29-May-2002 : Version 1 (code moved from SerialDate class) (DG);
40   *
41   */
42
43  package org.jfree.date;
44
45  /**
46   * Useful constants for months.  Note that these are NOT equivalent to the
47   * constants defined by java.util.Calendar (where JANUARY=0 and DECEMBER=11).
48   * <P>
49   * Used by the SerialDate and RegularTimePeriod classes.
50   *
51   * @author David Gilbert
52   */
53  public interface MonthConstants {
54
55      /** Constant for January. */
56      public static final int JANUARY = 1;
57
58      /** Constant for February. */
59      public static final int FEBRUARY = 2;
60
61      /** Constant for March. */
62      public static final int MARCH = 3;
63
64      /** Constant for April. */
65      public static final int APRIL = 4;
66
67      /** Constant for May. */
68      public static final int MAY = 5;
69
70      /** Constant for June. */
```

```
71    public static final int JUNE = 6;
72
73    /** Constant for July. */
74    public static final int JULY = 7;
75
76    /** Constant for August. */
77    public static final int AUGUST = 8;
78
79    /** Constant for September. */
80    public static final int SEPTEMBER = 9;
81
82    /** Constant for October. */
83    public static final int OCTOBER = 10;
84
85    /** Constant for November. */
86    public static final int NOVEMBER = 11;
87
88    /** Constant for December. */
89    public static final int DECEMBER = 12;
90
91 }
```

代码清单 B-4　BobsSerialDateTest.java

```
 1 package org.jfree.date.junit;
 2
 3 import junit.framework.TestCase;
 4 import org.jfree.date.*;
 5 import static org.jfree.date.SerialDate.*;
 6
 7 import java.util.*;
 8
 9 public class BobsSerialDateTest extends TestCase {
10
11   public void testIsValidWeekdayCode() throws Exception {
12     for (int day = 1; day <= 7; day++)
13       assertTrue(isValidWeekdayCode(day));
14     assertFalse(isValidWeekdayCode(0));
15     assertFalse(isValidWeekdayCode(8));
16   }
17
18   public void testStringToWeekdayCode() throws Exception {
19
20     assertEquals(-1, stringToWeekdayCode("Hello"));
21     assertEquals(MONDAY, stringToWeekdayCode("Monday"));
22     assertEquals(MONDAY, stringToWeekdayCode("Mon"));
23 //todo    assertEquals(MONDAY,stringToWeekdayCode("monday"));
24 //     assertEquals(MONDAY,stringToWeekdayCode("MONDAY"));
25 //     assertEquals(MONDAY, stringToWeekdayCode("mon"));
26
27     assertEquals(TUESDAY, stringToWeekdayCode("Tuesday"));
28     assertEquals(TUESDAY, stringToWeekdayCode("Tue"));
29 //     assertEquals(TUESDAY,stringToWeekdayCode("tuesday"));
30 //     assertEquals(TUESDAY,stringToWeekdayCode("TUESDAY"));
```

```
31 //    assertEquals(TUESDAY, stringToWeekdayCode("tue"));
32 //    assertEquals(TUESDAY, stringToWeekdayCode("tues"));
33
34       assertEquals(WEDNESDAY, stringToWeekdayCode("Wednesday"));
35       assertEquals(WEDNESDAY, stringToWeekdayCode("Wed"));
36 //    assertEquals(WEDNESDAY,stringToWeekdayCode("wednesday"));
37 //    assertEquals(WEDNESDAY,stringToWeekdayCode("WEDNESDAY"));
38 //    assertEquals(WEDNESDAY, stringToWeekdayCode("wed"));
39
40       assertEquals(THURSDAY, stringToWeekdayCode("Thursday"));
41       assertEquals(THURSDAY, stringToWeekdayCode("Thu"));
42 //    assertEquals(THURSDAY,stringToWeekdayCode("thursday"));
43 //    assertEquals(THURSDAY,stringToWeekdayCode("THURSDAY"));
44 //    assertEquals(THURSDAY, stringToWeekdayCode("thu"));
45 //    assertEquals(THURSDAY, stringToWeekdayCode("thurs"));
46
47       assertEquals(FRIDAY, stringToWeekdayCode("Friday"));
48       assertEquals(FRIDAY, stringToWeekdayCode("Fri"));
49 //    assertEquals(FRIDAY,stringToWeekdayCode("friday"));
50 //    assertEquals(FRIDAY,stringToWeekdayCode("FRIDAY"));
51 //    assertEquals(FRIDAY, stringToWeekdayCode("fri"));
52
53       assertEquals(SATURDAY, stringToWeekdayCode("Saturday"));
54       assertEquals(SATURDAY, stringToWeekdayCode("Sat"));
55 //    assertEquals(SATURDAY,stringToWeekdayCode("saturday"));
56 //    assertEquals(SATURDAY,stringToWeekdayCode("SATURDAY"));
57 //    assertEquals(SATURDAY, stringToWeekdayCode("sat"));
58
59       assertEquals(SUNDAY, stringToWeekdayCode("Sunday"));
60       assertEquals(SUNDAY, stringToWeekdayCode("Sun"));
61 //    assertEquals(SUNDAY,stringToWeekdayCode("sunday"));
62 //    assertEquals(SUNDAY,stringToWeekdayCode("SUNDAY"));
63 //    assertEquals(SUNDAY, stringToWeekdayCode("sun"));
64    }
65
66    public void testWeekdayCodeToString() throws Exception {
67       assertEquals("Sunday", weekdayCodeToString(SUNDAY));
68       assertEquals("Monday", weekdayCodeToString(MONDAY));
69       assertEquals("Tuesday", weekdayCodeToString(TUESDAY));
70       assertEquals("Wednesday", weekdayCodeToString(WEDNESDAY));
71       assertEquals("Thursday", weekdayCodeToString(THURSDAY));
72       assertEquals("Friday", weekdayCodeToString(FRIDAY));
73       assertEquals("Saturday", weekdayCodeToString(SATURDAY));
74    }
75
76    public void testIsValidMonthCode() throws Exception {
77       for (int i = 1; i <= 12; i++)
78          assertTrue(isValidMonthCode(i));
79       assertFalse(isValidMonthCode(0));
80       assertFalse(isValidMonthCode(13));
81    }
82
83    public void testMonthToQuarter() throws Exception {
84       assertEquals(1, monthCodeToQuarter(JANUARY));
```

```
 85      assertEquals(1, monthCodeToQuarter(FEBRUARY));
 86      assertEquals(1, monthCodeToQuarter(MARCH));
 87      assertEquals(2, monthCodeToQuarter(APRIL));
 88      assertEquals(2, monthCodeToQuarter(MAY));
 89      assertEquals(2, monthCodeToQuarter(JUNE));
 90      assertEquals(3, monthCodeToQuarter(JULY));
 91      assertEquals(3, monthCodeToQuarter(AUGUST));
 92      assertEquals(3, monthCodeToQuarter(SEPTEMBER));
 93      assertEquals(4, monthCodeToQuarter(OCTOBER));
 94      assertEquals(4, monthCodeToQuarter(NOVEMBER));
 95      assertEquals(4, monthCodeToQuarter(DECEMBER));
 96
 97      try {
 98        monthCodeToQuarter(-1);
 99        fail("Invalid Month Code should throw exception");
100      } catch (IllegalArgumentException e) {
101      }
102    }
103
104    public void testMonthCodeToString() throws Exception {
105      assertEquals("January", monthCodeToString(JANUARY));
106      assertEquals("February", monthCodeToString(FEBRUARY));
107      assertEquals("March", monthCodeToString(MARCH));
108      assertEquals("April", monthCodeToString(APRIL));
109      assertEquals("May", monthCodeToString(MAY));
110      assertEquals("June", monthCodeToString(JUNE));
111      assertEquals("July", monthCodeToString(JULY));
112      assertEquals("August", monthCodeToString(AUGUST));
113      assertEquals("September", monthCodeToString(SEPTEMBER));
114      assertEquals("October", monthCodeToString(OCTOBER));
115      assertEquals("November", monthCodeToString(NOVEMBER));
116      assertEquals("December", monthCodeToString(DECEMBER));
117
118      assertEquals("Jan", monthCodeToString(JANUARY, true));
119      assertEquals("Feb", monthCodeToString(FEBRUARY, true));
120      assertEquals("Mar", monthCodeToString(MARCH, true));
121      assertEquals("Apr", monthCodeToString(APRIL, true));
122      assertEquals("May", monthCodeToString(MAY, true));
123      assertEquals("Jun", monthCodeToString(JUNE, true));
124      assertEquals("Jul", monthCodeToString(JULY, true));
125      assertEquals("Aug", monthCodeToString(AUGUST, true));
126      assertEquals("Sep", monthCodeToString(SEPTEMBER, true));
127      assertEquals("Oct", monthCodeToString(OCTOBER, true));
128      assertEquals("Nov", monthCodeToString(NOVEMBER, true));
129      assertEquals("Dec", monthCodeToString(DECEMBER, true));
130
131      try {
132        monthCodeToString(-1);
133        fail("Invalid month code should throw exception");
134      } catch (IllegalArgumentException e) {
135      }
136
137    }
138
```

```
139    public void testStringToMonthCode() throws Exception {
140        assertEquals(JANUARY,stringToMonthCode("1"));
141        assertEquals(FEBRUARY,stringToMonthCode("2"));
142        assertEquals(MARCH,stringToMonthCode("3"));
143        assertEquals(APRIL,stringToMonthCode("4"));
144        assertEquals(MAY,stringToMonthCode("5"));
145        assertEquals(JUNE,stringToMonthCode("6"));
146        assertEquals(JULY,stringToMonthCode("7"));
147        assertEquals(AUGUST,stringToMonthCode("8"));
148        assertEquals(SEPTEMBER,stringToMonthCode("9"));
149        assertEquals(OCTOBER,stringToMonthCode("10"));
150        assertEquals(NOVEMBER, stringToMonthCode("11"));
151        assertEquals(DECEMBER,stringToMonthCode("12"));
152
153 //todo    assertEquals(-1, stringToMonthCode("0"));
154 //       assertEquals(-1, stringToMonthCode("13"));
155
156        assertEquals(-1,stringToMonthCode("Hello"));
157
158        for (int m = 1; m <= 12; m++) {
159          assertEquals(m, stringToMonthCode(monthCodeToString(m, false)));
160          assertEquals(m, stringToMonthCode(monthCodeToString(m, true)));
161        }
162
163 //     assertEquals(1,stringToMonthCode("jan"));
164 //     assertEquals(2,stringToMonthCode("feb"));
165 //     assertEquals(3,stringToMonthCode("mar"));
166 //     assertEquals(4,stringToMonthCode("apr"));
167 //     assertEquals(5,stringToMonthCode("may"));
168 //     assertEquals(6,stringToMonthCode("jun"));
169 //     assertEquals(7,stringToMonthCode("jul"));
170 //     assertEquals(8,stringToMonthCode("aug"));
171 //     assertEquals(9,stringToMonthCode("sep"));
172 //     assertEquals(10,stringToMonthCode("oct"));
173 //     assertEquals(11,stringToMonthCode("nov"));
174 //     assertEquals(12,stringToMonthCode("dec"));
175
176 //     assertEquals(1,stringToMonthCode("JAN"));
177 //     assertEquals(2,stringToMonthCode("FEB"));
178 //     assertEquals(3,stringToMonthCode("MAR"));
179 //     assertEquals(4,stringToMonthCode("APR"));
180 //     assertEquals(5,stringToMonthCode("MAY"));
181 //     assertEquals(6,stringToMonthCode("JUN"));
182 //     assertEquals(7,stringToMonthCode("JUL"));
183 //     assertEquals(8,stringToMonthCode("AUG"));
184 //     assertEquals(9,stringToMonthCode("SEP"));
185 //     assertEquals(10,stringToMonthCode("OCT"));
186 //     assertEquals(11,stringToMonthCode("NOV"));
187 //     assertEquals(12,stringToMonthCode("DEC"));
188
189 //     assertEquals(1,stringToMonthCode("january"));
190 //     assertEquals(2,stringToMonthCode("february"));
191 //     assertEquals(3,stringToMonthCode("march"));
192 //     assertEquals(4,stringToMonthCode("april"));
```

```
193 //      assertEquals(5,stringToMonthCode("may"));
194 //      assertEquals(6,stringToMonthCode("june"));
195 //      assertEquals(7,stringToMonthCode("july"));
196 //      assertEquals(8,stringToMonthCode("august"));
197 //      assertEquals(9,stringToMonthCode("september"));
198 //      assertEquals(10,stringToMonthCode("october"));
199 //      assertEquals(11,stringToMonthCode("november"));
200 //      assertEquals(12,stringToMonthCode("december"));
201
202 //      assertEquals(1,stringToMonthCode("JANUARY"));
203 //      assertEquals(2,stringToMonthCode("FEBRUARY"));
204 //      assertEquals(3,stringToMonthCode("MAR"));
205 //      assertEquals(4,stringToMonthCode("APRIL"));
206 //      assertEquals(5,stringToMonthCode("MAY"));
207 //      assertEquals(6,stringToMonthCode("JUNE"));
208 //      assertEquals(7,stringToMonthCode("JULY"));
209 //      assertEquals(8,stringToMonthCode("AUGUST"));
210 //      assertEquals(9,stringToMonthCode("SEPTEMBER"));
211 //      assertEquals(10,stringToMonthCode("OCTOBER"));
212 //      assertEquals(11,stringToMonthCode("NOVEMBER"));
213 //      assertEquals(12,stringToMonthCode("DECEMBER"));
214     }
215
216     public void testIsValidWeekInMonthCode() throws Exception {
217       for (int w = 0; w <= 4; w++) {
218         assertTrue(isValidWeekInMonthCode(w));
219       }
220       assertFalse(isValidWeekInMonthCode(5));
221     }
222
223     public void testIsLeapYear() throws Exception {
224       assertFalse(isLeapYear(1900));
225       assertFalse(isLeapYear(1901));
226       assertFalse(isLeapYear(1902));
227       assertFalse(isLeapYear(1903));
228       assertTrue(isLeapYear(1904));
229       assertTrue(isLeapYear(1908));
230       assertFalse(isLeapYear(1955));
231       assertTrue(isLeapYear(1964));
232       assertTrue(isLeapYear(1980));
233       assertTrue(isLeapYear(2000));
234       assertFalse(isLeapYear(2001));
235       assertFalse(isLeapYear(2100));
236     }
237
238     public void testLeapYearCount() throws Exception {
239       assertEquals(0, leapYearCount(1900));
240       assertEquals(0, leapYearCount(1901));
241       assertEquals(0, leapYearCount(1902));
242       assertEquals(0, leapYearCount(1903));
243       assertEquals(1, leapYearCount(1904));
244       assertEquals(1, leapYearCount(1905));
245       assertEquals(1, leapYearCount(1906));
246       assertEquals(1, leapYearCount(1907));
```

```
247         assertEquals(2, leapYearCount(1908));
248         assertEquals(24, leapYearCount(1999));
249         assertEquals(25, leapYearCount(2001));
250         assertEquals(49, leapYearCount(2101));
251         assertEquals(73, leapYearCount(2201));
252         assertEquals(97, leapYearCount(2301));
253         assertEquals(122, leapYearCount(2401));
254     }
255
256     public void testLastDayOfMonth() throws Exception {
257         assertEquals(31, lastDayOfMonth(JANUARY, 1901));
258         assertEquals(28, lastDayOfMonth(FEBRUARY, 1901));
259         assertEquals(31, lastDayOfMonth(MARCH, 1901));
260         assertEquals(30, lastDayOfMonth(APRIL, 1901));
261         assertEquals(31, lastDayOfMonth(MAY, 1901));
262         assertEquals(30, lastDayOfMonth(JUNE, 1901));
263         assertEquals(31, lastDayOfMonth(JULY, 1901));
264         assertEquals(31, lastDayOfMonth(AUGUST, 1901));
265         assertEquals(30, lastDayOfMonth(SEPTEMBER, 1901));
266         assertEquals(31, lastDayOfMonth(OCTOBER, 1901));
267         assertEquals(30, lastDayOfMonth(NOVEMBER, 1901));
268         assertEquals(31, lastDayOfMonth(DECEMBER, 1901));
269         assertEquals(29, lastDayOfMonth(FEBRUARY, 1904));
270     }
271
272     public void testAddDays() throws Exception {
273         SerialDate newYears = d(1, JANUARY, 1900);
274         assertEquals(d(2, JANUARY, 1900), addDays(1, newYears));
275         assertEquals(d(1, FEBRUARY, 1900), addDays(31, newYears));
276         assertEquals(d(1, JANUARY, 1901), addDays(365, newYears));
277         assertEquals(d(31, DECEMBER, 1904), addDays(5 * 365, newYears));
278     }
279
280     private static SpreadsheetDate d(int day, int month, int year) {return new
            SpreadsheetDate(day, month, year);}
281
282     public void testAddMonths() throws Exception {
283         assertEquals(d(1, FEBRUARY, 1900), addMonths(1, d(1, JANUARY, 1900)));
284         assertEquals(d(28, FEBRUARY, 1900), addMonths(1, d(31, JANUARY, 1900)));
285         assertEquals(d(28, FEBRUARY, 1900), addMonths(1, d(30, JANUARY, 1900)));
286         assertEquals(d(28, FEBRUARY, 1900), addMonths(1, d(29, JANUARY, 1900)));
287         assertEquals(d(28, FEBRUARY, 1900), addMonths(1, d(28, JANUARY, 1900)));
288         assertEquals(d(27, FEBRUARY, 1900), addMonths(1, d(27, JANUARY, 1900)));
289
290         assertEquals(d(30, JUNE, 1900), addMonths(5, d(31, JANUARY, 1900)));
291         assertEquals(d(30, JUNE, 1901), addMonths(17, d(31, JANUARY, 1900)));
292
293         assertEquals(d(29, FEBRUARY, 1904), addMonths(49, d(31, JANUARY, 1900)));
294
295     }
296
297     public void testAddYears() throws Exception {
298         assertEquals(d(1, JANUARY, 1901), addYears(1, d(1, JANUARY, 1900)));
299         assertEquals(d(28, FEBRUARY, 1905), addYears(1, d(29, FEBRUARY, 1904)));
```

```
300        assertEquals(d(28, FEBRUARY, 1905), addYears(1, d(28, FEBRUARY, 1904)));
301        assertEquals(d(28, FEBRUARY, 1904), addYears(1, d(28, FEBRUARY, 1903)));
302    }
303
304    public void testGetPreviousDayOfWeek() throws Exception {
305        assertEquals(d(24, FEBRUARY, 2006), getPreviousDayOfWeek(FRIDAY,
                d(1, MARCH, 2006)));
306        assertEquals(d(22, FEBRUARY, 2006), getPreviousDayOfWeek(WEDNESDAY,
                d(1, MARCH, 2006)));
307        assertEquals(d(29, FEBRUARY, 2004), getPreviousDayOfWeek(SUNDAY,
                d(3, MARCH, 2004)));
308        assertEquals(d(29, DECEMBER, 2004), getPreviousDayOfWeek(WEDNESDAY,
                d(5, JANUARY, 2005)));
309
310        try {
311            getPreviousDayOfWeek(-1, d(1, JANUARY, 2006));
312            fail("Invalid day of week code should throw exception");
313        } catch (IllegalArgumentException e) {
314        }
315    }
316
317    public void testGetFollowingDayOfWeek() throws Exception {
318 //     assertEquals(d(1, JANUARY, 2005),getFollowingDayOfWeek(SATURDAY,
                d(25, DECEMBER, 2004)));
319        assertEquals(d(1, JANUARY, 2005), getFollowingDayOfWeek(SATURDAY,
                d(26, DECEMBER, 2004)));
320        assertEquals(d(3, MARCH, 2004), getFollowingDayOfWeek(WEDNESDAY,
                d(28, FEBRUARY, 2004)));
321
322        try {
323            getFollowingDayOfWeek(-1, d(1, JANUARY, 2006));
324            fail("Invalid day of week code should throw exception");
325        } catch (IllegalArgumentException e) {
326        }
327    }
328
329    public void testGetNearestDayOfWeek() throws Exception {
330        assertEquals(d(16, APRIL, 2006), getNearestDayOfWeek(SUNDAY, d(16, APRIL,
                2006)));
331        assertEquals(d(16, APRIL, 2006), getNearestDayOfWeek(SUNDAY, d(17, APRIL,
                2006)));
332        assertEquals(d(16, APRIL, 2006), getNearestDayOfWeek(SUNDAY, d(18, APRIL,
                2006)));
333        assertEquals(d(16, APRIL, 2006), getNearestDayOfWeek(SUNDAY, d(19, APRIL,
                2006)));
334        assertEquals(d(23, APRIL, 2006), getNearestDayOfWeek(SUNDAY, d(20, APRIL,
                2006)));
335        assertEquals(d(23, APRIL, 2006), getNearestDayOfWeek(SUNDAY, d(21, APRIL,
                2006)));
336        assertEquals(d(23, APRIL, 2006), getNearestDayOfWeek(SUNDAY, d(22, APRIL,
                2006)));
337
338 //todo    assertEquals(d(17, APRIL, 2006), getNearestDayOfWeek(MONDAY,
                d(16, APRIL, 2006)));
```

```
339     assertEquals(d(17, APRIL, 2006), getNearestDayOfWeek(MONDAY, d(17, APRIL,
            2006)));
340     assertEquals(d(17, APRIL, 2006), getNearestDayOfWeek(MONDAY, d(18, APRIL,
            2006)));
341     assertEquals(d(17, APRIL, 2006), getNearestDayOfWeek(MONDAY, d(19, APRIL,
            2006)));
342     assertEquals(d(17, APRIL, 2006), getNearestDayOfWeek(MONDAY, d(20, APRIL,
            2006)));
343     assertEquals(d(24, APRIL, 2006), getNearestDayOfWeek(MONDAY, d(21, APRIL,
            2006)));
344     assertEquals(d(24, APRIL, 2006), getNearestDayOfWeek(MONDAY, d(22, APRIL,
            2006)));
345
346 //    assertEquals(d(18, APRIL, 2006), getNearestDayOfWeek(TUESDAY,
            d(16, APRIL, 2006)));
347 //    assertEquals(d(18, APRIL, 2006), getNearestDayOfWeek(TUESDAY,
            d(17, APRIL, 2006)));
348     assertEquals(d(18, APRIL, 2006), getNearestDayOfWeek(TUESDAY, d(18, APRIL,
            2006)));
349     assertEquals(d(18, APRIL, 2006), getNearestDayOfWeek(TUESDAY, d(19, APRIL,
            2006)));
350     assertEquals(d(18, APRIL, 2006), getNearestDayOfWeek(TUESDAY, d(20, APRIL,
            2006)));
351     assertEquals(d(18, APRIL, 2006), getNearestDayOfWeek(TUESDAY, d(21, APRIL,
            2006)));
352     assertEquals(d(25, APRIL, 2006), getNearestDayOfWeek(TUESDAY, d(22, APRIL,
            2006)));
353
354 //    assertEquals(d(19, APRIL, 2006), getNearestDayOfWeek(WEDNESDAY
            d(16, APRIL, 2006)));
355 //    assertEquals(d(19, APRIL, 2006), getNearestDayOfWeek(WEDNESDAY,
            d(17, APRIL, 2006)));
356 //    assertEquals(d(19, APRIL, 2006), getNearestDayOfWeek(WEDNESDAY,
            d(18, APRIL, 2006)));
357     assertEquals(d(19, APRIL, 2006), getNearestDayOfWeek(WEDNESDAY,
            d(19, APRIL, 2006)));
358     assertEquals(d(19, APRIL, 2006), getNearestDayOfWeek(WEDNESDAY,
            d(20, APRIL, 2006)));
359     assertEquals(d(19, APRIL, 2006), getNearestDayOfWeek(WEDNESDAY,
            d(21, APRIL, 2006)));
360     assertEquals(d(19, APRIL, 2006), getNearestDayOfWeek(WEDNESDAY,
            d(22, APRIL, 2006)));
361
362 //    assertEquals(d(13, APRIL, 2006), getNearestDayOfWeek(THURSDAY,
            d(16, APRIL, 2006)));
363 //    assertEquals(d(20, APRIL, 2006), getNearestDayOfWeek(THURSDAY,
            d(17, APRIL, 2006)));
364 //    assertEquals(d(20, APRIL, 2006), getNearestDayOfWeek(THURSDAY,
            d(18, APRIL, 2006)));
365 //    assertEquals(d(20, APRIL, 2006), getNearestDayOfWeek(THURSDAY,
            d(19, APRIL, 2006)));
366     assertEquals(d(20, APRIL, 2006), getNearestDayOfWeek(THURSDAY,
            d(20, APRIL, 2006)));
367     assertEquals(d(20, APRIL, 2006), getNearestDayOfWeek(THURSDAY,
```

```
               d(21, APRIL, 2006)));
368       assertEquals(d(20, APRIL, 2006), getNearestDayOfWeek(THURSDAY,
               d(22, APRIL, 2006)));
369
370 //    assertEquals(d(14, APRIL, 2006), getNearestDayOfWeek(FRIDAY, d(16, APRIL,
               2006)));
371 //    assertEquals(d(14, APRIL, 2006), getNearestDayOfWeek(FRIDAY, d(17, APRIL,
               2006)));
372 //    assertEquals(d(21, APRIL, 2006), getNearestDayOfWeek(FRIDAY, d(18, APRIL,
               2006)));
373 //    assertEquals(d(21, APRIL, 2006), getNearestDayOfWeek(FRIDAY, d(19, APRIL,
               2006)));
374 //    assertEquals(d(21, APRIL, 2006), getNearestDayOfWeek(FRIDAY, d(20, APRIL,
               2006)));
375       assertEquals(d(21, APRIL, 2006), getNearestDayOfWeek(FRIDAY, d(21, APRIL,
               2006)));
376       assertEquals(d(21, APRIL, 2006), getNearestDayOfWeek(FRIDAY, d(22, APRIL,
               2006)));
377
378 //    assertEquals(d(15, APRIL, 2006), getNearestDayOfWeek(SATURDAY,
               d(16, APRIL, 2006)));
379 //    assertEquals(d(15, APRIL, 2006), getNearestDayOfWeek(SATURDAY,
               d(17, APRIL, 2006)));
380 //    assertEquals(d(15, APRIL, 2006), getNearestDayOfWeek(SATURDAY,
               d(18, APRIL, 2006)));
381 //    assertEquals(d(22, APRIL, 2006), getNearestDayOfWeek(SATURDAY,
               d(19, APRIL, 2006)));
382 //    assertEquals(d(22, APRIL, 2006), getNearestDayOfWeek(SATURDAY,
               d(20, APRIL, 2006)));
383 //    assertEquals(d(22, APRIL, 2006), getNearestDayOfWeek(SATURDAY,
               d(21, APRIL, 2006)));
384       assertEquals(d(22, APRIL, 2006), getNearestDayOfWeek(SATURDAY,
               d(22, APRIL, 2006)));
385
386       try {
387         getNearestDayOfWeek(-1, d(1, JANUARY, 2006));
388         fail("Invalid day of week code should throw exception");
389       } catch (IllegalArgumentException e) {
390       }
391     }
392
393     public void testEndOfCurrentMonth() throws Exception {
394       SerialDate d = SerialDate.createInstance(2);
395       assertEquals(d(31, JANUARY, 2006), d.getEndOfCurrentMonth(d(1, JANUARY, 2006)));
396       assertEquals(d(28, FEBRUARY, 2006), d.getEndOfCurrentMonth(d(1, FEBRUARY, 2006)));
397       assertEquals(d(31, MARCH, 2006), d.getEndOfCurrentMonth(d(1, MARCH, 2006)));
398       assertEquals(d(30, APRIL, 2006), d.getEndOfCurrentMonth(d(1, APRIL, 2006)));
399       assertEquals(d(31, MAY, 2006), d.getEndOfCurrentMonth(d(1, MAY, 2006)));
400       assertEquals(d(30, JUNE, 2006), d.getEndOfCurrentMonth(d(1, JUNE, 2006)));
401       assertEquals(d(31, JULY, 2006), d.getEndOfCurrentMonth(d(1, JULY, 2006)));
402       assertEquals(d(31, AUGUST, 2006), d.getEndOfCurrentMonth(d(1, AUGUST, 2006)));
403       assertEquals(d(30, SEPTEMBER, 2006), d.getEndOfCurrentMonth
               (d(1, SEPTEMBER, 2006)));
404       assertEquals(d(31, OCTOBER, 2006), d.getEndOfCurrentMonth(d(1, OCTOBER, 2006)));
```

```
405      assertEquals(d(30, NOVEMBER, 2006), d.getEndOfCurrentMonth(d(1, NOVEMBER, 2006)));
406      assertEquals(d(31, DECEMBER, 2006), d.getEndOfCurrentMonth(d(1, DECEMBER, 2006)));
407      assertEquals(d(29, FEBRUARY, 2008), d.getEndOfCurrentMonth(d(1, FEBRUARY, 2008)));
408    }
409
410    public void testWeekInMonthToString() throws Exception {
411      assertEquals("First",weekInMonthToString(FIRST_WEEK_IN_MONTH));
412      assertEquals("Second",weekInMonthToString(SECOND_WEEK_IN_MONTH));
413      assertEquals("Third",weekInMonthToString(THIRD_WEEK_IN_MONTH));
414      assertEquals("Fourth",weekInMonthToString(FOURTH_WEEK_IN_MONTH));
415      assertEquals("Last",weekInMonthToString(LAST_WEEK_IN_MONTH));
416
417 //todo    try {
418 //        weekInMonthToString(-1);
419 //        fail("Invalid week code should throw exception");
420 //      } catch (IllegalArgumentException e) {
421 //      }
422    }
423
424    public void testRelativeToString() throws Exception {
425      assertEquals("Preceding",relativeToString(PRECEDING));
426      assertEquals("Nearest",relativeToString(NEAREST));
427      assertEquals("Following",relativeToString(FOLLOWING));
428
429 //todo    try {
430 //        relativeToString(-1000);
431 //        fail("Invalid relative code should throw exception");
432 //      } catch (IllegalArgumentException e) {
433 //      }
434    }
435
436    public void testCreateInstanceFromDDMMYYY() throws Exception {
437      SerialDate date = createInstance(1, JANUARY, 1900);
438      assertEquals(1,date.getDayOfMonth());
439      assertEquals(JANUARY,date.getMonth());
440      assertEquals(1900,date.getYYYY());
441      assertEquals(2,date.toSerial());
442    }
443
444    public void testCreateInstanceFromSerial() throws Exception {
445      assertEquals(d(1, JANUARY, 1900),createInstance(2));
446      assertEquals(d(1, JANUARY, 1901), createInstance(367));
447    }
448
449    public void testCreateInstanceFromJavaDate() throws Exception {
450      assertEquals(d(1, JANUARY, 1900),
                     createInstance(new GregorianCalendar(1900,0,1).getTime()));
451      assertEquals(d(1, JANUARY, 2006),
                     createInstance(new GregorianCalendar(2006,0,1).getTime()));
452    }
453
454    public static void main(String[] args) {
455      junit.textui.TestRunner.run(BobsSerialDateTest.class);
456    }
```

```
457 }
```

代码清单 B-5　SpreadsheetDate.java

```
 1 /* ========================================================================
 2  * JCommon : a free general purpose class library for the Java(tm) platform
 3  * ========================================================================
 4  *
 5  * (C) Copyright 2000-2005, by Object Refinery Limited and Contributors.
 6  *
 7  *
 8  *
 9  * This library is free software; you can redistribute it and/or modify it
10  * under the terms of the GNU Lesser General Public License as published by
11  * the Free Software Foundation; either version 2.1 of the License, or
12  * (at your option) any later version.
13  *
14  * This library is distributed in the hope that it will be useful, but
15  * WITHOUT ANY WARRANTY; without even the implied warranty of MERCHANTABILITY
16  * or FITNESS FOR A PARTICULAR PURPOSE. See the GNU Lesser General Public
17  * License for more details.
18  *
19  * You should have received a copy of the GNU Lesser General Public
20  * License along with this library; if not, write to the Free Software
21  * Foundation, Inc., 51 Franklin Street, Fifth Floor, Boston, MA  02110-1301,
22  * USA.
23  *
24  * [Java is a trademark or registered trademark of Sun Microsystems, Inc.
25  * in the United States and other countries.]
26  *
27  * --------------------
28  * SpreadsheetDate.java
29  * --------------------
30  * (C) Copyright 2000-2005, by Object Refinery Limited and Contributors.
31  *
32  * Original Author:  David Gilbert (for Object Refinery Limited);
33  * Contributor(s):   -;
34  *
35  * $Id: SpreadsheetDate.java,v 1.8 2005/11/03 09:25:39 mungady Exp $
36  *
37  * Changes
38  * -------
39  * 11-Oct-2001 : Version 1 (DG);
40  * 05-Nov-2001 : Added getDescription() and setDescription() methods (DG);
41  * 12-Nov-2001 : Changed name from ExcelDate.java to SpreadsheetDate.java (DG);
42  *               Fixed a bug in calculating day, month and year from serial
43  *               number (DG);
44  * 24-Jan-2002 : Fixed a bug in calculating the serial number from the day,
45  *               month and year.  Thanks to Trevor Hills for the report (DG);
46  * 29-May-2002 : Added equals(Object) method (SourceForge ID 558850) (DG);
47  * 03-Oct-2002 : Fixed errors reported by Checkstyle (DG);
48  * 13-Mar-2003 : Implemented Serializable (DG);
49  * 04-Sep-2003 : Completed isInRange() methods (DG);
50  * 05-Sep-2003 : Implemented Comparable (DG);
51  * 21-Oct-2003 : Added hashCode() method (DG);
```

```java
 52  *
 53  */
 54
 55 package org.jfree.date;
 56
 57 import java.util.Calendar;
 58 import java.util.Date;
 59
 60 /**
 61  * Represents a date using an integer, in a similar fashion to the
 62  * implementation in Microsoft Excel.  The range of dates supported is
 63  * 1-Jan-1900 to 31-Dec-9999.
 64  * <P>
 65  * Be aware that there is a deliberate bug in Excel that recognises the year
 66  * 1900 as a leap year when in fact it is not a leap year. You can find more
 67  * information on the Microsoft website in article Q181370:
 68  * <P>
 69  *
 70  * <P>
 71  * Excel uses the convention that 1-Jan-1900 = 1.  This class uses the
 72  * convention 1-Jan-1900 = 2.
 73  * The result is that the day number in this class will be different to the
 74  * Excel figure for January and February 1900...but then Excel adds in an extra
 75  * day (29-Feb-1900 which does not actually exist!) and from that point forward
 76  * the day numbers will match.
 77  *
 78  * @author David Gilbert
 79  */
 80 public class SpreadsheetDate extends SerialDate {
 81
 82   /** For serialization. */
 83   private static final long serialVersionUID = -2039586705374454461L;
 84
 85   /**
 86    * The day number (1-Jan-1900 = 2, 2-Jan-1900 = 3, ..., 31-Dec-9999 =
 87    * 2958465).
 88    */
 89   private int serial;
 90
 91   /** The day of the month (1 to 28, 29, 30 or 31 depending on the month). */
 92   private int day;
 93
 94   /** The month of the year (1 to 12). */
 95   private int month;
 96
 97   /** The year (1900 to 9999). */
 98   private int year;
 99
100   /** An optional description for the date. */
101   private String description;
102
103   /**
104    * Creates a new date instance.
105    *
```

```java
106      * @param day   the day (in the range 1 to 28/29/30/31).
107      * @param month  the month (in the range 1 to 12).
108      * @param year   the year (in the range 1900 to 9999).
109      */
110     public SpreadsheetDate(final int day, final int month, final int year) {
111
112         if ((year >= 1900) && (year <= 9999)) {
113             this.year = year;
114         }
115         else {
116             throw new IllegalArgumentException(
117                 "The 'year' argument must be in range 1900 to 9999."
118             );
119         }
120
121         if ((month >= MonthConstants.JANUARY)
122                 && (month <= MonthConstants.DECEMBER)) {
123             this.month = month;
124         }
125         else {
126             throw new IllegalArgumentException(
127                 "The 'month' argument must be in the range 1 to 12."
128             );
129         }
130
131         if ((day >= 1) && (day <= SerialDate.lastDayOfMonth(month, year))) {
132             this.day = day;
133         }
134         else {
135             throw new IllegalArgumentException("Invalid 'day' argument.");
136         }
137
138         // the serial number needs to be synchronised with the day-month-year...
139         this.serial = calcSerial(day, month, year);
140
141         this.description = null;
142
143     }
144
145     /**
146      * Standard constructor - creates a new date object representing the
147      * specified day number (which should be in the range 2 to 2958465.
148      *
149      * @param serial  the serial number for the day (range: 2 to 2958465).
150      */
151     public SpreadsheetDate(final int serial) {
152
153         if ((serial >= SERIAL_LOWER_BOUND) && (serial <= SERIAL_UPPER_BOUND)) {
154             this.serial = serial;
155         }
156         else {
157             throw new IllegalArgumentException(
158                 "SpreadsheetDate: Serial must be in range 2 to 2958465.");
159         }
```

```
160
161        // the day-month-year needs to be synchronised with the serial number...
162        calcDayMonthYear();
163
164    }
165
166    /**
167     * Returns the description that is attached to the date.  It is not
168     * required that a date have a description, but for some applications it
169     * is useful.
170     *
171     * @return The description that is attached to the date.
172     */
173    public String getDescription() {
174        return this.description;
175    }
176
177    /**
178     * Sets the description for the date.
179     *
180     * @param description the description for this date (<code>null</code>
181     *                    permitted).
182     */
183    public void setDescription(final String description) {
184        this.description = description;
185    }
186
187    /**
188     * Returns the serial number for the date, where 1 January 1900 = 2
189     * (this corresponds, almost, to the numbering system used in Microsoft
190     * Excel for Windows and Lotus 1-2-3).
191     *
192     * @return The serial number of this date.
193     */
194    public int toSerial() {
195        return this.serial;
196    }
197
198    /**
199     * Returns a <code>java.util.Date</code> equivalent to this date.
200     *
201     * @return The date.
202     */
203    public Date toDate() {
204        final Calendar calendar = Calendar.getInstance();
205        calendar.set(getYYYY(), getMonth() - 1, getDayOfMonth(), 0, 0, 0);
206        return calendar.getTime();
207    }
208
209    /**
210     * Returns the year (assume a valid range of 1900 to 9999).
211     *
212     * @return The year.
213     */
```

```
214    public int getYYYY() {
215        return this.year;
216    }
217
218    /**
219     * Returns the month (January = 1, February = 2, March = 3).
220     *
221     * @return The month of the year.
222     */
223    public int getMonth() {
224        return this.month;
225    }
226
227    /**
228     * Returns the day of the month.
229     *
230     * @return The day of the month.
231     */
232    public int getDayOfMonth() {
233        return this.day;
234    }
235
236    /**
237     * Returns a code representing the day of the week.
238     * <P>
239     * The codes are defined in the {@link SerialDate} class as:
240     * <code>SUNDAY</code>, <code>MONDAY</code>, <code>TUESDAY</code>,
241     * <code>WEDNESDAY</code>, <code>THURSDAY</code>, <code>FRIDAY</code>, and
242     * <code>SATURDAY</code>.
243     *
244     * @return A code representing the day of the week.
245     */
246    public int getDayOfWeek() {
247        return (this.serial + 6) % 7 + 1;
248    }
249
250    /**
251     * Tests the equality of this date with an arbitrary object.
252     * <P>
253     * This method will return true ONLY if the object is an instance of the
254     * {@link SerialDate} base class, and it represents the same day as this
255     * {@link SpreadsheetDate}.
256     *
257     * @param object  the object to compare (<code>null</code> permitted).
258     *
259     * @return A boolean.
260     */
261    public boolean equals(final Object object) {
262
263        if (object instanceof SerialDate) {
264            final SerialDate s = (SerialDate) object;
265            return (s.toSerial() == this.toSerial());
266        }
```

```
267        else {
268            return false;
269        }
270
271    }
272
273    /**
274     * Returns a hash code for this object instance.
275     *
276     * @return A hash code.
277     */
278    public int hashCode() {
279        return toSerial();
280    }
281
282    /**
283     * Returns the difference (in days) between this date and the specified
284     * 'other' date.
285     *
286     * @param other  the date being compared to.
287     *
288     * @return The difference (in days) between this date and the specified
289     *         'other' date.
290     */
291    public int compare(final SerialDate other) {
292        return this.serial - other.toSerial();
293    }
294
295    /**
296     * Implements the method required by the Comparable interface.
297     *
298     * @param other  the other object (usually another SerialDate).
299     *
300     * @return A negative integer, zero, or a positive integer as this object
301     *         is less than, equal to, or greater than the specified object.
302     */
303    public int compareTo(final Object other) {
304        return compare((SerialDate) other);
305    }
306
307    /**
308     * Returns true if this SerialDate represents the same date as the
309     * specified SerialDate.
310     *
311     * @param other  the date being compared to.
312     *
313     * @return <code>true</code> if this SerialDate represents the same date as
314     *         the specified SerialDate.
315     */
316    public boolean isOn(final SerialDate other) {
317        return (this.serial == other.toSerial());
318    }
319
320    /**
```

```
321      s* Returns true if this SerialDate represents an earlier date compared to
322       * the specified SerialDate.
323       *
324       * @param other  the date being compared to.
325       *
326       * @return <code>true</code> if this SerialDate represents an earlier date
327       *         compared to the specified SerialDate.
328       */
329      public boolean isBefore(final SerialDate other) {
330          return (this.serial < other.toSerial());
331      }
332
333      /**
334       * Returns true if this SerialDate represents the same date as the
335       * specified SerialDate.
336       *
337       * @param other  the date being compared to.
338       *
339       * @return <code>true</code> if this SerialDate represents the same date
340       *         as the specified SerialDate.
341       */
342      public boolean isOnOrBefore(final SerialDate other) {
343          return (this.serial <= other.toSerial());
344      }
345
346      /**
347       * Returns true if this SerialDate represents the same date as the
348       * specified SerialDate.
349       *
350       * @param other  the date being compared to.
351       *
352       * @return <code>true</code> if this SerialDate represents the same date
353       *         as the specified SerialDate.
354       */
355      public boolean isAfter(final SerialDate other) {
356          return (this.serial > other.toSerial());
357      }
358
359      /**
360       * Returns true if this SerialDate represents the same date as the
361       * specified SerialDate.
362       *
363       * @param other  the date being compared to.
364       *
365       * @return <code>true</code> if this SerialDate represents the same date as
366       *         the specified SerialDate.
367       */
368      public boolean isOnOrAfter(final SerialDate other) {
369          return (this.serial >= other.toSerial());
370      }
371
372      /**
373       * Returns <code>true</code> if this {@link SerialDate} is within the
374       * specified range (INCLUSIVE).  The date order of d1 and d2 is not
```

```java
     * important.
     *
     * @param d1  a boundary date for the range.
     * @param d2  the other boundary date for the range.
     *
     * @return A boolean.
     */
    public boolean isInRange(final SerialDate d1, final SerialDate d2) {
        return isInRange(d1, d2, SerialDate.INCLUDE_BOTH);
    }

    /**
     * Returns true if this SerialDate is within the specified range (caller
     * specifies whether or not the end-points are included).  The order of d1
     * and d2 is not important.
     *
     * @param d1  one boundary date for the range.
     * @param d2  a second boundary date for the range.
     * @param include  a code that controls whether or not the start and end
     *                 dates are included in the range.
     *
     * @return <code>true</code> if this SerialDate is within the specified
     *         range.
     */
    public boolean isInRange(final SerialDate d1, final SerialDate d2,
                             final int include) {
        final int s1 = d1.toSerial();
        final int s2 = d2.toSerial();
        final int start = Math.min(s1, s2);
        final int end = Math.max(s1, s2);
        
        final int s = toSerial();
        if (include == SerialDate.INCLUDE_BOTH) {
            return (s >= start && s <= end);
        }
        else if (include == SerialDate.INCLUDE_FIRST) {
            return (s >= start && s < end);
        }
        else if (include == SerialDate.INCLUDE_SECOND) {
            return (s > start && s <= end);
        }
        else {
            return (s > start && s < end);
        }
    }

    /**
     * Calculate the serial number from the day, month and year.
     * <P>
     * 1-Jan-1900 = 2.
     *
     * @param d  the day.
     * @param m  the month.
     * @param y  the year.
```

```java
429     *
430     * @return the serial number from the day, month and year.
431     */
432    private int calcSerial(final int d, final int m, final int y) {
433       final int yy = ((y - 1900) * 365) + SerialDate.leapYearCount(y - 1);
434       int mm = SerialDate.AGGREGATE_DAYS_TO_END_OF_PRECEDING_MONTH[m];
435       if (m > MonthConstants.FEBRUARY) {
436          if (SerialDate.isLeapYear(y)) {
437             mm = mm + 1;
438          }
439       }
440       final int dd = d;
441       return yy + mm + dd + 1;
442    }
443
444    /**
445     * Calculate the day, month and year from the serial number.
446     */
447    private void calcDayMonthYear() {
448
449       // get the year from the serial date
450       final int days = this.serial - SERIAL_LOWER_BOUND;
451       // overestimated because we ignored leap days
452       final int overestimatedYYYY = 1900 + (days / 365);
453       final int leaps = SerialDate.leapYearCount(overestimatedYYYY);
454       final int nonleapdays = days - leaps;
455       // underestimated because we overestimated years
456       int underestimatedYYYY = 1900 + (nonleapdays / 365);
457
458       if (underestimatedYYYY == overestimatedYYYY) {
459          this.year = underestimatedYYYY;
460       }
461       else {
462          int ss1 = calcSerial(1, 1, underestimatedYYYY);
463          while (ss1 <= this.serial) {
464             underestimatedYYYY = underestimatedYYYY + 1;
465             ss1 = calcSerial(1, 1, underestimatedYYYY);
466          }
467          this.year = underestimatedYYYY - 1;
468       }
469
470       final int ss2 = calcSerial(1, 1, this.year);
471
472       int[] daysToEndOfPrecedingMonth
473          = AGGREGATE_DAYS_TO_END_OF_PRECEDING_MONTH;
474
475       if (isLeapYear(this.year)) {
476          daysToEndOfPrecedingMonth
477             = LEAP_YEAR_AGGREGATE_DAYS_TO_END_OF_PRECEDING_MONTH;
478       }
479
480       // get the month from the serial date
481       int mm = 1;
482       int sss = ss2 + daysToEndOfPrecedingMonth[mm] - 1;
```

```
483            while (sss < this.serial) {
484                mm = mm + 1;
485                sss = ss2 + daysToEndOfPrecedingMonth[mm] - 1;
486            }
487            this.month = mm - 1;
488
489            // what's left is d(+1);
490            this.day = this.serial - ss2
491                     - daysToEndOfPrecedingMonth[this.month] + 1;
492
493        }
494
495 }
```

代码清单 B-6　RelativeDayOfWeekRule.java

```
 1 /* ========================================================================
 2  * JCommon : a free general purpose class library for the Java(tm) platform
 3  * ========================================================================
 4  *
 5  * (C) Copyright 2000-2005, by Object Refinery Limited and Contributors.
 6  *
 7  *
 8  *
 9  * This library is free software; you can redistribute it and/or modify it
10  * under the terms of the GNU Lesser General Public License as published by
11  * the Free Software Foundation; either version 2.1 of the License, or
12  * (at your option) any later version.
13  *
14  * This library is distributed in the hope that it will be useful, but
15  * WITHOUT ANY WARRANTY; without even the implied warranty of MERCHANTABILITY
16  * or FITNESS FOR A PARTICULAR PURPOSE. See the GNU Lesser General Public
17  * License for more details.
18  *
19  * You should have received a copy of the GNU Lesser General Public
20  * License along with this library; if not, write to the Free Software
21  * Foundation, Inc., 51 Franklin Street, Fifth Floor, Boston, MA  02110-1301,
22  * USA.
23  *
24  * [Java is a trademark or registered trademark of Sun Microsystems, Inc.
25  * in the United States and other countries.]
26  *
27  * -------------------------
28  * RelativeDayOfWeekRule.java
29  * -------------------------
30  * (C) Copyright 2000-2003, by Object Refinery Limited and Contributors.
31  *
32  * Original Author:  David Gilbert (for Object Refinery Limited);
33  * Contributor(s):   -;
34  *
35  * $Id: RelativeDayOfWeekRule.java,v 1.6 2005/11/16 15:58:40 taqua Exp $
36  *
37  * Changes (from 26-Oct-2001)
38  * --------------------------
```

```
39   * 26-Oct-2001 : Changed package to com.jrefinery.date.*;
40   * 03-Oct-2002 : Fixed errors reported by Checkstyle (DG);
41   *
42   */
43
44  package org.jfree.date;
45
46  /**
47   * An annual date rule that returns a date for each year based on (a) a
48   * reference rule; (b) a day of the week; and (c) a selection parameter
49   * (SerialDate.PRECEDING, SerialDate.NEAREST, SerialDate.FOLLOWING).
50   * <P>
51   * For example, Good Friday can be specified as 'the Friday PRECEDING Easter
52   * Sunday'.
53   *
54   * @author David Gilbert
55   */
56  public class RelativeDayOfWeekRule extends AnnualDateRule {
57
58      /** A reference to the annual date rule on which this rule is based. */
59      private AnnualDateRule subrule;
60
61      /**
62       * The day of the week (SerialDate.MONDAY, SerialDate.TUESDAY, and so on).
63       */
64      private int dayOfWeek;
65
66      /** Specifies which day of the week (PRECEDING, NEAREST or FOLLOWING). */
67      private int relative;
68
69      /**
70       * Default constructor - builds a rule for the Monday following 1 January.
71       */
72      public RelativeDayOfWeekRule() {
73          this(new DayAndMonthRule(), SerialDate.MONDAY, SerialDate.FOLLOWING);
74      }
75
76      /**
77       * Standard constructor - builds rule based on the supplied sub-rule.
78       *
79       * @param subrule  the rule that determines the reference date.
80       * @param dayOfWeek  the day-of-the-week relative to the reference date.
81       * @param relative  indicates *which* day-of-the-week (preceding, nearest
82       *                  or following).
83       */
84      public RelativeDayOfWeekRule(final AnnualDateRule subrule,
85          final int dayOfWeek, final int relative) {
86          this.subrule = subrule;
87          this.dayOfWeek = dayOfWeek;
88          this.relative = relative;
89      }
90
91      /**
92       * Returns the sub-rule (also called the reference rule).
```

```
 93      *
 94      * @return The annual date rule that determines the reference date for this
 95      *         rule.
 96      */
 97     public AnnualDateRule getSubrule() {
 98         return this.subrule;
 99     }
100
101     /**
102      * Sets the sub-rule.
103      *
104      * @param subrule   the annual date rule that determines the reference date
105      *                  for this rule.
106      */
107     public void setSubrule(final AnnualDateRule subrule) {
108         this.subrule = subrule;
109     }
110
111     /**
112      * Returns the day-of-the-week for this rule.
113      *
114      * @return the day-of-the-week for this rule.
115      */
116     public int getDayOfWeek() {
117         return this.dayOfWeek;
118     }
119
120     /**
121      * Sets the day-of-the-week for this rule.
122      *
123      * @param dayOfWeek the day-of-the-week (SerialDate.MONDAY,
124      *                  SerialDate.TUESDAY, and so on).
125      */
126     public void setDayOfWeek(final int dayOfWeek) {
127         this.dayOfWeek = dayOfWeek;
128     }
129
130     /**
131      * Returns the 'relative' attribute, that determines *which*
132      * day-of-the-week we are interested in (SerialDate.PRECEDING,
133      * SerialDate.NEAREST or SerialDate.FOLLOWING).
134      *
135      * @return The 'relative' attribute.
136      */
137     public int getRelative() {
138         return this.relative;
139     }
140
141     /**
142      * Sets the 'relative' attribute (SerialDate.PRECEDING, SerialDate.NEAREST,
143      * SerialDate.FOLLOWING).
144      *
145      * @param relative  determines *which* day-of-the-week is selected by this
146      *                  rule.
```

```
147      */
148     public void setRelative(final int relative) {
149       this.relative = relative;
150     }
151
152     /**
153      * Creates a clone of this rule.
154      *
155      * @return a clone of this rule.
156      *
157      * @throws CloneNotSupportedException this should never happen.
158      */
159     public Object clone() throws CloneNotSupportedException {
160       final RelativeDayOfWeekRule duplicate
161           = (RelativeDayOfWeekRule) super.clone();
162       duplicate.subrule = (AnnualDateRule) duplicate.getSubrule().clone();
163       return duplicate;
164     }
165
166     /**
167      * Returns the date generated by this rule, for the specified year.
168      *
169      * @param year  the year (1900 &lt;= year &lt;= 9999).
170      *
171      * @return The date generated by the rule for the given year (possibly
172      *         <code>null</code>).
173      */
174     public SerialDate getDate(final int year) {
175
176       // check argument...
177       if ((year < SerialDate.MINIMUM_YEAR_SUPPORTED)
178         || (year > SerialDate.MAXIMUM_YEAR_SUPPORTED)) {
179         throw new IllegalArgumentException(
180             "RelativeDayOfWeekRule.getDate(): year outside valid range.");
181       }
182
183       // calculate the date...
184       SerialDate result = null;
185       final SerialDate base = this.subrule.getDate(year);
186
187       if (base != null) {
188         switch (this.relative) {
189           case(SerialDate.PRECEDING):
190             result = SerialDate.getPreviousDayOfWeek(this.dayOfWeek, base);
191
192             break;
193           case(SerialDate.NEAREST):
194             result = SerialDate.getNearestDayOfWeek(this.dayOfWeek, base);
195
196             break;
197           case(SerialDate.FOLLOWING):
198             result = SerialDate.getFollowingDayOfWeek(this.dayOfWeek, base);
199
200             break;
```

```
201            default:
202                break;
203            }
204        }
205        return result;
206
207    }
208
209 }
```

代码清单 B-7　DayDate.java（最终版本）

```
1  /* ========================================================================
2   * JCommon : a free general purpose class library for the Java(tm) platform
3   * ========================================================================
4   *
5   * (C) Copyright 2000-2005, by Object Refinery Limited and Contributors.
...
36   */
37  package org.jfree.date;
38
39  import java.io.Serializable;
40  import java.util.*;
41
42  /**
43   * An abstract class that represents immutable dates with a precision of
44   * one day.  The implementation will map each date to an integer that
45   * represents an ordinal number of days from some fixed origin.
46   *
47   * Why not just use java.util.Date?  We will, when it makes sense.  At times,
48   * java.util.Date can be *too* precise - it represents an instant in time,
49   * accurate to 1/1000th of a second (with the date itself depending on the
50   * time-zone).  Sometimes we just want to represent a particular day (e.g. 21
51   * January 2015) without concerning ourselves about the time of day, or the
52   * time-zone, or anything else.  That's what we've defined DayDate for.
53   *
54   * Use DayDateFactory.makeDate to create an instance.
55   *
56   * @author David Gilbert
57   * @author Robert C. Martin did a lot of refactoring.
58   */
59
60  public abstract class DayDate implements Comparable, Serializable {
61      public abstract int getOrdinalDay();
62      public abstract int getYear();
63      public abstract Month getMonth();
64      public abstract int getDayOfMonth();
65
66      protected abstract Day getDayOfWeekForOrdinalZero();
67
68      public DayDate plusDays(int days) {
69          return DayDateFactory.makeDate(getOrdinalDay() + days);
70      }
71
```

```java
 72    public DayDate plusMonths(int months) {
 73      int thisMonthAsOrdinal = getMonth().toInt() - Month.JANUARY.toInt();
 74      int thisMonthAndYearAsOrdinal = 12 * getYear() + thisMonthAsOrdinal;
 75      int resultMonthAndYearAsOrdinal = thisMonthAndYearAsOrdinal + months;
 76      int resultYear = resultMonthAndYearAsOrdinal / 12;
 77      int resultMonthAsOrdinal = resultMonthAndYearAsOrdinal % 12 + Month.JANUARY.toInt();
 78      Month resultMonth = Month.fromInt(resultMonthAsOrdinal);
 79      int resultDay = correctLastDayOfMonth(getDayOfMonth(), resultMonth, resultYear);
 80      return DayDateFactory.makeDate(resultDay, resultMonth, resultYear);
 81    }
 82
 83    public DayDate plusYears(int years) {
 84      int resultYear = getYear() + years;
 85      int resultDay = correctLastDayOfMonth(getDayOfMonth(), getMonth(), resultYear);
 86      return DayDateFactory.makeDate(resultDay, getMonth(), resultYear);
 87    }
 88
 89    private int correctLastDayOfMonth(int day, Month month, int year) {
 90      int lastDayOfMonth = DateUtil.lastDayOfMonth(month, year);
 91      if (day > lastDayOfMonth)
 92        day = lastDayOfMonth;
 93      return day;
 94    }
 95
 96    public DayDate getPreviousDayOfWeek(Day targetDayOfWeek) {
 97      int offsetToTarget = targetDayOfWeek.toInt() - getDayOfWeek().toInt();
 98      if (offsetToTarget >= 0)
 99        offsetToTarget -= 7;
100      return plusDays(offsetToTarget);
101    }
102
103    public DayDate getFollowingDayOfWeek(Day targetDayOfWeek) {
104      int offsetToTarget = targetDayOfWeek.toInt() - getDayOfWeek().toInt();
105      if (offsetToTarget <= 0)
106        offsetToTarget += 7;
107      return plusDays(offsetToTarget);
108    }
109
110    public DayDate getNearestDayOfWeek(Day targetDayOfWeek) {
111      int offsetToThisWeeksTarget = targetDayOfWeek.toInt() - getDayOfWeek().toInt();
112      int offsetToFutureTarget = (offsetToThisWeeksTarget + 7) % 7;
113      int offsetToPreviousTarget = offsetToFutureTarget - 7;
114
115      if (offsetToFutureTarget > 3)
116        return plusDays(offsetToPreviousTarget);
117      else
118        return plusDays(offsetToFutureTarget);
119    }
120
121    public DayDate getEndOfMonth() {
122      Month month = getMonth();
123      int year = getYear();
124      int lastDay = DateUtil.lastDayOfMonth(month, year);
125      return DayDateFactory.makeDate(lastDay, month, year);
```

```
126      }
127
128      public Date toDate() {
129        final Calendar calendar = Calendar.getInstance();
130        int ordinalMonth = getMonth().toInt() - Month.JANUARY.toInt();
131        calendar.set(getYear(), ordinalMonth, getDayOfMonth(), 0, 0, 0);
132        return calendar.getTime();
133      }
134
135      public String toString() {
136        return String.format("%02d-%s-%d", getDayOfMonth(), getMonth(), getYear());
137      }
138
139      public Day getDayOfWeek() {
140        Day startingDay = getDayOfWeekForOrdinalZero();
141        int startingOffset = startingDay.toInt() - Day.SUNDAY.toInt();
142        int ordinalOfDayOfWeek = (getOrdinalDay() + startingOffset) % 7;
143        return Day.fromInt(ordinalOfDayOfWeek + Day.SUNDAY.toInt());
144      }
145
146      public int daysSince(DayDate date) {
147        return getOrdinalDay() - date.getOrdinalDay();
148      }
149
150      public boolean isOn(DayDate other) {
151        return getOrdinalDay() == other.getOrdinalDay();
152      }
153
154      public boolean isBefore(DayDate other) {
155        return getOrdinalDay() < other.getOrdinalDay();
156      }
157
158      public boolean isOnOrBefore(DayDate other) {
159        return getOrdinalDay() <= other.getOrdinalDay();
160      }
161
162      public boolean isAfter(DayDate other) {
163        return getOrdinalDay() > other.getOrdinalDay();
164      }
165
166      public boolean isOnOrAfter(DayDate other) {
167        return getOrdinalDay() >= other.getOrdinalDay();
168      }
169
170      public boolean isInRange(DayDate d1, DayDate d2) {
171        return isInRange(d1, d2, DateInterval.CLOSED);
172      }
173
174      public boolean isInRange(DayDate d1, DayDate d2, DateInterval interval) {
175        int left = Math.min(d1.getOrdinalDay(), d2.getOrdinalDay());
176        int right = Math.max(d1.getOrdinalDay(), d2.getOrdinalDay());
177        return interval.isIn(getOrdinalDay(), left, right);
178      }
179    }
```

代码清单 B-8　Month.java（最终版本）

```java
package org.jfree.date;

import java.text.DateFormatSymbols;

public enum Month {
  JANUARY(1), FEBRUARY(2), MARCH(3),
  APRIL(4),   MAY(5),      JUNE(6),
  JULY(7),    AUGUST(8),   SEPTEMBER(9),
  OCTOBER(10),NOVEMBER(11),DECEMBER(12);
  private static DateFormatSymbols dateFormatSymbols = new DateFormatSymbols();
  private static final int[] LAST_DAY_OF_MONTH =
    {0, 31, 28, 31, 30, 31, 30, 31, 31, 30, 31, 30, 31};

  private int index;

  Month(int index) {
    this.index = index;
  }

  public static Month fromInt(int monthIndex) {
    for (Month m : Month.values()) {
      if (m.index == monthIndex)
        return m;
    }
    throw new IllegalArgumentException("Invalid month index " + monthIndex);
  }

  public int lastDay() {
    return LAST_DAY_OF_MONTH[index];
  }

  public int quarter() {
    return 1 + (index - 1) / 3;
  }

  public String toString() {
    return dateFormatSymbols.getMonths()[index - 1];
  }

  public String toShortString() {
    return dateFormatSymbols.getShortMonths()[index - 1];
  }

  public static Month parse(String s) {
    s = s.trim();
    for (Month m : Month.values())
      if (m.matches(s))
        return m;

    try {
      return fromInt(Integer.parseInt(s));
    }
    catch (NumberFormatException e) {}
```

```java
54         throw new IllegalArgumentException("Invalid month " + s);
55     }
56
57     private boolean matches(String s) {
58         return s.equalsIgnoreCase(toString()) ||
59                 s.equalsIgnoreCase(toShortString());
60     }
61
62     public int toInt() {
63         return index;
64     }
65 }
```

代码清单 B-9 Day.java（最终版本）

```java
 1 package org.jfree.date;
 2
 3 import java.util.Calendar;
 4 import java.text.DateFormatSymbols;
 5
 6 public enum Day {
 7     MONDAY(Calendar.MONDAY),
 8     TUESDAY(Calendar.TUESDAY),
 9     WEDNESDAY(Calendar.WEDNESDAY),
10     THURSDAY(Calendar.THURSDAY),
11     FRIDAY(Calendar.FRIDAY),
12     SATURDAY(Calendar.SATURDAY),
13     SUNDAY(Calendar.SUNDAY);
14
15     private final int index;
16     private static DateFormatSymbols dateSymbols = new DateFormatSymbols();
17
18     Day(int day) {
19         index = day;
20     }
21
22     public static Day fromInt(int index) throws IllegalArgumentException {
23         for (Day d : Day.values())
24             if (d.index == index)
25                 return d;
26         throw new IllegalArgumentException(
27             String.format("Illegal day index: %d.", index));
28     }
29
30     public static Day parse(String s) throws IllegalArgumentException {
31         String[] shortWeekdayNames =
32             dateSymbols.getShortWeekdays();
33         String[] weekDayNames =
34             dateSymbols.getWeekdays();
35
36         s = s.trim();
37         for (Day day : Day.values()) {
38             if (s.equalsIgnoreCase(shortWeekdayNames[day.index]) ||
39                     s.equalsIgnoreCase(weekDayNames[day.index])) {
```

```
40        return day;
41      }
42    }
43    throw new IllegalArgumentException(
44      String.format("%s is not a valid weekday string", s));
45  }
46
47  public String toString() {
48    return dateSymbols.getWeekdays()[index];
49  }
50
51  public int toInt() {
52    return index;
53  }
54 }
```

代码清单 B-10　DateInterval.java（最终版本）

```
1 package org.jfree.date;
2
3 public enum DateInterval {
4   OPEN {
5     public boolean isIn(int d, int left, int right) {
6       return d > left && d < right;
7     }
8   },
9   CLOSED_LEFT {
10    public boolean isIn(int d, int left, int right) {
11      return d >= left && d < right;
12    }
13  },
14  CLOSED_RIGHT {
15    public boolean isIn(int d, int left, int right) {
16      return d > left && d <= right;
17    }
18  },
19  CLOSED {
20    public boolean isIn(int d, int left, int right) {
21      return d >= left && d <= right;
22    }
23  };
24
25  public abstract boolean isIn(int d, int left, int right);
26 }
```

代码清单 B-11　WeekInMonth.java（最终版本）

```
1 package org.jfree.date;
2
3 public enum WeekInMonth {
4   FIRST(1), SECOND(2), THIRD(3), FOURTH(4), LAST(0);
5   private final int index;
6
```

```
 7    WeekInMonth(int index) {
 8      this.index = index;
 9    }
10
11    public int toInt() {
12      return index;
13    }
14 }
```

代码清单 B-12　WeekdayRange.java（最终版本）

```
1 package org.jfree.date;
2
3 public enum WeekdayRange {
4   LAST, NEAREST, NEXT
5 }
```

代码清单 B-13　DateUtil.java（最终版本）

```
 1 package org.jfree.date;
 2
 3 import java.text.DateFormatSymbols;
 4
 5 public class DateUtil {
 6   private static DateFormatSymbols dateFormatSymbols = new DateFormatSymbols();
 7
 8   public static String[] getMonthNames() {
 9     return dateFormatSymbols.getMonths();
10   }
11
12   public static boolean isLeapYear(int year) {
13     boolean fourth = year % 4 == 0;
14     boolean hundredth = year % 100 == 0;
15     boolean fourHundredth = year % 400 == 0;
16     return fourth && (!hundredth || fourHundredth);
17   }
18
19   public static int lastDayOfMonth(Month month, int year) {
20     if (month == Month.FEBRUARY && isLeapYear(year))
21       return month.lastDay() + 1;
22     else
23       return month.lastDay();
24   }
25
26   public static int leapYearCount(int year) {
27     int leap4 = (year - 1896) / 4;
28     int leap100 = (year - 1800) / 100;
29     int leap400 = (year - 1600) / 400;
30     return leap4 - leap100 + leap400;
31   }
32 }
```

附录 B org.jfree.date.SerialDate 383

代码清单 B-14　DayDateFactory.java（最终版本）

```java
 1 package org.jfree.date;
 2
 3 public abstract class DayDateFactory {
 4   private static DayDateFactory factory = new SpreadsheetDateFactory();
 5   public static void setInstance(DayDateFactory factory) {
 6     DayDateFactory.factory = factory;
 7   }
 8
 9   protected abstract DayDate _makeDate(int ordinal);
10   protected abstract DayDate _makeDate(int day, Month month, int year);
11   protected abstract DayDate _makeDate(int day, int month, int year);
12   protected abstract DayDate _makeDate(java.util.Date date);
13   protected abstract int _getMinimumYear();
14   protected abstract int _getMaximumYear();
15
16   public static DayDate makeDate(int ordinal) {
17     return factory._makeDate(ordinal);
18   }
19
20   public static DayDate makeDate(int day, Month month, int year) {
21     return factory._makeDate(day, month, year);
22   }
23
24   public static DayDate makeDate(int day, int month, int year) {
25     return factory._makeDate(day, month, year);
26   }
27
28   public static DayDate makeDate(java.util.Date date) {
29     return factory._makeDate(date);
30   }
31
32   public static int getMinimumYear() {
33     return factory._getMinimumYear();
34   }
35
36   public static int getMaximumYear() {
37     return factory._getMaximumYear();
38   }
39 }
```

代码清单 B-15　SpreadsheetDateFactory.java（最终版本）

```java
 1 package org.jfree.date;
 2
 3 import java.util.*;
 4
 5 public class SpreadsheetDateFactory extends DayDateFactory {
 6   public DayDate _makeDate(int ordinal) {
 7     return new SpreadsheetDate(ordinal);
 8   }
 9
10   public DayDate _makeDate(int day, Month month, int year) {
```

```
11       return new SpreadsheetDate(day, month, year);
12     }
13
14     public DayDate _makeDate(int day, int month, int year) {
15       return new SpreadsheetDate(day, month, year);
16     }
17
18     public DayDate _makeDate(Date date) {
19       final GregorianCalendar calendar = new GregorianCalendar();
20       calendar.setTime(date);
21       return new SpreadsheetDate(
22         calendar.get(Calendar.DATE),
23         Month.fromInt(calendar.get(Calendar.MONTH) + 1),
24         calendar.get(Calendar.YEAR));
25     }
26
27     protected int _getMinimumYear() {
28       return SpreadsheetDate.MINIMUM_YEAR_SUPPORTED;
29     }
30
31     protected int _getMaximumYear() {
32       return SpreadsheetDate.MAXIMUM_YEAR_SUPPORTED;
33     }
34 }
```

代码清单 B-16   SpreadsheetDate.java（最终版本）

```
 1  /* ========================================================================
 2   * JCommon : a free general purpose class library for the Java(tm) platform
 3   * ========================================================================
 4   *
 5   * (C) Copyright 2000-2005, by Object Refinery Limited and Contributors.
 6   *
...
52   *
53   */
54
55  package org.jfree.date;
56
57  import static org.jfree.date.Month.FEBRUARY;
58
59  import java.util.*;
60
61  /**
62   * Represents a date using an integer, in a similar fashion to the
63   * implementation in Microsoft Excel.  The range of dates supported is
64   * 1-Jan-1900 to 31-Dec-9999.
65   * <p/>
66   * Be aware that there is a deliberate bug in Excel that recognises the year
67   * 1900 as a leap year when in fact it is not a leap year.  You can find more
68   * information on the Microsoft website in article Q181370:
69   * <p/>
70   *
71   * <p/>
```

```
72  * Excel uses the convention that 1-Jan-1900 = 1.  This class uses the
73  * convention 1-Jan-1900 = 2.
74  * The result is that the day number in this class will be different to the
75  * Excel figure for January and February 1900...but then Excel adds in an extra
76  * day (29-Feb-1900 which does not actually exist!) and from that point forward
77  * the day numbers will match.
78  *
79  * @author David Gilbert
80  */
81 public class SpreadsheetDate extends DayDate {
82   public static final int EARLIEST_DATE_ORDINAL = 2;      // 1/1/1900
83   public static final int LATEST_DATE_ORDINAL = 2958465;  // 12/31/9999
84   public static final int MINIMUM_YEAR_SUPPORTED = 1900;
85   public static final int MAXIMUM_YEAR_SUPPORTED = 9999;
86   static final int[] AGGREGATE_DAYS_TO_END_OF_PRECEDING_MONTH =
87     {0, 0, 31, 59, 90, 120, 151, 181, 212, 243, 273, 304, 334, 365};
88   static final int[] LEAP_YEAR_AGGREGATE_DAYS_TO_END_OF_PRECEDING_MONTH =
89     {0, 0, 31, 60, 91, 121, 152, 182, 213, 244, 274, 305, 335, 366};
90
91   private int ordinalDay;
92   private int day;
93   private Month month;
94   private int year;
95
96   public SpreadsheetDate(int day, Month month, int year) {
97     if (year < MINIMUM_YEAR_SUPPORTED || year > MAXIMUM_YEAR_SUPPORTED)
98       throw new IllegalArgumentException(
99         "The 'year' argument must be in range " +
100         MINIMUM_YEAR_SUPPORTED + " to " + MAXIMUM_YEAR_SUPPORTED + ".");
101     if (day < 1 || day > DateUtil.lastDayOfMonth(month, year))
102       throw new IllegalArgumentException("Invalid 'day' argument.");
103
104     this.year = year;
105     this.month = month;
106     this.day = day;
107     ordinalDay = calcOrdinal(day, month, year);
108   }
109
110   public SpreadsheetDate(int day, int month, int year) {
111     this(day, Month.fromInt(month), year);
112   }
113
114   public SpreadsheetDate(int serial) {
115     if (serial < EARLIEST_DATE_ORDINAL || serial > LATEST_DATE_ORDINAL)
116       throw new IllegalArgumentException(
117         "SpreadsheetDate: Serial must be in range 2 to 2958465.");
118
119     ordinalDay = serial;
120     calcDayMonthYear();
121   }
122
123   public int getOrdinalDay() {
124     return ordinalDay;
125   }
```

```java
126
127   public int getYear() {
128     return year;
129   }
130
131   public Month getMonth() {
132     return month;
133   }
134
135   public int getDayOfMonth() {
136     return day;
137   }
138
139   protected Day getDayOfWeekForOrdinalZero() {return Day.SATURDAY;}
140
141   public boolean equals(Object object) {
142     if (!(object instanceof DayDate))
143       return false;
144
145     DayDate date = (DayDate) object;
146     return date.getOrdinalDay() == getOrdinalDay();
147   }
148
149   public int hashCode() {
150     return getOrdinalDay();
151   }
152
153   public int compareTo(Object other) {
154     return daysSince((DayDate) other);
155   }
156
157   private int calcOrdinal(int day, Month month, int year) {
158     int leapDaysForYear = DateUtil.leapYearCount(year - 1);
159     int daysUpToYear = (year - MINIMUM_YEAR_SUPPORTED) * 365 + leapDaysForYear;
160     int daysUpToMonth = AGGREGATE_DAYS_TO_END_OF_PRECEDING_MONTH[month.toInt()];
161     if (DateUtil.isLeapYear(year) && month.toInt() > FEBRUARY.toInt())
162       daysUpToMonth++;
163     int daysInMonth = day - 1;
164     return daysUpToYear + daysUpToMonth + daysInMonth + EARLIEST_DATE_ORDINAL;
165   }
166
167   private void calcDayMonthYear() {
168     int days = ordinalDay - EARLIEST_DATE_ORDINAL;
169     int overestimatedYear = MINIMUM_YEAR_SUPPORTED + days / 365;
170     int nonleapdays = days - DateUtil.leapYearCount(overestimatedYear);
171     int underestimatedYear = MINIMUM_YEAR_SUPPORTED + nonleapdays / 365;
172
173     year = huntForYearContaining(ordinalDay, underestimatedYear);
174     int firstOrdinalOfYear = firstOrdinalOfYear(year);
175     month = huntForMonthContaining(ordinalDay, firstOrdinalOfYear);
176     day = ordinalDay - firstOrdinalOfYear - daysBeforeThisMonth(month.toInt());
177   }
178
179   private Month huntForMonthContaining(int anOrdinal, int firstOrdinalOfYear) {
```

```
180      int daysIntoThisYear = anOrdinal - firstOrdinalOfYear;
181      int aMonth = 1;
182      while (daysBeforeThisMonth(aMonth) < daysIntoThisYear)
183        aMonth++;
184
185      return Month.fromInt(aMonth - 1);
186    }
187
188    private int daysBeforeThisMonth(int aMonth) {
189      if (DateUtil.isLeapYear(year))
190        return LEAP_YEAR_AGGREGATE_DAYS_TO_END_OF_PRECEDING_MONTH[aMonth] - 1;
191      else
192        return AGGREGATE_DAYS_TO_END_OF_PRECEDING_MONTH[aMonth] - 1;
193    }
194
195    private int huntForYearContaining(int anOrdinalDay, int startingYear) {
196      int aYear = startingYear;
197      while (firstOrdinalOfYear(aYear) <= anOrdinalDay)
198        aYear++;
199
200      return aYear - 1;
201    }
202
203    private int firstOrdinalOfYear(int year) {
204      return calcOrdinal(1, Month.JANUARY, year);
205    }
206
207    public static DayDate createInstance(Date date) {
208      GregorianCalendar calendar = new GregorianCalendar();
209      calendar.setTime(date);
210      return new SpreadsheetDate(calendar.get(Calendar.DATE),
211                                 Month.fromInt(calendar.get(Calendar.MONTH) + 1),
212                                 calendar.get(Calendar.YEAR));
213
214    }
215 }
```

结束语

2005 年，在参加于丹佛举行的敏捷大会时，Elisabeth Hedrickson 递给我一条绿色腕带，类似 Lance Armstrong 带热那款①。这条腕带上面写着"沉迷测试"（Test Obsessed）的字样。我高兴地戴上，并自豪地一直系着。自从 1999 年从 Kent Beck 那儿学到 TDD 以来，我的确迷上了测试驱动开发。

不过跟着就发生了些奇事。我发现自己无法取下腕带，不仅因为腕带很紧，而且那也是条精神上的紧箍咒。那条腕带就是我职业道德的宣告，也是我承诺尽己所能写出最好代码的提示。取下它，仿佛就是违背了这些宣告和承诺似的。

所以它还在我的手腕上。在写代码时，我用余光就可以瞟见它。它一直提醒我，我做了写出整洁代码的承诺。

---

① 美国自行车运动员 Lance Armstrong 于 1997 年被诊断出罹患睾丸癌后，成立基金会，委托耐克公司生产了数百万条黄色腕带，为基金会筹款。这款腕带很快成为一种时尚。——译者注